Advances in Electronic and Computer Engineering

Advances in Electronic and Computer Engineering

Edited by **Annie Kent**

WILLFORD PRESS
New York

Published by Willford Press,
118-35 Queens Blvd., Suite 400,
Forest Hills, NY 11375, USA
www.willfordpress.com

Advances in Electronic and Computer Engineering
Edited by Annie Kent

International Standard Book Number: 978-1-68285-004-6 (Hardback)

Printed in the United States of America.

Contents

Preface

Over the recent decade, advancements and applications have progressed exponentially. This has led to the increased interest in this field and projects are being conducted to enhance knowledge. The main objective of this book is to present some of the critical challenges and provide insights into possible solutions. This book will answer the varied questions that arise in the field and also provide an increased scope for furthering studies.

In this age of rapid technological progress, the scope of computer and electronic engineering studies are vast. This book integrates the complex applications of computer engineering and electronics in the most comprehensible manner. Algorithms, semiconductors, circuits, mobile computing, etc. are some of the topics that have been delved into. This book is a complete source of knowledge on the significant concepts and latest advancements of these disciplines. This book will be an indispensable source of reference for the students and professionals in the field of electronics and computer engineering.

I hope that this book, with its visionary approach, will be a valuable addition and will promote interest among readers. Each of the authors has provided their extraordinary competence in their specific fields by providing different perspectives as they come from diverse nations and regions. I thank them for their contributions.

Editor

1

Global approach of mean service satisfaction assessment

Ahmed Dooguy Kora[1], Ibrahima CISSE[1], Jean-Pierre Cances[2]

[1]*Department of Telecommunications, ESMT, Dakar, Senegal*
[2]*Parc d'Ester, Ecole Nationale Superieure d'Ingenieurs de Limoges, Limoges Cedex, France*
E-mail: koraahmed@yahoo.fr

Abstract: A theoretical expression for mobile service satisfaction assessment has been proposed. Mobile networks users' satisfaction is a major concern for the operators and regulators. Therefore a certain level of network qualification is required to be offered to consumers by operators thanks to the decisions initiated by the regulation authority. For the assessment of the level of satisfaction, several methodologies and tools (measuring and monitoring) have emerged. Ranking in two broad categories, namely the objective and subjective methods, both have advantages as well as disadvantages. This Letter has proposed a unified approach to evaluate more objectively users' level of satisfaction of a service based on the most common network key performance indicators (KPIs) rate following the different methods. This approach's main advantage is that it has taken advantages of the different positive aspects of the existing methods and outperformed their limitations thanks to the introduced concept of global KPI. In addition, the size of samples according to each method has been considered. A mean service satisfaction theoretical expression has been proposed to regulation authority, consumers association and operators as common base of service satisfaction assessment.

1 Introduction

Today's generation mobile networks are designed for multimedia communication with higher data rate access to information and services on public and private networks. The deployment of the network is followed by its commercial launch once it is fully operational. The operator monitors its quality of service (QoS) for multiple purposes because it enables to optimise traffic and investigate critical areas where the network performance could require adequate changes.

Indeed, these advanced technologies have absolutely changed the human life through different services offered to the user.

However, this important progress does not prevent the advent of innovative ideas that mainly affect the target user satisfaction assessment. These new services suitable to the competition between mobile operators contribute to the improvement of the QoS and quality of experience (QoE) in an area. Thus, the tendency to respond to these new needs of the users has pushed these operators to improve the QoS and QoE which are subject of many papers [1–7]. This quality implies the continuous monitoring of the operating state of the network and all the key performance indicators (KPIs). To perform such task, engineering tools could help a lot. Competition between telecommunications operators is very dynamic because the only way to maintain potential customers and attract others is to offer constantly new services with the highest possible quality. However, there are many approaches to estimate a service satisfaction level for mobile networks and every solution has its pros and cons [8]. In this Letter, we have investigated the effectiveness and limitations of each method in order to propose a unified approach combining the different possible methods which are objectives methods [drive test (DT), counters from operating and maintenance centre (OMC) and controllers of protocols] and subjective ones (the complaints of customers, customer surveys etc.). This approach's main advantage is that it combines the positive aspects of all the above enumerated methods based on the size of the considered samples. This approach could be used by regulation authority and operators as common base of service satisfaction assessment.

The rest of this Letter is organised as follows: Section 2 presents the actors and existing methodologies used to perform QoS assessment, Section 3 is dedicated to the proposed unified approach. Section 4 ends this Letter with a conclusion.

2 Common QoS and QoE assessment methods

Three actors are involved in service satisfaction assessment.

Regulators: ensure compliance by network operators through the specifications of their license conditions including the service quality.

Operators: follow and optimise their networks in order to meet these specifications. They satisfy their subscribers and maximise returns on their investments.

End users: constitute the third party implementing tools to objectively evaluate the QoS provided by network operators [9].

To better identify all aspects of service quality assessment, operators as well as regulators use several techniques or tools to assess the performance of a mobile network [10, 11]. These techniques and appropriate tools may be specific to a given interface or nodes in a mobile network. Thus, the approaches differ according to the relevance of their results in time and in space. It is important to note that up to now no tool or technique permits to really obtain a three-dimensional (full time, full space with sharp location and subjective) assessment of the KPI for a mobile network [12] as illustrated in Fig. 1.

2.1 Drive test

The DT is one key approach in the performance measurement process, with the goal of collecting measurement data as a function of selected location and time. The DT allows the mobile network to be tested through the use of team technicians who behave as users in order to investigate on the network performance. Data collected at the end of the DT provide information mainly on the technical characteristics of the network and not necessarily on the real needs of users. They are presented in a percentage format after collecting a number of statistical samples. The DT is difficult to apply to the total area in the network. It requires a good car able to access any way at any time. Once the data have been collected over the selected itinerary, engineers can use this data together with post-processing software to identify any problems in the network for trouble

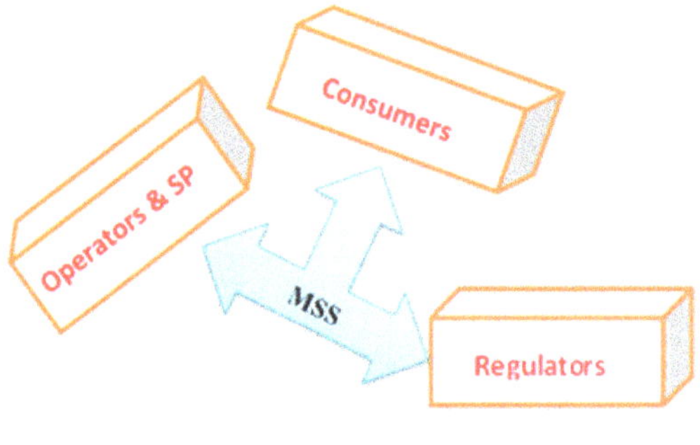

Fig. 1 *Main actors of MSS assessment*

shooting purpose. The DT is performed in order to carry out all different scenarios to test for mobile communication including outdoor, indoors, in car etc. Most of the time, a DT investigates on the coverage, the difficulty to establish a communication, the call drop, the quality of the communication, handover failures etc.

Fig. 2 presents in addition to the base transceiver station (BTS)/ node B, the kit used for a DT. The kit usually contains at least one mobile phone, a global positioning system (GPS) receiver and a computer. The mobile(s) phone(s) could be equipped with special software used to display radio measurements data. The mobile is connected to the data acquisition computer. The mobile assess allowed radio parameters which are transferred to the computer. The GPS receiver is used to record the exact location of the geographical position of each point of measurement. It is essential to identify the collection points that have problems. The laptop computer processes and records the data received from the mobiles and the GPS receiver thanks to an acquisition application. It displays the state of the network on the screen.

The DT kit is set up in a vehicle or indoors and long communications as well as short communications are initiated. A long communication could be initiated and terminated by the user himself.

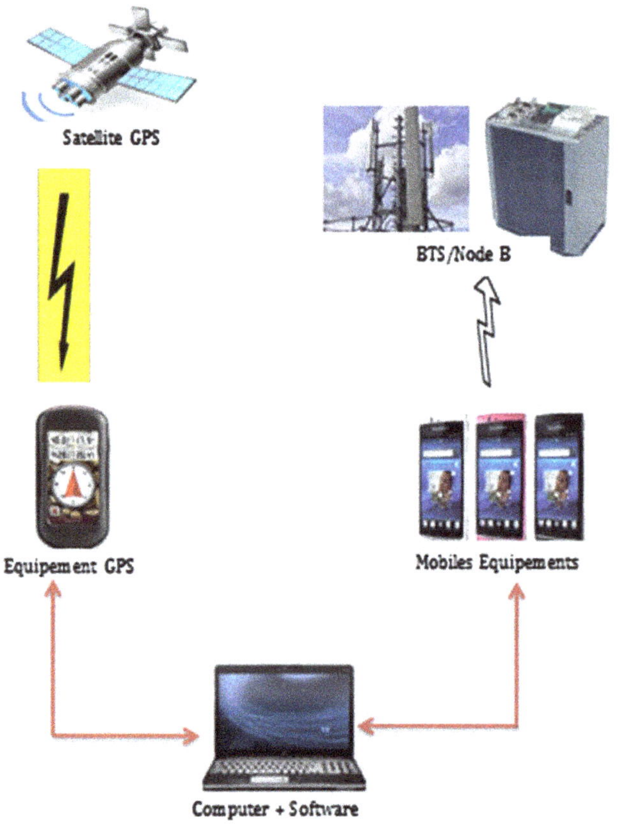

Fig. 2 *Tools of DT*

A short communication is usually automatically generated by the system for equal or less to 3 min. A long call can be used to measure the handover success rate (HOSR) as well as the Rx quality, whereas communications setup success rate is measured on a short call. Other parameters as received signal level, interference on broadcast control channel (BCCH), nearest sector detection etc. are also observed. The report of the DT could be depicted in a graphical form with a bar chart and conclusions are derived based on the statistics. A DT can detect several problems as no – working sites/sectors or TRXs, no-active radio network features like frequency hopping, coverage overlaps, C/I and C/A analysis, high-interference spots, call setup failure, drop calls, capacity problems, handovers failure, accessibility and retain ability of the network, equipment performance, faulty installations etc.

The KPI most considered by the actors depends on the objective of the DT. KPI's for operators include and not limited to call drop rate (CDR), call setup success rate (CSSR), traffic channel (TCH) usage, standalone dedicated control channel (SDCCH) usage, handover statistics and connection establishment.

KPIs data are generally collected from the network OMC, DT statistics using mobile test equipment, protocol analyser statistic or QoE campaign. KPIs values are estimated using formulas and compilation of various data. Different KPIs are involved in call setup testing, the continuation of the call and the volume of traffic, and the QoS across the network.

A threshold level is determined for each KPI, if the threshold is exceeded an alarm is sent to the supervision to indicate the presence of trouble ticket related to the functions that could affect that KPI. The common KPIs and their thresholds [13, 14] are given as follows:

CSSR (>95%)

$$\mathrm{CSSR} = \frac{\text{successfully completed call setups}}{\text{call setups attempts}} \tag{1}$$

Drop call rate (CDR) (<2%)

$$\mathrm{CDR} = \frac{\text{dropped calls due to RF loss and HO failure}}{\text{successfully call established}} \tag{2}$$

HOSR (>98%)

$$\mathrm{HOSR} = \frac{\text{successful handovers}}{\text{handover attempts}} \tag{3}$$

Handover failure rate (HOFR) (<3%)

$$\mathrm{HOFR} = \frac{\text{handover failures with dropped calls}}{\text{handover attempts}} \tag{4}$$

TCH blocking (<2%)

$$\mathrm{TCH}_{\text{congestion}} = \frac{\text{blocked TCH assignments}}{\text{TCH assignments attempts}} \tag{5}$$

SDCCH channel blocking (<3%)

$$\mathrm{SDCCH}_{\text{congestion}} = \frac{\text{blocked SDCCH assignments}}{\text{immediate assignment attempts}} \tag{6}$$

SDCCH access success rate (>97%)

$$\mathrm{TASRR} = \frac{\text{successfully immediate assignments}}{\text{immediate assignment attempts}} \tag{7}$$

Note: TASR: Traffic assignments success rate.

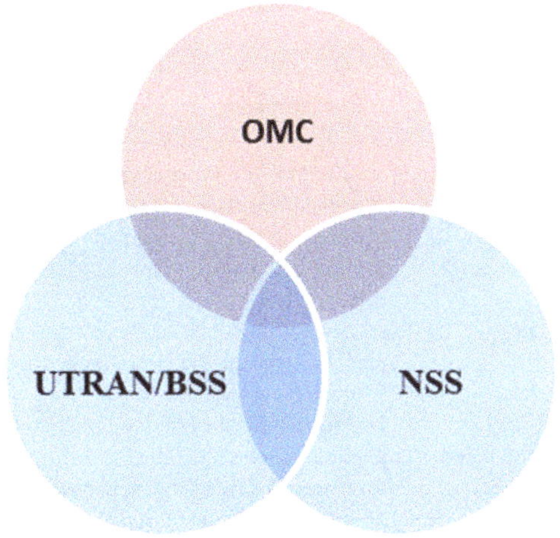

Fig. 3 *OMC in a mobile network*

2.2 Operation and maintenance centre

The OMC is connected to all managed elements in the switching system and in the base station sub system (BSS). The OMC functions include fault management (maintenance tasks), administration and commercial operation (subscription, charging, statistics etc.), security management, network configuration, operation and performance management.

The operation and maintenance functions are based on the concepts of the telecommunication management network which is standardised in the ITU-T series M.30.

Fig. 3 depicts how OMC system covers all the elements of a mobile network.

In operation and maintenance centres, counters are used for various events to provide the operator specific information about the state and the quality of the network. OMC equipment's manage the mobile network equipment's. There are two sections in the OMC, the OMC–R which is in charge of the radio part (BSS) and the OMC-S dealing with the switching part. According to the DT data, we have especially focused on the OMC–R. The OMC–R is responsible for radio performance management, measures are based on the collection of the counters calculated by the sensors. The OMC measures are used in several departments as statistics, network planning and optimisation, investigation in case of problems on the network, real-time network analysis. The OMC data are provided in the form of raw data, and so they are usable, they are turned into performance indicators to allow operators to obtain information about the QoS (Fig. 4) offered by their network and decide which optimisation strategy to apply. The OMC data can be combined with other sources which can enable getting the network KPI as in Figs. 4 and 5.

These indicators allow localisation of network anomalies, and therefore the identification and diagnosis of the causes of these problems in order to react with appropriate corrective actions. The processing of data in the OMC is an assessment method with three advantages:

- Economic (facilities centred at a network level allow information on the entire network).
- Automatic (collecting information is dynamic and does not require manual complexity).
- Exhaustive [occurs permanently in time with the same (Fig. 5) KPIs].

The following parameters can quantify network quality in OMC [15]: Dropped call rate (CDR) which indicates the number of premature call terminations (threshold generally between 0 and 5%).

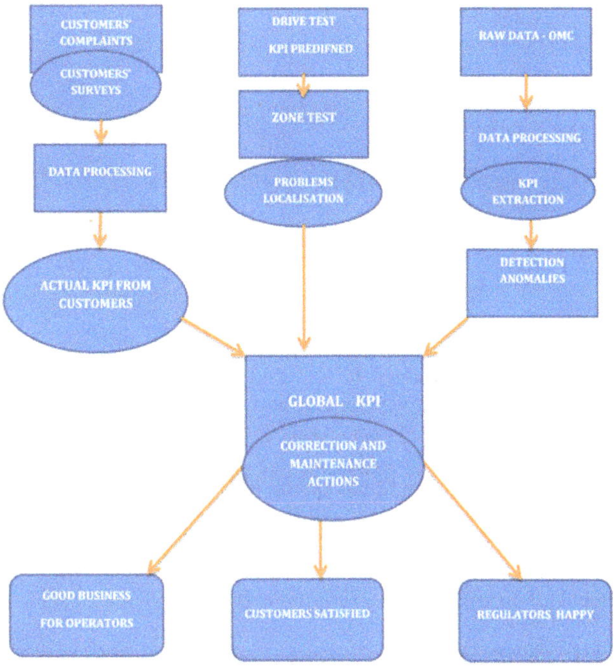

Fig. 4 *QoS management*

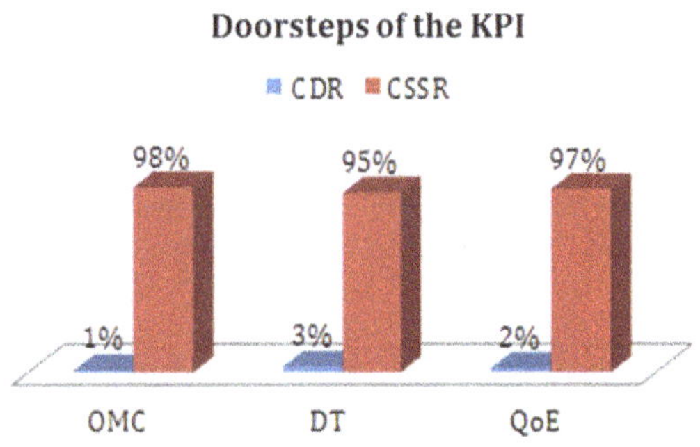

Fig. 5 *Graphic representation of KPIs*

Blocking (grade of service) which indicates number of failed call establishments because of network problems, primarily congestion (threshold generally between 2 and 5%)

$$\mathrm{GOS} = \frac{\text{number of call attempts} - \text{number of call success}}{\text{number of call attempts}} \tag{8}$$

Traffic throughput in Erlangs measures the amount of traffic throughput across the network. It can be expressed as the sum of the duration of all the calls divided by the measurement time

$$\text{traffic for one suscriber} = \frac{\text{number of calls per hour per suscriber} \times \text{MHT}}{3600} \tag{9}$$

Note:

- MHT means (mean holding time) generally 90 s = 2,5 mE/ subscriber.
- Erlangs indicates the amount of traffic carried by the network.

TCH assignment failure rate (< 2%)

$$\text{TAFR} = \frac{\text{TCH assignment failures}}{\text{TCH assignment attempts}} \tag{10}$$

radio resource control (RRC) connection setup success rate (> 97%)

$$\text{RNC} - \text{CSSR} = \frac{\text{successfully completed call setups}}{\text{valid call setups attempts}} \tag{11}$$

Other KPIs like inter-frequency hard HOSR, soft HOSR, inter-systems hard/soft HOSR could be also estimated. Some periodic counters are daily, weekly or monthly basis as cell traffic level, rapid fault identification, call setup rate, TCH/SDCCH congestion, TCH drop call rate etc.

Despite all these advantages, this approach compared with DT is mainly limited by the difficulty to identify with more precision the specific local coordinates of a detected event in the cell. In addition to this, the OMC data transferred to the regulation authority could be corrupted if appropriate strategy is not established.

2.3 *Protocols analysers data*

A protocol analyser is a high-impedance measuring device that can be intrusively or non-intrusively connected to a live transmission link in order to monitor the activity on that link without disrupting the information flow. Its primary purpose is to monitor control traffic rather than user data, that is, the bidirectional passage protocol information [16].

A protocol analyser can be connected to the nodes of the mobile network over a period of time. During this time, the transmission paths are constantly measured and these measurements are logged in the form of trace files for subsequent analysis. A typical protocol analyser generally has the following characteristics:

- Hardware capable of attaching to the transmission medium.
- Software capable of capturing and decoding the raw protocol data passing along it.
- Software capable of filtering and/or searching for specific protocol data and parameters.
- A screen to display measurement results.
- Facilities for remote access.
- A storage device to record measurements for subsequent processing.

The use of protocol analysers enables reduced and less sophisticated OMC statistical measurements instead of wide network measurements. In the case of a service satisfaction assessment for the whole network or a given area, OMC data can be taken rather than the restricted view offered by expensive analysers.

2.4 *Benefits and limitations of drive testing*

The limitations and benefits of a DT [17] are summarised, respectively, in Tables 1 and 2 as follows.

2.5 *Subjective methods*

Subjective methods are mainly based on the perception of the user about the QoS. The way the final user needs is satisfied. The results of a subjective campaign must be specified with objective measures so that significant results provided by users are much more relevant. The quality perceived by user does not necessarily indicate the technical characteristics of a network [18]. To specify the technical results of subjective measures, trained technicians perform field investigations in conducting quantitative studies to collect the point of view of users. In these types of surveys for a large number of users in a specific geographic area, it is closed specific questions. The approach can be done based on a list of questions, telephone calls, by internet, mail etc. The results are gathered from a number of tests and statistics are elaborated in order to establish a synthesis.

Customers complaints are also a vital source of the feedback on the QoS, and should not be ignored because they are proofs of those parameters that have not been supported by quantitative and objective measures. These complaints have to specify in details trouble areas for troubleshooting purpose. They also help regulators to point out to the operator's coverage holes. The result facilitates the choice of KPI (Table 3) for the future assessment performance of a network [19].

Table 1 DT limitations

Limitations	
replicating subscriber usage profiles	even if mobile test can simulate a standard mobile station (MS) from a functional viewpoint; it is less simple to accurately replicate subscribers usage profiles in terms of frequency of received calls and types of services used
restricted access area	restrictions on accessing certain areas may preclude accurate DT measurements in a specific geographical area
limited wide network performance measurement	Owing to limited resources, it is not possible to obtain a global and sharp assessment of the whole network performance within an acceptable time period because of having continuously redeployment test teams. The field investigations can be done only if field test engineers can physically access the area. However, not all areas are easy to access. Alternative investigation methods could be used
based mostly on down-link analysis	measurements generally transmitted by the MS to the PC are on the down-link. Partial information is gathered during DTs regarding up-link behaviour

Table 2 DT benefits

Benefits	
replicate subscriber conditions	DT is carried out using an MS. The mobile used is the common one in the network for objective RF performance. However, software is added to enable performance data to be measured and exported to a recording device
comparative operator performance	in the increasingly competitive cellular market, competitor's strengths and weaknesses can be key factor to gaining and/or maintaining their market. DT enables operators to compare the performance of their own network with those of their competitors for optimisation purpose
focusing on particular parameter set and/or regions	DT can be a resource and time-intensive activity. By selecting a suitable route and parameter, the DT team can focus on specific areas instead of wide network testing

Table 3 Common KPIs thresholds

Global KPI from OMC/DT	Threshold, %
CSSR	>97%
dropped call rate	<1%
HOFR	<2%
congestion SDCCH/TCH	<3%
RRC connection setup success rate	98%

3 Proposed global approach

The previous section has shown that objective and subjective approaches to assess users' level of satisfaction converge towards the same target. The choice of KPIs depends on actors (regulators, operators and customers) (see Table 3). It aims to satisfy the end user needs.

Combining subjective and objective methods allows consolidating the results and taking into account more aspects of user's satisfaction [20, 21]. Then in order to mitigate this assessment based on the diversity of the methods with their own strength and weakness, a global approach which has been introduced in this Letter overcomes the previous enumerated limitations.

Objective measures from the OMC (Table 4) counters provide oriented KPI for optimisation, maintenance and supervision. The DT is also based on oriented KPIs for investigation on network performance, benchmarking, optimisation and service satisfaction assessment etc.

The main KPIs of DT as well as those of the OMC are converging as shown in the following Table 4 [22].

Our idea is to introduce an overall KPI whose assessment is based on the combining of the subjective and objective investigations. The objective is that this approach of calculation is more representative and reliable. Formula (12) has been introduced to evaluate the weight to allocate to each KPI value for a given method [23] in the proposed case

$$\alpha^{i}_{\text{KPI}} = \frac{\text{number of } i \text{ samples}}{\text{aggregated samples number of considered methods}} \tag{12}$$

This way of calculations of a KPI coefficient supposes:

- The doorsteps of confidence and the margins of mistake accepted might be identical.
- The rounded probabilistic has the same degree.
- The size of the samples could vary from a method to another.
- The approximations are at thousandth near and the methods of probabilistic samplings are used.
- The methods of calculations of a KPI at the OMC might not vary from one operator to another.

The main parameters in each case include, but not limited to:

- The geographical area (the zone of cover considered).
- The number of raised measures.
- The type of measures which have been performed.
- The precision to consider.

Table 4 Example of KPIs for different methods

	CDR	CSSR
OMC	(15/1500) = 1%	(1470/1500) = 98%
DT	(30/1000) = 3%	(950/1000) = 95%
QoE	(16/800) = 2%	(776/800) = 97%

According to different ways of user satisfaction level of assessment [19–25], let us denote:

N_{DT}: number of sample of the DT method.
KPI_{DT}: percentage qualifying the performance of the network according to the DT method.
N_{QoE}: number of sample of the QoE method.
KPI_{QoE}: percentage qualifying the performance of the network according to the QoE method.
N_{OMC}: number of sample of the OMC counters.
KPI_{OMC}: percentage qualifying the performance of the network according to the OMC counters.
N: total number of sample when considering the three methods.

$$N = N_{\text{DT}} + N_{\text{OMC}} + N_{\text{QoE}} \tag{13}$$

For example, the global KPI percentage $\text{KPI}_{\text{global}}$ according to the combination of the three methods (DT, QoE and OMC) could be expressed as

$$\text{KPI}_{\text{global}} = \frac{N_{\text{DT}}}{N} \times \text{KPI}_{\text{DT}} + \frac{N_{\text{OMC}}}{N} \times \text{KPI}_{\text{OMC}} + \frac{N_{\text{QoE}}}{N} \times \text{KPI}_{\text{QoE}} \tag{14}$$

The total number of samples is given by:
'$N = N_{\text{DT}} + N_{\text{OMC}} + N_{\text{QoE}} = 1000 + 800 + 1500 = 3300$ samples'

Based on QoE, DT and OMC, one can rewrite (14) as

$$\text{KPI}_{\text{global}} = \frac{1}{N} \times \sum_{i=1}^{3} N_i \times \text{KPI}_i \tag{15}$$

where N_i is the number of samples according to method i, and 'i' takes values as DT, QoE and OMC.

The resulted calculations of global call setup success rate (Fig. 6) and global dropped call rate (Fig. 7) applied to previous numerical

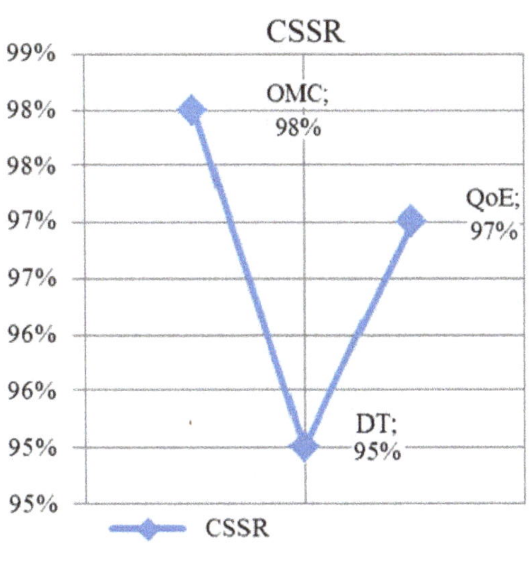

Fig. 6 *CSSR of different methods*

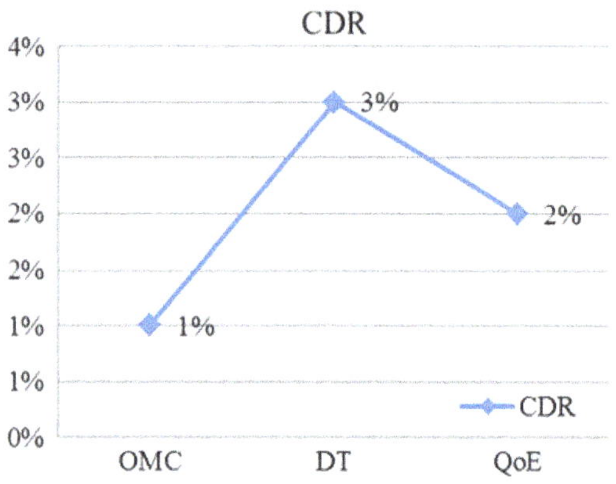

Fig. 7 *CDR of different methods*

example could be expressed as follows

$$\begin{aligned} \mathrm{CSSR_{global}} &= \frac{1000}{3300}\times 95\% + \frac{800}{3300}\times 97\% + \frac{1500}{3300}\times 98\% \\ \mathrm{CSSR_{global}} &= 96.848\% \\ \mathrm{CDR_{global}} &= \frac{1000}{3300}\times 3\% + \frac{800}{3300}\times 2\% + \frac{1500}{3300}\times 1\% \\ \mathrm{CDR_{global}} &= 1.848\% \end{aligned} \tag{16}$$

For regulation authority, the collection of OMC data from the operators could present some limitations because it could be corrupted by them, then it could be better to proceed as

$$\begin{aligned} \mathrm{KPI_{global}} &= \frac{1}{N}\times\sum_{i=1}^{2} N_i \times \mathrm{KPI}_i \\ &= \frac{N_{\mathrm{DT}}}{N}\times \mathrm{KPI_{DT}} + \frac{N_{\mathrm{QoE}}}{N}\times \mathrm{KPI_{QoE}} \end{aligned} \tag{17}$$

N_i is number of samples based on method i describing DT and QoE. The $\mathrm{KPI_{global}}$ expression could be extended to KPI values built from the protocol analyser.

The overall assessment of the quality of the communication depending on for QoE or signal quality indicator given by DT or OMC could be classified at least in two cases (bad and good). In our model, let us introduce three cases (bad, good and very good) corresponding, respectively, to 35, 70 and 100%.

The overall mean service satisfaction (MSS) could be assessed as follows

$$\mathrm{MSSA} = \frac{(N_{i\mathrm{bad}} \times 0.35 + N_{i\mathrm{good}} \times 0.7 + N_{i\mathrm{vgood}})}{N \times (1 - \mathrm{CDR_{global}})} \tag{18}$$

where $N_{i\mathrm{bad}}$ corresponds to the number of bad communications, $N_{i\mathrm{good}}$ corresponds to the number of good communications and $N_{i\mathrm{vgood}}$ corresponds to the number of excellent communications, N corresponds to the total number of communications initiated as defined in (13). $(N_{i\mathrm{bad}} + N_{i\mathrm{good}} + N_{i\mathrm{vgood}})$ corresponds to the total number of started communications which have ended.

4 Conclusion

Service satisfaction assessment has been a major preoccupation after the phase of commercial launch of a mobile network. This Letter has shown that the different methodologies used have their own limitations. Thus, while being based on the advantages and limits of the different methods, a global KPI assessment approach has been introduced. Taking into account the size of collected information according to each method, appropriate evaluated coefficient values are allocated to KPIs emanated from the existing methods. This has enabled the formulated expression of MSS which would be useful for regulators, network operators and consumer associations.

5 References

[1] Jin D., Wang K., Feng L.: 'End-to-end QoS performance analysis of wireless mesh network on RM-AODV routing protocol'. Int. Conf. Automatic Control and Artificial Intelligence (ACAI 2012), 2012, pp. 987–990

[2] Park J.-T., Chun S.-M.: 'QoS-guaranteed IP mobility management for fast-moving vehicles with multiple network interfaces ', *IET Commun.*, 2012, **6**, (15), pp. 2287–2295

[3] Zhang X., Phillips C.: 'A novel heuristic for overlay mapping with enhanced resilience and QoS'. IET Int. Conf. Communication Technology and Application (ICCTA 2011), 2011, pp. 540–545

[4] Ma M., Lu J., Fu C.P.: 'Hierarchical scheduling framework for QoS service in WiMAX point-to-multi-point networks', *IET Commun.*, 2010, **4**, (9), pp. 1073–1082

[5] Sun Z., He D., Liang L., Cruickshank H.: 'Internet QoS and traffic modelling', *IEE Proc., Softw.*, 2004, **151**, (5), pp. 248–255

[6] Wei G., Min W., Chao Y., Bingyao C., Jiaying D.: 'Users/applications based QoS system under Linux'. IET Int. Communication Conf. Wireless Mobile & Computing (CCWMC 2009), 2009, pp. 385–388

[7] Carter S.F., Macfadyen N.W., Martin G.A., Southgate R.L.: 'Techniques for the study of QoS in IP networks' in ed. by R. Ackerley, *Telecommun. Perform. Eng.*, 2004, pp. 67–96

[8] International Telecommunication Union: 'Méthode d'évaluation de la qualité de transmission', Recommendation UIT-T P.800

[9] Available at http://www.qosforum.com, accessed July 1999

[10] ANSI: 'Quality of service for business multimedia conferencing', ANSI T1.522, 2000

[11] UIT-T: 'Connections retain ability objective for the international telephone service', UIT-T recommendation E.850

[12] Available at http://www.fratel.org/, accessed March–June 2013

[13] Available at http://www.3gpp.org/, accessed March–July 2013

[14] Available at http://www.wapiti.telecom_lille1.eu/commun/ens/peda/options/ST/RIO/pub/exposes/exposesser2010_ttnfa2011/coquerel_bourguin/mqos_en_place5.html, accessed April 2013

[15] Available at http://www.fr.scribd.com/doc/91188121/Alcatel-Lucent-W-CDMA-UTRAN-Parameters-Description, accessed May 2013

[16] Available at http://www.astellia.com/fr/sondes, accessed April 2013

[17] Available at http://www.wapiti.telecomlille1.eu/commun/ens/peda/options/st/rio/pub/exposes/exposesrio2008-ttnfa2009/Belhachemi-Arab/g1000.htm, accessed April 2013

[18] Available at http://www.telecomsource.net/showthread.php?586-Huawei-kpi-definations-with-counter-id#.UejII7odcvw, accessed June 2013

[19] ITU-T: 'Amendment 1 (01/07): new appendix I – definition of quality of experience (QoE)', ITU-T recommendation P.10/G.100, 2007

[20] ITU-T: 'Terms and definitions related to quality of service and network performance including dependability', ITU-T recommendation E-800

[21] ETSI: 'Digital cellular telecommunications system (phase 2 +); telecommunication management; performance management (PM); performance measurements – GSM', (3GPP TS 52.402 version 9.0.0 Release 9), ETSI TS 152 402, V9.0.0 (2010-02)

[22] Available at http://www.artp.sn/, accessed June 2013

[23] ITU-T: 'Measurements of the performance of common channel signaling networks', ITU-T recommendation E.505

[24] Ameigeiras P., Ramos-Munoz J., Navaros-Ortiz J., Mogensen P., Lopez-Soler J.M.: 'QoE oriented cross-layer design of a resource allocation algorithm in beyond 3G systems', *Comput. Commun.*, 2010, **33**, (5), pp. 571–582 Available at http://www.dx.doi.org/10.4304/jcm.4.9. 669-680

[25] Thakolsri S., Khan S., Steinbach E., Kellerer W.: 'QoE-driven cross-layer optimization for high speed downlink packet access', *J. Commun.*, 2009, **4**, (9), pp. 669–680

2

Accessibility of dynamic web applications with emphasis on visually impaired users

Kingsley Okoye, Hossein Jahankhani, Abdel-Rahman H. Tawil

School of Architecture Computing and Engineering, University of East London, London E16 2RD, UK
E-mail: u0926644@uel.ac.uk

Abstract: As the internet is fast migrating from static web pages to dynamic web pages, the users with visual impairment find it confusing and challenging when accessing the contents on the web. There is evidence that dynamic web applications pose accessibility challenges for the visually impaired users. This study shows that a difference can be made through the basic understanding of the technical requirement of users with visual impairment and addresses a number of issues pertinent to the accessibility needs for such users. We propose that only by designing a framework that is structurally flexible, by removing unnecessary extras and thereby making every bit useful (fit-for-purpose), will visually impaired users be given an increased capacity to intuitively access e-contents. This theory is implemented in a dynamic website for the visually impaired designed in this study. Designers should be aware of how the screen reading software works to enable them make reasonable adjustments or provide alternative content that still corresponds to the objective content to increase the possibility of offering faultless service to such users. The result of our research reveals that materials can be added to a content repository or re-used from existing ones by identifying the content types and then transforming them into a flexible and accessible one that fits the requirements of the visually impaired through our method (no-frill + agile methodology) rather than computing in advance or designing according to a given specification.

1 Introduction

A greater number of, if not all, individuals now use the internet as a means to communicate, and to a large extent work and collaborate with each other. However, these innovations in information and communication technology utilisation have paved the way for improvement as well as challenges for both the users and the developers. These challenges do not exclude the disabled users. Recently there have been a couple of questions arising as to how best can the use of this fast changing technology be used to intuitively address the needs of the users with disability as well as what is needed in future to better support these users, especially the visually impaired users, since the internet is fast migrating from static web pages to dynamic web pages; because of the increasing rate at which demand for rich internet application and multimedia content is growing. The software developers have failed to consider what kind of support could be built for such users that do not really want to deal with all the technical details that come along with the dynamic web applications, focusing only on the increasing demand for rich internet applications. Garrigós *et al.* [1] mention that because of the growing demand for web applications offering a rich user experience, user-centric web applications are being replaced by the so-called rich internet applications (RIAs), which provide an interface, interaction and functionality capabilities similar to desktop applications. RIA development has new requirements and concerns that come into play [2], complicating the task of the software developers, such as impose limited screen size, more difficult interaction and poorer multimedia support. The dynamic web developer's community is well-aware of these challenging difficulties [3, 4], because these approaches do not yet cover all design concerns usually encountered in state-of-the-art applications such as the dynamic web. One of the major unsupported aspects is the personalisation of content and requirement to the specific user and his/her context, specifically for users with disabilities, such as the visually impaired.

1.1 Scope of study

This paper has focused on investigating some of the challenging features found in recent web applications, and techniques to measure and improve the ability and performance of visually impaired users in using and accessing the contents embodied in them; as a step towards bridging the digital gap between dynamic web applications and the visually impaired users. In this paper, we presented the hypothesis guiding this research followed by a review of existing literatures. Next, the data collection, analysis and results were presented. Subsequently, the implementation and practical element and the last section concludes this research work with discussion of findings, contribution of this research work, limitations and recommendation for future research.

1.2 Objective

This research presents a novel framework for practice capable of enabling a deeper understanding of accessibility requirements for dynamic web applications centred around users with visual impairment, through evaluation of collected facts and analysis of results from face-to-face and online survey of the user group mentioned in this paper. We then propose and implement solutions to the dynamic web accessibility issues by designing and deploying a software application and its implementation for best practice, which then informs software developers on how best to significantly and effectively approach the design of dynamic web application contents with the visually impaired users in mind.

2 Statement of problem

Various studies claim that the visually impaired users are the category of disabled users of web applications that need the most assistance because of the level of challenge they face in using and accessing the e-contents [5–8]. Some of these challenges have been identified as decreasing ability to focus on near tasks, changing colour perception and sensitivity, pupil shrinkage and decreasing contrast sensitivity [3]. This research paper aims at capturing the experiences of past researchers in the context of web application development as well as visually impaired persons themselves; as an indication and means to revealing the areas where support is lacking and thereby discuss a road map for further improvement and reviews.

2.1 Accessibility issues for the visually impaired ICT users

Visual impairment refers to someone who is blind or partially sighted [9]. The Royal National Institute for the Blind (RNIB)

[10] describes persons with visual impairment as 'people with irretrievable loss of sight'. According to [9], about 314 million people are visually impaired globally, of which ~15% (45 million) are blind. These figures were justified by [11] who also points out that 'globally there are over 314 million visually impaired people: 45 million of them are totally blind'. Visual impairment is a worldwide disability problem, which has been seen as a 'global public health problem'. However, Mulloy *et al.* [12] argue that the use of assistive technology (AT) for users with visual impairments and blindness has the potential to improve interactivity and accessibility outcome via enhancement of existing sight abilities and/or engagement of other senses. Visual impairment exists in various groups of individuals including children, unskilled, disabled and also the elderly as they face serious problems in using information and communications technology (ICT) tools. In [13], Adetoro *et al.* confirm that individuals with visual impairment, like every other person or group, need information to minimise ambiguity, identify and resolve problems and eventually enhance their performance and interactivity especially in using ICT. In [14], Thylefors stated that the number of visually impaired persons is expected to rise by 2020 as the total number of elderly persons is estimated to double in number, reaching two billion worldwide. According to [15], the incidence of visual impairment will continue to increase at the rate of ($p = 0.01$) with greater age. This increase, according to them, is not associated with gender, environment or level of experience of such person. Our paper focuses on providing a solution for an accessible platform for the visually impaired users of ICT regardless of age and sex. Despite the fact there have been other available means through which the visually impaired persons interact with the web applications, a large number of them still need assistance to intuitively interact with the contents embodied in these applications [6, 12]. According to Petrie *et al.* [5], the new flash content of RIAs and colour contrast levels are causing new accessibility problems for the visually impaired users. The screen reader reads static hypertext markup language (HTML) pages by analysing the HTML tag structures and allow users to navigate through pages using a keyboard with shortcuts and key combinations, as a result of this most dynamic content is not accessible through the use of screen reader since such contents are designed and built with DHTML tag structures [16]. The subjective observation by these researchers also stresses that because the visually impaired person uses the screen reader in combination with keyboard, they find it difficult to interact with the graphics and moving objects since most often the view of such contents changes dynamically in response to mouse operation. The findings of [17] further suggest that there are various measures that it would be possible to take towards alleviating the situation, in the form of further improvements to retrieval systems, to search interfaces and to text-to-speech screen readers. However, Freire *et al.* [18] state that it is a very challenging accessibility issue since even texts are treated as images, which explains why the challenge increasingly lies not in the visual impairment but in the design of the technology that mediates their access to and use of the dynamic web application. That is why new evidence is needed to help us plan; both because of the increase in demand and because of the changing lifestyles.

2.2 *Dynamic web application for the visually impaired*

Web-based applications are those applications which are directly accessible using any available browser and which do not need to be installed on the user's computer, while standalone applications include the 'downloadable' applications from the web, which cannot be accessed directly from the browser but need to be installed locally on the user's computer [19]. Following these, Avila *et al.* [20] refer to dynamic web application as 'a multimedia platform that incorporates animation, sound, flash, video and interactivity into a standalone product or onto a web page'. According to them, the application uses various multimedia channels ranging from simple web page decorations and banner advertisements to fully interactive training and electronic forms. A fully accessible website is one that is designed to make use of the latest web technologies, such as multimedia, while at the same time accommodating the needs of those who have difficulty with or are unable to use these technologies, such as the visually impaired [21]. According to Adetoro *et al.* [13], dynamic web applications for visually impaired users should be provided with the intention of meeting their accessibility and information needs by transcribing the application into an alternative format in order to increase the performance of the visually impaired users in using such technologies. This is a multi-faceted problem that requires research into technical challenges; from user modelling to context analysis [22]. In [23], font sizes should be made larger for the visually impaired because of the decline in visual acuity. This shows that colour and text combination is critical when designing web application for the visually impaired users. Reference [24] requires content developers to make electronic and information technology accessible to users with disabilities, including blindness, colour blindness and visual impairment. The law includes standards for software applications, operating systems, web-based applications and multimedia. The technical criteria of section 508 [24] pointed out two design approaches for making web applications accessible to users with disabilities which include:

- Provision of multiple ways to operate the technology and retrieve information so that users can choose alternatives based on their physical capabilities.
- Provide support for the AT being used by the users with disabilities to ensure improved accessibility.

3 Research methodology

We focused on the target user group 'the visually impaired ICT users' as it is very important to know the users and what their previous and present experience is. Understanding the users' technical skill is as important as knowing their expectation [25].

3.1 *Participants*

Research participants comprised various categories of blind and partially sighted people who use the services and are registered with the RNIB, as well as professionals at the organisation, in London, UK. The categories of the participants involved both older and adult persons, trying as much as possible to cross the range of sex, ethnicity, age, skills and experience. The internet-mediated research was also implemented as a simple online-based methodology, whereby participants were contacted and sent an experimental text-based questionnaire via e-mail and focus groups.

3.2 *Research design*

To gather the necessary information with which to attempt to provide answers to the research questions. A holistic design method was used to gather the information for this research, which draws on both quantitative and qualitative methods of collecting data [26]. With the visually impaired persons as the unit of analysis, the research utilised the combination of cluster and random sampling for its primary data collection, trying as much as possible to represent every unit. The case study sampling was used to examine the problem of accessibility and what is missing in the design of dynamic web contents for the visually impaired, based on the information we obtained from the survey, interviews and questionnaire, and to test the implications of the result against the current literature and present a representation of the results using both quantitative and qualitative methods of data analysis.

3.2.1 Design procedure: Evidence was collected via face-to-face interaction as well as online interviews and reviews with a number of key informants. These were the visually impaired persons themselves, professionals, careers, advisors as well as service providers who work with the visually impaired persons. Documentary evidence was also obtained in the form of journals, written articles and policies on visual disabilities from the various participants. Where possible, more information was gathered from online forums, e-mail discussions and comments by visually impaired users as well as other professionals who work with such persons. The internet-assisted means of data collection was chosen with the aim of improving the reliability and validity of the research, as it was perceived to help eliminate the natural human error and lack of control over the research. Consequently, structured questionnaires were distributed via the internet to some of these key informants and were set to collect responses via the internet, enhancing the anonymity of the researcher and the respondent. The outcome was positive as it helped the researcher overcome the barriers of sex, ethnicity and age and also encouraged honesty and an increase in response rate without violating the ethical issues in relation to the participants.

3.3 Data presentation

The face-to-face and interview information was collected through a visit to the RNIB, an organisation that has been providing solutions for people who are visually impaired for over 30 years. One of the UK's leading providers of access technology working with and for people with sight problems, providing over 60 services; many of which are designed to support professionals in their work and the almost two million people with sight loss [27]. Initially, there were over 12 000 people registered to the 60 services being provided by RNIB. The visit to the organisation presented participants with two simple tasks to complete, trying as much as possible not to cross boundaries considering the ethical issues involved in the research. The procedure was to ask the visually impaired persons, who use the screen reader, to access the web (particularly the dynamic websites) followed by a questionnaire which was handed to them to complete. This resulted in a sample of 25 respondents out of the 30 screen reader users that were interviewed. However, since this method did not generate many responses and proved time-consuming, it was decided that an internet-mediated approach might prove beneficial in collecting more effective and less time-consuming data. To this end, a number of participants were sent a questionnaire via the internet and within 3 weeks of sending the requests 61 responses were obtained. A copy of the link to the questionnaire was attached into the body of the e-mail message being sent to the participants to complete [28] and the HTML code to the questionnaire was also copied and pasted to add the web link to any webpage (<a href='http://www.surveymonkey.com/s/HQLW7ND'>Click here to take survey). These methods allowed the researcher to reach a vast and diverse number of potential participants, as well as providing the respondents with easy access to the questionnaire thereby increasing the time and cost efficiency of the research.

Table 1 Table showing the percentage face-to-face response by the screen reader users

Responses	Number of completers	Percentage, %
confusing – finds it difficult to access dynamic websites	18	72
not very confusing	5	20
not confusing	2	8

3.3.1 Data analysis: A review of the responses we obtained from the participants was conducted and two results lists were generated based on the responses: one of persons who use the screen reader programmes to access the computer, and another sample of visually impaired individuals, designers, professionals and advisers who responded through the online survey. This resulted in a sample of 86 respondents and 14 no-responses from the participants on the overall web accessibility issues which were raised. A total of 100 questionnaires were distributed, both by face-to-face interview and through e-mail request and online forums.

3.4 Face-to-face interview with screen reader users

About 30 questionnaires were distributed to the participants who use the screen reading programmes to access the web at the RNIB research centre. Out of the 30 questionnaires that were distributed face-to-face 25 responded to the survey, representing about 83.3% of the distributed questionnaires.

From Table 1, our sample shows that 72% of the group finds it difficult to access dynamic web pages, 20% says it is not very confusing, and only 8% agree they can easily access the dynamic websites. Fig. 1 shows a pie chart representation of our face-to-face interview with the visually impaired users at RNIB.

3.5 Online survey

The online survey was highly successful in generating a large sample of participants in less time and cost. Messages requesting the participants to complete the questionnaire were sent to a handful of professionals, groups and individuals; yet in just 3 weeks 61 responses were received. Fig. 2 shows a screenshot of the response we gathered through our online survey using survey monkey [28].

The collection of data was a random one and no duplication was found in the data collected, as respondents provided their submissions online via the web-based survey system. A total of 70 questionnaires were sent out to the participants via e-mail attachment and web links. Out of the 70 questionnaires that were distributed to the participants 61 responded, representing about 87% of the distributed questionnaires. 68.9% of the respondents admitted that dynamic websites are confusing and inaccessible to the visually impaired, 21.3% say it is not very confusing while 9.8% believe they are not confusing. Fig. 3 shows a pie chart representation of the online survey in percentage.

Following the result in Fig. 2, the various data that we collected were organised in both tables and charts pertinent to the research findings and outcomes, as we have seen above. This was to ensure consistency and to enable comparison, to see if the method works and to identify factors that might lead to a bottleneck in the entire research process. In the next section, we analyse

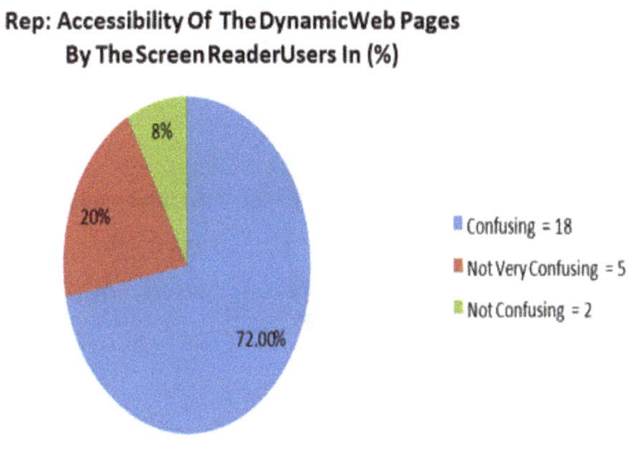

Fig. 1 *Result of the face-to-face respondents in percentage*

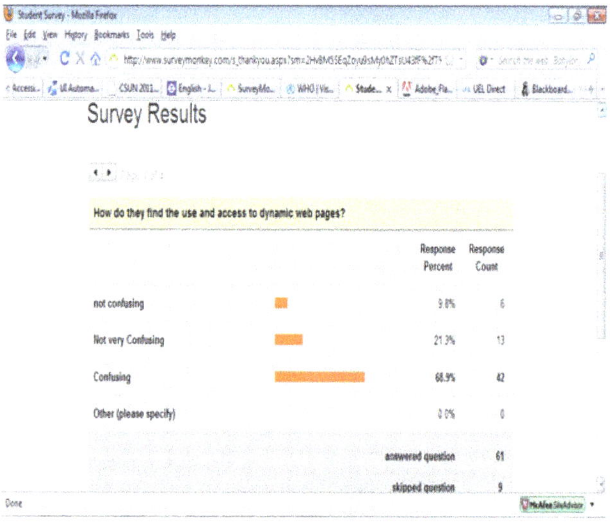

Fig. 2 *Screenshot of the online survey respondents and result*

the results of the face-to-face interview in relation to the online survey.

3.6 Resulting analysis

The results of the face-to-face and online survey were represented in a table as well as bar chart to enable us compare the result of the data we collected (see Fig. 4).

The results were calculated to help determine the overall mean (%) of responses which we represented in Table 2.

The result at the end of the research which provided a sample of 86 respondents in all, clearly shows that:
60 respondents = who says that dynamic web application are 'confusing' to the visually impaired.
18 respondents = says they are 'Not Very Confusing'.
8 respondents = says 'Not Confusing'.

In summary, Fig. 5 presented the results of the data analysis which shows 60 out of 86 participants say that dynamic web applications are confusing for the visually impaired users; which is more than 100% greater than the Not Very Confusing & Not Confusing respondents. This is evidence that dynamic web applications are to some extent inaccessible and confusing to the visually impaired users.

3.7 Research findings

Even though the visually impaired users were already used to navigating through web pages and other e-contents using a screen

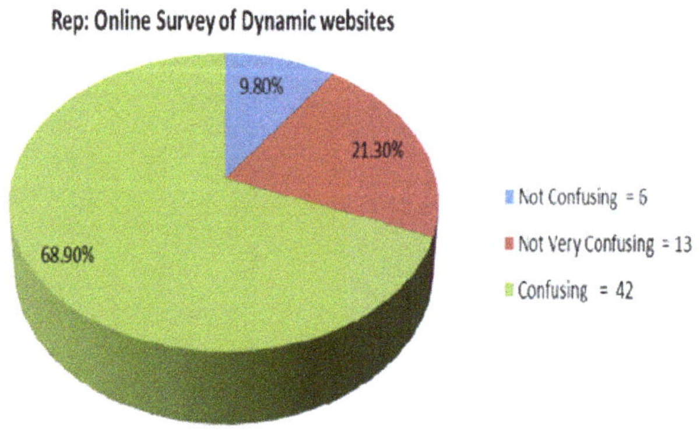

Fig. 3 *Result of the online survey respondents in percentage*

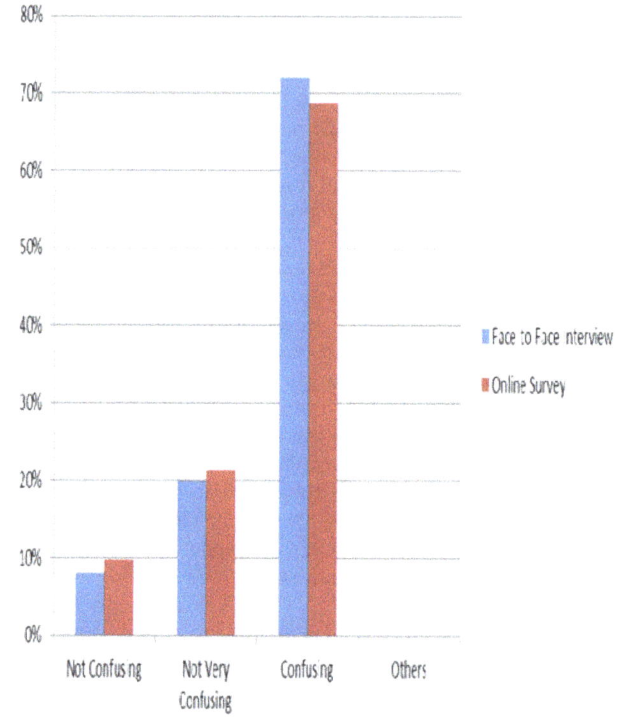

Fig. 4 *Bar chart representing the face-to-face plotted against the online survey results*

reader, being able to navigate through the content on the dynamic web pages was a completely novel experience for them; ranging from lack of further contextual information to limited control over the interaction and the amount of information that was being displayed, which on several occasions was confusing to them. These results indicate that the dynamic web pages are causing impediments to the visually impaired users when accessing the contents.

3.8 Implementation of solution

From the previous sections, it was seen that dynamic web applications are confusing and to some extent inaccessible to the visually impaired users. However, in this section, the research paper, as part of its contribution towards enhancing the use and access to the dynamic web applications by the visually impaired users, draws prototypes as well as implementation of a technical design approach intended towards the creation of an accessible dynamic website for the visually impaired, ensuring keyboard accessibility and providing accessible user interface control over the font size and page colours as well as voice over programmes.

3.8.1 Proposed website design: Developing accessible dynamic applications is not only feasible but increasingly practical. Creating dynamic websites and allowing the visually impaired users to access its pages requires two different approaches, which will be tackled in this section.

Table 2 Mean (%) of the face-to-face against online survey result analysis

	Not confusing, %	Not very confusing, %	Confusing, %	Others, %
face-to-face	8	20	72	0
online survey	9.8	21.3	68.9	0
mean, %	8.9	20.65	70.45	0

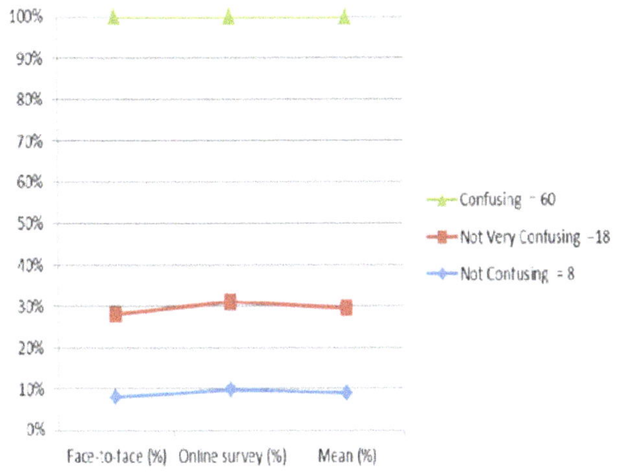

Fig. 5 *Line graph showing the mean (%) of the responses*

- The developers need an authoring tool to create the DHTML (dynamic) content and to package it into the product the visually impaired user will see.
- Secondly, the users generally need an assistive device such as screen reader to run or access the end product on their computers.

3.8.2 Technical high-level specification of our proposed website: To focus on applying the principles of accessibility to the dynamic website and to make sure the proposed new system is efficient, this paper chooses to use the following application software which is found to be more consistent and reliable than other authoring tools and works more readily with 'ActionScript' which supports testing for accessibility. Table 3 shows the software requirements for our proposed website and their functions in the design phase.

3.8.3 Structure of the web application in diagrammatic form and the homepage screenshot: Figs. 6 and 7 show the different pages as well as the enhanced legibility criteria embodied in our new website and the screenshot of the homepage, respectively.

Table 3 Software requirement for the new proposed website

Software requirements	Functions
Adobe Dreamweaver CS4 Professional	for designing the website graphical user interface, buttons and tables
Adobe Flash CS4 Professional	for designing the dynamic contents, moving images and pages
Adobe Fireworks CS5 Professional	firework is used for complete solution for creating, optimising and integrating web graphics, videos, sounds and rich texts
Adobe-Photo-shop CS4	Adobe Photo-shop was used for the design of navigation buttons and the top bar and also many other graphics that have been used in the design
ActionScript, version 3.0	coding and programming
Adobe Flash Player, version 10	for running and testing the App which is expected to run on any browser
JAWS 4.0	a screen reading programme for testing and implementing the whole application
Mozilla Firefox or Internet Explorer 8.0	since the browsers feed information about the web contents to the screen reading programme using MSAA. It is expected that any browser that supports MSAA; such as the Mozilla Firefox or Internet Explorer will be able to run the programme

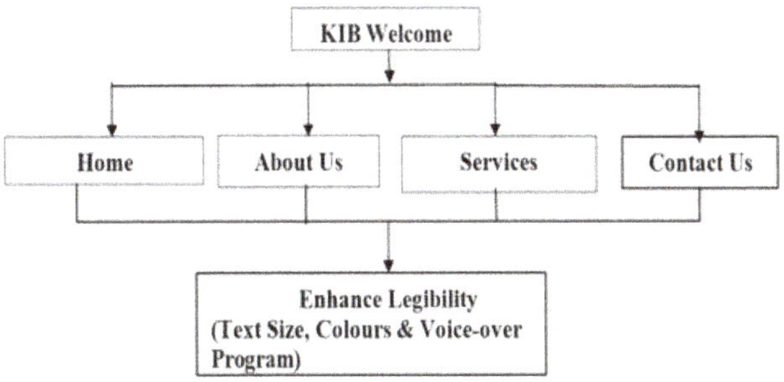

Fig. 6 *Control navigation of pages in our website named (KIB)*

3.9 Implementation test

By default text +9 objects in dynamic websites are already read and can also identify objects with a text label using the screen reader. It is up to the developer to add text equivalent to images and movies, which is a good design consideration when designing for the visually impaired user to provide further context. Setting the Tab index also allows the designer to control the reading order of the elements in the flash movie, which allows users to navigate using only the Tab and Enter keys on the keyboard. Information about the list of the available shortcuts as text equivalent was also provided or made available for the screen reader users through the help area button. In Adobe Flash CS4, combo and list boxes are already accessible, we only have to enable the accessibility object by using the command **enable.Accessibility().**

In Fig. 8 Text equivalent and Tab index number can be added using the Accessibility Panel or Action Script. To detect whether the screen reader is on, use the function Accessibility.isActive().

Flash CS4 Professional also offers a new component to display captions that are either contained in a World Wide Web consortium timed text, XML file (distribution format exchange profile (DFXP)) or integrated with Flash video (FLV) file as cue points. Custom screen magnifiers and an option to change the text size through font formatting also support the visually impaired user in accessing the dynamic contents.

3.9.1 Research contribution: This research paper, as part of its contributions towards the current debate on how best to design web applications for the visually impaired users, affirms that if designers would apply the combination of the no-frills and agile

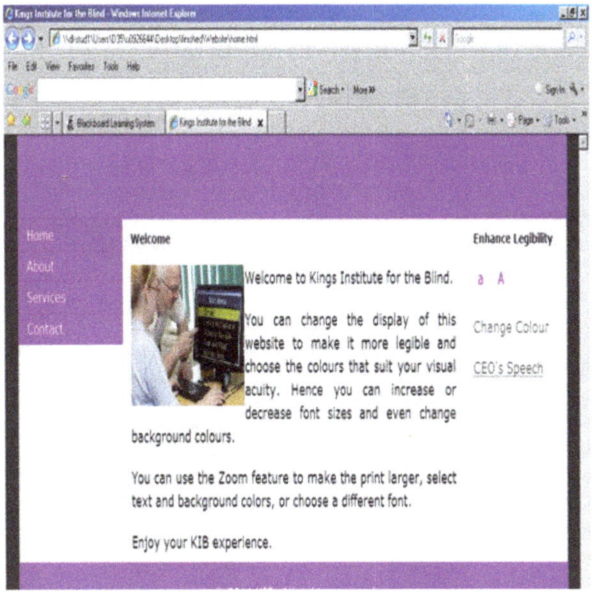

Fig. 7 *Screenshot of the new website homepage*

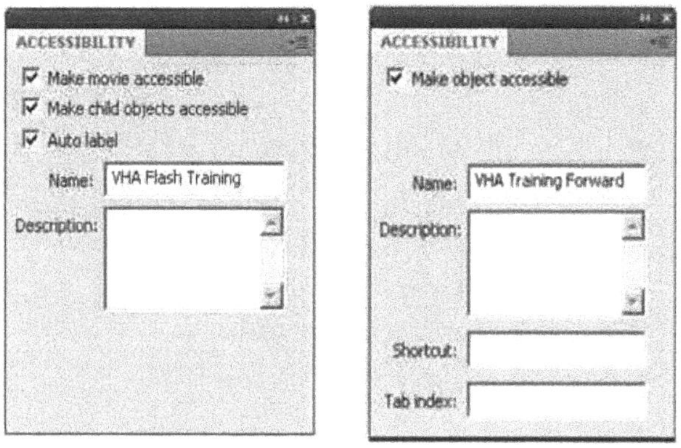

Fig. 8 *Accessibility panel*

methodology in correspondence to the article of the section 508 main requirement for software application development it will go a long way in solving the problem of accessing dynamic web pages by the visually impaired ICT users.

The research attempts to validate such theoretical impact by suggesting two types of methodology. One that suggests the removal of unnecessary and non-essential contents, new materials to be added to a content repository and re-use of already existing contents (the no-frills methodology), and another which identifies the potential users as well as the content type to create prototypes fitting the expected user's need and finally evaluates the developed product with test or heuristics to analyse its impact on the users (the agile methodology). Thus this research work affirms that:

> Only by designing a framework that is structurally flexible, by removing unnecessary extras thereby making every bit useful (fit-for-purpose) for the visually impaired users, will such persons be given an increased capacity to intuitively access e-contents.

The audio of a streaming video which automatically starts playing when a particular website is opened seems to interfere with the synthesised assistive voice generated from the screen reading software, which is confusing for the user with visual impairment [5, 11, 21, 22, 29]. Consequently, this study affirms that:

> Designers should be aware of how the screen reading software works to enable them make reasonable adjustments or provide alternative content that still corresponds to the objective content to increase the possibility of offering faultless service to the users with visual impairment.

Dynamic HTMLs such as DHTML and XML metadata have the ability of simplifying or reorganising contents, thereby providing or enabling additional data or navigation methods to be manually added to the contents, which may improve the accessibility of the dynamic web or multimedia applications made available for the visually impaired users. Thus, this paper proposes this last bit of its contribution in relation to this topic, that:

> Materials can be added to a content repository or re-used from an already existing one by identifying the content types and transforming them into a flexible and accessible one for the visually impaired (no-frill + agile methodology), rather than computing in advance or designing according to a given specification (Moore's model).

Content developers should have a basic understanding of how the screen readers obtain information through an application programming interface such as Microsoft active accessibility (MSAA) when accessing an application and then convey that information to the user. Although MSAA recognises and automatically conveys some types of information to the user, it is required in most cases that the content developers must perform specific steps to expose the information to MSAA so that it can be passed along to the user:

- The web application must have support for MSAA: the interface through which accessibility information is conveyed to AT.
- Provide scalable graphics and magnification capability for users with visual impairment.
- Ensure accessible video playback controls that allow users to stop, forward, rewind and pause presentations.
- Provide customised colour swatches that allow developers to design for users who are colour blind
- Mouse-free navigation that allows users to navigate via the keyboard.

3.9.2 From our practical section: The implementation shows some practical suggestions for creating accessible web contents that are exciting, engaging and usable for the visually impaired using the relevant software and authoring tools. Visually impaired users can access the best experience on the web, if developers will step outside of their frame of reference and keep all the technical user requirements of the visually impaired in mind; considering the fact that such users might not be able to use the mouse, might not see the fonts clearly or maybe not at all and may view contents in different colours or colourless.

4 Conclusion

First, this paper attempts to revitalise the research on providing accessible dynamic web applications for the visually impaired users. In doing so, many additional studies to the current debate and revelation of what is missing in the literature were sparked. Many dynamic web applications appear to be confusing as well as inaccessible, in general, to the users with visual impairment. Since the contents are dynamic and time-based, it can be challenging to make it accessible to AT users, such as screen readers. Careful developer involvement is required to ensure that the moving and flash contents are exposed in a logical manner. On the other hand, many of the steps required to achieve accessibility are not technically difficult, but one of the biggest challenges for dynamic website and software developers in general, is to understand the perspective of the visually impaired users as well as how the various screen reading programmes work and take that perspective into account during the early stages of development. The second major contribution of this paper is in its findings and the contribution to the body of knowledge towards provision of accessible web contents for such users with visual impairment. Results of this paper are consistent with prior literature about the impact of dynamic web pages on the visually impaired users, with an attempt to validate its proposal that if designers would apply the combination of the no-frills and agile methodology in correspondence to the article of the section 508 main requirement for software application development by designing a framework that is structurally flexible, removal of unnecessary extras thereby making every bit useful (fit-for-purpose), the visually impaired users will be given an increased capacity to intuitively access the dynamic web applications.

4.1 Future work

There are many areas where further research can be applied; the scope of the survey could focus on a slightly larger sample of dynamic web users with visual impairment to provide more authenticity. Research could also be applied to the various categories of dynamic web users with visual impairment, as the level of visual acuity in different users may also correlate to their level of access to dynamic web contents. Therefore, this could be another area of further research. Another option for future work is to evaluate dynamic web applications with other types of web users with different disabilities to determine their accessibility in that aspect.

5 Acknowledgments

The authors would like to thank the anonymous reviewers for their valuable comments.

6 References

[1] Garrigós I., Melia S., Casteleyn S.: 'Personalizing the interface in rich Internet applications', *J. Web Inf. Syst. Eng., WISE 2009. Lect. Notes Comput. Sci.*, 2009, **5802**, pp. 365–378

[2] Wright J.M., Dietrich J.B.: 'Requirements for rich Internet application design methodologies'. Proc. of the Ninth Int. Conf. on Web Information Systems Engineering, (WISE), 2008

[3] Saldaño V., Martin A., Gaetán G., Vilte D.: 'Web accessibility for older users: a southern argentinean view'. Proc. of the Eighth Int. Conf. on Software Engineering Advances ICSEA, 2013

[4] Ali L., Jahankhani H., Jahankhani H.: 'E-accessibility of higher education websites'. Presented at the ECEL Seventh European Conf. on e-Learning, Agia Napa Cyprus, 6–7 November 2008

[5] Petrie H., Power C., Velasco C.A., Boticario J.G.: 'Accessibility of blended and E-learning for mature age and disabled students and staff: introduction to the special thematic session', *Lect. Notes in Comput. Sci.*, 2010, **6179/2010**, pp. 484–485

[6] Swallow D., Petrie H., Power C.: 'Understanding and supporting the needs of educational professionals working with students with disabilities and mature age students'. Proc. of the Computers Helping People with Special Needs, Lecture Notes in Computer Science, 2010, vol. 6179/2010, pp. 486–491

[7] Santos O.C., Boticario J.G., delViso A.F., de laCámara S.P., Sánchez C.R., Restrepo E.G.: 'Basic skills training to disabled and adult learners through an accessible e-learning platform'. Proc. of the Universal Access in Human Computer Interaction. Applications and Services, Lecture Notes in Computer Science, 2007, vol. 4556/2007, pp. 796–805

[8] Ali L., Jahankhani H., Jahankhani H.: 'Accessibility evolution tools comparison'. Presented at the Second Annual Conf. on Advances in Computing and Technology, Proc. of the AC&T, UK, 2007

[9] NHS: 'Visual impairment: your health, your choices'. 2009. [Online] Retrieved from http://www.nhs.uk/conditions/Visual-impairment/Pages/Introduction.aspx, accessed 23rd March 2013

[10] RNIB: 'New literature on sight problems: changing the way we think about blindness'. Proc. of Supporting Blind and Partially sighted People, 2010, vol. 2010/84

[11] Veal D., Maj S.P.: 'A graphical user interface for the visually impaired – an evaluation', *J. Mod. Appl. Sci.*, 2010, **4**, (12), pp. 83–89

[12] Mulloy A.M., Gevarter C., Hopkins M., Sutherland K.S., Ramdoss S.T.: 'Assistive technology for students with visual impairments and blindness', In Lancioni G., Singh N. (ed.). *Assistive technologies for people with diverse abilities Psychopathology series*, 1st edn., (Springer, New York, 2014), pp. 113–156

[13] Adetoro N.: 'Reading interest and information needs of persons with visual impairment in Nigeria', *South Afr. J. Libr. Inf. Sci.*, 2010, **76**, (1), pp. 49–56

[14] Thylefors B.: 'A mission for vision'. Proc. of Lancet 354 Suppl, 1999, p. SIV44

[15] Sheng Y.Q., Liang X., Yang H., Wang Y.X., Jonas J.B.: 'Five-year incidence of visual impairment and blindness in adult Chinese', *Beijing Eye Study J. Ophthalmol.*, 2010, **118**, (6), pp. 1069–1075

[16] Asakawa C., Itoh T., Takagi H., Miyashita H.: 'Accessibility evaluation for multimedia content', *Lect. Notes Comput. Sci.*, 2007, **4556/2007**, pp. 11–19

[17] Hunsucker R.: 'Making life easier for the visually impaired web searcher: it is now clearer how this should and can be done, but implementation lags', *J. Evidence Based Libr. Inf. Pract.*, 2013, **8**, (1), pp. 90–93

[18] Freire A.P., Linhalis F., Bianchini S.L., Fortes R.P.M., Pimental M.C.: 'Revealing the whiteboard to blind students: an inclusive approach to provide mediation in synchronous e-learning activities', *J. Comput. Educ.*, 2010, **54**, (4), pp. 866–876

[19] Bocconi S., Dini S., Ferlino L., Martinoli C., Ott M.: 'ICT educational tools and visually impaired students: different answers to different accessibility needs'. Proc. of the Universal Access in Human–Computer Interaction. Applications and Services, Lecture notes in Computer science, 2007, vol. 4556/2007, pp. 491–500

[20] Avila J., Crowe E., Mendez M.L.: 'Creating dynamic, interactive, accessible flash'. Proc. of the 26th Annual Int. Technology & Persons with Disabilities Conf. Centre on disabilities, Cunningham Lindsey, Birmingham, UK, March 2011

[21] RNIB: 'Making your Teaching inclusive: what is visual impairment?', 2010. [Online] Available at http://www.open.ac.uk/inclusiveteaching/pages/inclusive-teaching/printed-materials.php, accessed 18 January 2014

[22] Syed T.A., SelCuk Candan K., Sangwoo H., Yan Qi.: 'Topic development pattern analysis-based adaptation of information spaces', *J. New Rev. Multimed.*, 2009, **15**, (1), pp. 73–96

[23] Zhao Z., Rau P., Zhang T., Salvendy G.: 'Visual search-based design and evaluation of screen magnifiers for older and visually impaired users', *Int. J. Hum.-Comput. Stud.*, 2009, **67**, (8), pp. 663–675

[24] Section 508 software design requirement. Available at http://www.pubbliaccesso.it/normative/DM080705-D-en.htm

[25] Kranjnc E., Feiner J., Schmidt S.: 'User centred interaction design for mobile applications: focused on visually impaired and blind people', *Lect. Notes Comput. Sci.*, 2010, **6389**, pp. 195–202

[26] Robson C.: 'Real world research: a resource for users of social research methods in applied settings' (Publication of John Wiley and Sons Ltd., West Sussex, UK, 2011, 3rd edn.)

[27] Web Design. Available at http://www.rnib.org.uk/professionals/webaccessibility/background/Pages/background.aspx

[28] Research Online Survey. Available at http://www.surveymonkey.com/s/HQLW7ND

[29] Asakawa C., Miyashita H., Sato D., Takagi H.: 'Aibrowser for multimedia: introducing multimedia content accessibility for visually impaired users'. Proc. of the Ninth Int. ACM SIGACCESS Conf. on Computers and Accessibility, Assets, New York, USA, 2007

3

New single-carrier transceiver scheme based on the discrete sine transform

Faisal Al-kamali

Department of Electrical, Faculty of Engineering and Architecture, IBB University, IBB, Yemen
E-mail: faisalalkamali@yahoo.com

Abstract: A discrete sine transform (DST)-based single-carrier transceiver scheme for broadband wireless communications is proposed and investigated. The proposed scheme uses a DST rather than the conventional discrete Fourier transform (DFT) as a basis function to implement the single-carrier system. The performance of the proposed scheme is studied and compared with the DFT-based single-carrier transceiver scheme and the discrete cosine transform based single-carrier transceiver scheme. Simulation results for single-carrier frequency division multiple access system are presented to demonstrate the effectiveness of the proposed scheme for broadband wireless communications.

1 Introduction

Wireless digital communication is rapidly expanding, resulting in a demand for wireless systems that are reliable and have a high spectral efficiency. In recent years, the single-carrier transmission technique such as single-carrier frequency division multiple access (SC-FDMA) has gained more and more attention when it comes to proposals for future wireless communication systems [1–4]. The main advantages of SC-FDMA system is the flexibility that arises from the fact that the total bandwidth is subdivided into many different subcarriers and the low peak-to-average power ratio (PAPR) as compared to the multicarrier systems [2]. This fact motivates the manufacturers to introduce this method in the uplink of long-term evolution (LTE), since reducing the sensitivity to non-linear amplification is of special relevance to mobile terminals [1]. There are two subcarrier mapping methods for the SC-FDMA system. The SC-FDMA system with blockwise subcarriers allocation is known as localised frequency division multiple access (LFDMA) [4]. The SC-FDMA system with regularly interleaved subcarriers allocation is also known as interleaved frequency division multiple access (IFDMA) [3].

In the literature, only the discrete Fourier transform (DFT) and the discrete cosine transform (DCT) were proposed to implement the SC-FDMA system [2–6]. The DFT- based SC-FDMA (DFT-SC-FDMA) system with the localised subcarrier mapping scheme has been adopted as the modulation and demodulation scheme of choice in the 3GPP LTE standard. In [5, 6], it was shown that the DCT-based SC-FDMA (DCT-SC-FDMA) system provides better performance than that of the DFT-SC-FDMA system, even in the presence of carrier frequency offset. Moreover, its PAPR is lower than that of the orthogonal frequency division multiple access (OFDMA). Up to now, the DST-based SC-FDMA (DST-SC-FDMA) system is not studied in the literature. DST uses only real arithmetics rather than the complex arithmetics used in the DFT. This reduces the signal processing complexity, and the in-phase/quadrature imbalance.

In this paper, we introduce a new single-carrier transceiver scheme based on the DST. The proposed transceiver uses a DST, rather than the DFT or the DCT, to implement the SC-FDMA system. The proposed scheme is described and its model is derived. Then, the bit error rate (BER) and the PAPR performances of the proposed DST-SC-FDMA scheme are studied and compared with the existing schemes. In contrast to the conventional DFT-SC-FDMA system, it is found that DST-SC-FDMA provides good BER performance and an acceptable PAPR performance, especially with the localised mapping scheme.

The rest of the paper is organised as follows. Section 2 introduces an overview about DST. Section 3 derives the system model of the proposed DST-SC-FDMA system. Section 4 briefly introduces the DFT-SC-FDMA system. Section 5 is about the DCT-SC-FDMA system. The PAPR problem is discussed in Section 6. Experimental results are given in Section 7. Finally, Section 8 concludes the paper.

2 Discrete sine transform (DST)

In mathematics, the DST is a Fourier-related transform similar to the DFT, but using a purely real matrix. It is equivalent to the imaginary parts of a DFT of roughly twice the length, operating on real data with odd symmetry. DSTs are widely employed in solving partial differential equations by spectral methods, where the different variants of the DST correspond to slightly different odd/even boundary conditions at the two ends of the array [7]. There are several types of the DST with slightly modified definitions. In this Letter, DST-I is considered. For simplicity, it is denoted by DST. The DST is given by

$$y(k) = \sum_{n=1}^{N} x(n)\sin\left(\pi\frac{kn}{N+1}\right), \quad k = 1, 2, \ldots, N \quad (1)$$

The inverse DST (IDST) is given by

$$x(n) = \frac{2}{N+1}\sum_{k=1}^{N} y(k)\sin\left(\pi\frac{kn}{N+1}\right), \quad n = 1, 2, \ldots, N \quad (2)$$

3 DST-SC-FDMA system

The complex exponential functions set are not the only orthogonal basis that can be used to construct baseband single-carrier signals. A single set of cosinusoidal and sinusoidal functions can be used as an orthogonal basis to implement the single-carrier scheme [5]. Fig. 1 describes the transceiver block diagram of the DST-SC-FDMA system. At the transmitter, the input data sequence of the uth user is encoded. The coded bits are mapped to multilevel symbols in one of modulation formats such as quadrature phase shift keying (QPSK), and 16-quadrature amplitude modulation (16QAM). The modulated symbols are grouped into blocks, each containing N symbols, followed by an N-point DST. The subcarrier mapping block assigns the DST outputs into M ($\geq N$) subcarriers that can be transmitted, and inserts zeros into any unused subcarriers. After performing an M-point IDST, a cyclic prefix (CP) is appended at the head of IDST outputs.

In matrix notation, the transmitted signal of the uth user (u = 1,2, …, U) can be formulated as follows

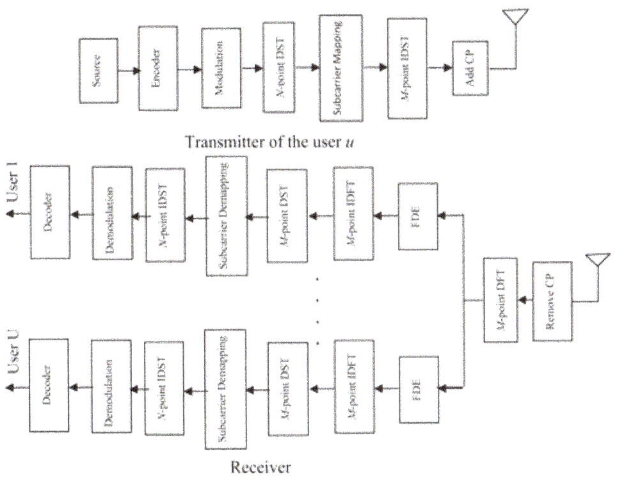

Fig. 1 *Structure of the DST-SC-FDMA system over a frequency selective channel*

$$\tilde{\boldsymbol{x}}_u = \boldsymbol{\Pi} \boldsymbol{S}_M^{-1} \boldsymbol{\Gamma}_u \boldsymbol{S}_N \boldsymbol{x}_u \tag{3}$$

where $\boldsymbol{x}_u$ is an $N \times 1$ vector containing the modulated symbols of the uth user. $\boldsymbol{S}_N$ is an $N \times N$ DST matrix. $\boldsymbol{S}_M^{-1}$ is an $M \times M$ IDST matrix. $\boldsymbol{\Gamma}_T^u$ is an $M \times N$ matrix describing the subcarriers mapping of the uth user. $M = Q \cdot N$, where Q is the maximum number of users that can transmit, simultaneously. $\boldsymbol{\Pi}$ is an $(M+N_C) \times M$ matrix, which adds a CP of length N_C.

The entries of $\boldsymbol{\Gamma}_u$ for both the localised and the interleaved systems are given in (4) and (5), respectively:

$$\boldsymbol{\Gamma}_u = \left[\boldsymbol{0}_{(u-1)N \times N}; \boldsymbol{I}_N; \boldsymbol{0}_{(M-uN)\times N}\right] \tag{4}$$

$$\boldsymbol{\Gamma}_u = \left[\boldsymbol{0}_{(u-1)\times N}; \boldsymbol{u}_1^{\mathrm{T}}; \boldsymbol{0}_{(Q-u)\times N}; \ldots; \boldsymbol{0}_{(u-1)\times N}; \boldsymbol{u}_N^{\mathrm{T}}; \boldsymbol{0}_{(Q-u)\times N}\right] \tag{5}$$

where $\boldsymbol{I}_N$ and $\boldsymbol{0}_{Q' \times N}$ matrices denote the $N \times N$ identity matrix and the $Q' \times N$ all-zero matrix. $\boldsymbol{u}_l$ ($l = 1, 2, \ldots, N$) denotes the unit column vector, of length N, with all zero entries except at l. $\boldsymbol{\Pi}$ can be represented as follows

$$\boldsymbol{\Pi} = [\boldsymbol{C}, \boldsymbol{I}_M]^{\mathrm{T}} \tag{6}$$

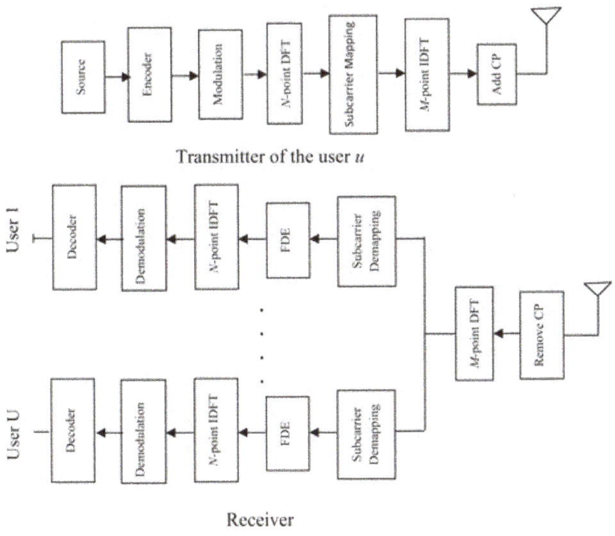

Fig. 2 *Structure of the DFT-SC-FDMA system over a frequency selective channel*

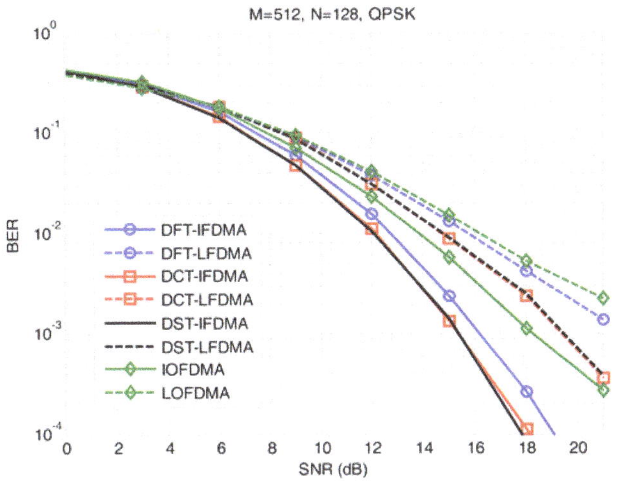

Fig. 3 *BER against the SNR for DFT-SC-FDMA, DCT-SC-FDMA, DST-SC-FDMA and OFDMA systems with different subcarrier mapping schemes and QPSK*

where

$$\boldsymbol{C} = \left[\boldsymbol{0}_{N_C \times (M-N_C)}, \boldsymbol{I}_{N_C}\right]^{\mathrm{T}} \tag{7}$$

At the receiver side, the CP is removed from the received signal and the received signal can be written as follows

$$\boldsymbol{r} = \sum_{u=1}^{U} \boldsymbol{H}_u \bar{\boldsymbol{x}}_u + \boldsymbol{n} \tag{8}$$

where $\bar{\boldsymbol{x}}_u$ is an $M \times 1$ vector representing the block of the transmitted symbols of the uth user. $\boldsymbol{H}_u$ is an $M \times M$ circulant matrix describing the multipath channel between the uth user and the base station. $\boldsymbol{n}$ is an $M \times 1$ vector describing the additive noise. Applying the DFT, we obtain

$$\boldsymbol{R} = \sum_{u=1}^{U} \boldsymbol{\Lambda}_u \boldsymbol{F}_M \bar{\boldsymbol{x}}_u + \boldsymbol{N} \tag{9}$$

where $\boldsymbol{\Lambda}_u$ is an $M \times M$ diagonal matrix containing the DFT of the circulant sequence of $\boldsymbol{H}_u$. $\boldsymbol{N}$ is the DFT of $\boldsymbol{n}$. $\boldsymbol{F}_M$ is an $M \times M$ DFT

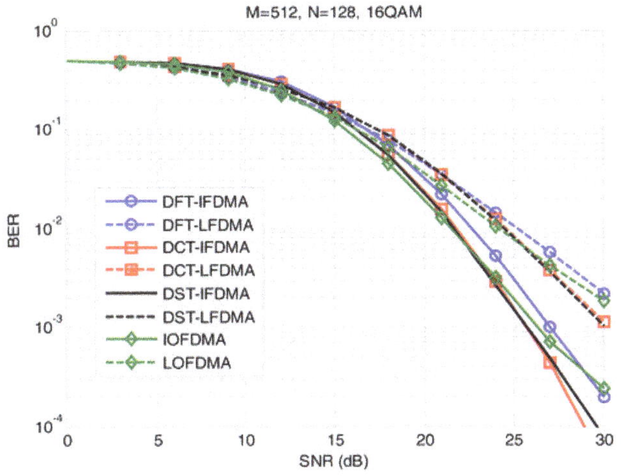

Fig. 4 *BER against the SNR for DFT-SC-FDMA, DCT-SC-FDMA, DST-SC-FDMA and OFDMA systems with different subcarrier mapping schemes and 16QAM*

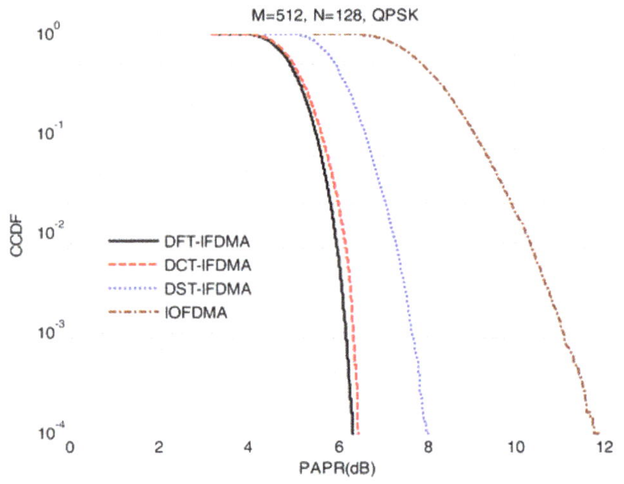

Fig. 5 *CCDFs of the PAPR for DFT-IFDMA, DCT-IFDMA, DST-IFDMA and IOFDMA systems with QPSK*

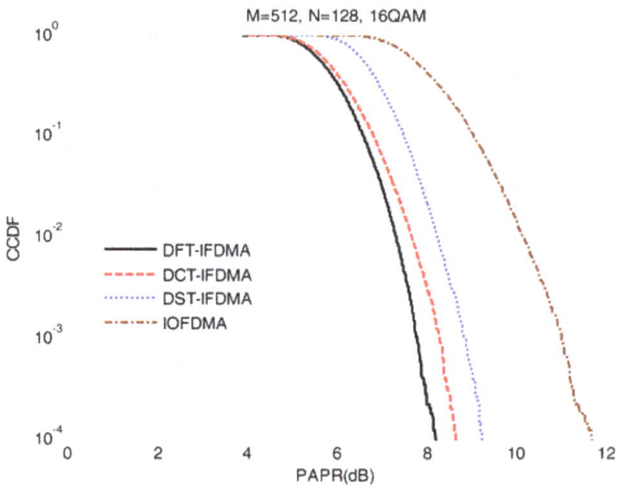

Fig. 7 *CCDFs of the PAPR for DFT-IFDMA, DCT-IFDMA, DST-IFDMA and IOFDMA systems with 16QAM*

matrix. The generic M-point DFT matrix has entries $[\boldsymbol{F}_M]_{p,\ q} = e^{-j2\pi(pq/M)}$, and its inverse DFT (IDFT) is $\boldsymbol{F}_M^{-1} = (1/M)\boldsymbol{F}_M^H$.

After that, the FDE, the M-point IDFT and the DST-SC-FDMA demodulation operations are performed to provide the estimate of the modulated symbols as follows

$$\hat{x}_u = \boldsymbol{S}_N^{-1}\boldsymbol{\Gamma}_u^{\mathrm{T}}\boldsymbol{S}_M\boldsymbol{F}_M^{-1}\boldsymbol{W}_u\boldsymbol{R} \qquad (10)$$

where $\boldsymbol{W}_u$ is the $M \times M$ FDE matrix of the uth user. Finally, the demodulation and the decoding processes are applied.

4 DFT-SC-FDMA system

A conventional DFT-SC-FDMA transceiver block diagram is shown in Fig. 2. There are U uplink users communicating with a base station through independent multipath fading channels. A total of M subcarriers is assumed and each user is assigned N subcarriers. In the DFT-SC-FDMA transmitter, the encoded signals are modulated and then transformed into the frequency domain via an N-point DFT. Then, the subcarriers are mapped in the frequency domain. After that, IDFT is performed, and a CP is added to the resulting signal. Finally, the resulting signal is transmitted through the wireless channel.

At the receiver, the CP is removed and the DFT is then applied. Finally, the subcarrier demapping, the equalisation, the IDFT, the demodulation and the decoding operations are performed.

5 DCT-SC-FDMA system

The structure of the DCT-SC-FDMA system is similar to that of the DST-SC-FDMA system in Section 3. The difference is that the DST and the IDST blocks at the transmitter and receiver are replaced by the DCT and the IDCT blocks, respectively. More details about the DCT-SC-FDMA system are found in [5].

In this paper, the DFT-based IFDMA is denoted by DFT-IFDMA, the DFT-based LFDMA is denoted by DFT-LFDMA, the DCT-based IFDMA is denoted by DCT-IFDMA, the DCT-based LFDMA is denoted by DCT-LFDMA, the DST-based IFDMA is denoted by DST-IFDMA, the DST-based LFDMA is denoted by DST-LFDMA, the DFT-based interleaved OFDMA is denoted by IOFDMA, and the DFT-based localised OFDMA is denoted by LOFDMA.

6 Peak power problem

The peak power problem causes the non-linear distortion in the power amplifier and reduces power efficiency [8]. The metric used to measure the impact of this problem is the PAPR. PAPR

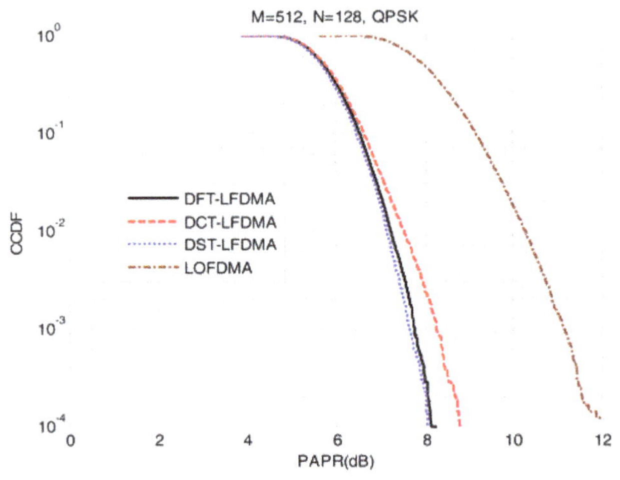

Fig. 6 *CCDFs of the PAPR for DFT-LFDMA, DCT-LFDMA, DST-LFDMA and LOFDMA systems with QPSK*

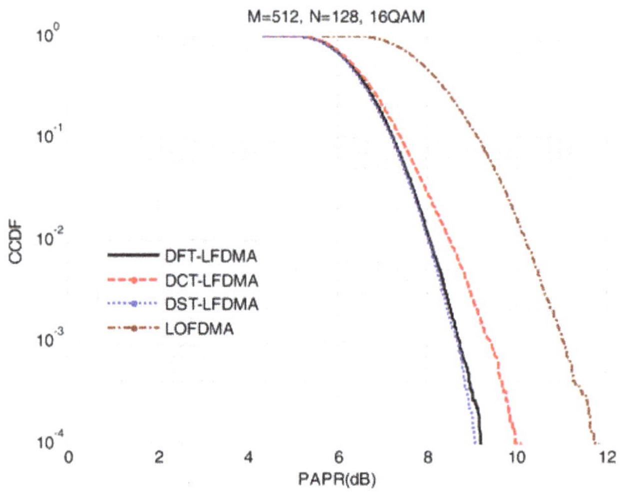

Fig. 8 *CCDFs of the PAPR for DFT-LFDMA, DCT-LFDMA, DST-LFDMA and LOFDMA systems with 16QAM*

Table 1 PAPR values at a CCDF = 10^{-4} for different systems

Modulation format, dB	DFT-IFDMA, dB	DFT-LFDMA, dB	DCT-IFDMA, dB	DCT-LFDMA, dB	DST-IFDMA, dB	DST-LFDMA, dB	IOFDMA and LOFDMA, dB
QPSK	6.5	8	6.6	8.8	8	8	11.75
16-QAM	8.25	9.25	8.75	10	9.25	9.25	11.75

is a commonly used measure of the range of a signal's amplitude. It is a reasonably good qualitative measure; signals with low PAPR generally require less power backoff and exhibit less performance sensitivity when amplified by a non-linear power amplifier than do signals with high PAPR [8]. An informative metric is the complementary cumulative distribution (CCDF) function of the signal amplitude measured over many samples. CCDF is the probability that the PAPR is higher than a certain PAPR value. The PAPR in dB can be expressed as [8]

$$\text{PAPR(dB)} = 10.\log_{10}\left(\frac{\max\left(|x(m)|^2\right)}{(1/M)\sum_{m=0}^{M-1}|x(m)|^2}\right) \quad (11)$$

where $x(m)$ is the mth symbol of the transmitted signal.

7 Simulation results

7.1 Simulation parameters

Monte Carlo simulation is used to evaluate the performance of the proposed DST-SC-FDMA system with different subcarriers mapping methods. For comparison purposes, the DFT-SC-FDMA, the DCT-SC-FDMA and the OFDMA systems are also simulated. In the simulated DST-SC-FDMA system, each user occupies 128 subcarriers. The total number of subcarriers $M = 512$ and the number of users $U = 4$. In each Monte Carlo realisation, all subcarriers are assigned among all users according to the subcarriers mapping method used. QPSK and 16QAM modulation schemes are used to generate a transmitted block for each user. The channel model used for simulations is the vehicular A model [9]. A convolutional code with memory length 7 and octal generator polynomials (133, 171) is chosen as the channel code.

7.2 BER performance

The BER performances for DST-SC-FDMA, DCT-SC-FDMA, DFT-SC-FDMA and OFDMA systems are simulated in Figs. 3 and 4 for the QPSK and the 16QAM, respectively. For each plot, it is clear that the performance of the DST-SC-FDMA system is better than that of the DFT-SC-FDMA and the OFDMA systems, especially with the QPSK. From the figures, it can be seen that the DST-SC-FDMA and the DCT-SC-FDMA systems provide the same performance with both the interleaved and the localised methods. At a BER = 10^{-3} with QPSK, the performance gain is about 2 dB for the DST-LFDMA system, and about 1 dB for the DCT-IFDMA system when compared to that of the DFT-LFDMA system and the DFT-IFDMA system, respectively. This is attributed to the energy concentration property of the DST.

7.3 PAPR performance

Figs. 5 and 6 illustrate the CCDFs of the PAPR for DST-SC-FDMA, DFT-SC-FDMA, DCT-SC-FDMA and OFDMA signals for the case of QPSK and for different subcarrier mapping schemes. Raised cosine pulse shaping filter with roll-off factor = 0.2, and 4 times oversampling is used. From Fig. 5, it is observed that the DST-IFDMA provide considerable gain in PAPR when compared to that of the IOFDMA system. However, it provides high PAPR when compared with that of the DFT-IFDMA and the DCT-IFDMA systems. From Fig. 6, it is shown that the PAPR of the DST-LFDMA signal is comparable to that of the conventional DFT-LFDMA and it is better than that of the DCT-LFDMA and the LOFDMA systems.

Figs. 7 and 8 plot the CCDF of the PAPR for DST-SC-FDMA, DFT-SC-FDMA, DCT-SC-FDMA and OFDMA signals for the case of 16QAM. Raised cosine pulse shaping filter with roll-off factor = 0.2, and 4 times oversampling is used. It is clearly seen that the PAPR of the DST-LFDMA is lower than that of the DCT-LFDMA and the LOFDMA systems. It is also noted that the PAPR of the DST-IFDMA is greater than that of the DFT-IFDMA and the DCT-IFDMA by about 1 and 0.5 dB, respectively.

Since the localised mapping is the subcarriers mapping method of the LTE standard and based on the previous figures, the proposed DST-LFDMA is more suitable for the future wireless communications since it provides better BER performance than that of the conventional DFT-LFDMA system and it has the same PAPR as that of the conventional DFT-LFDMA system.

A full comparison of PAPR at a CCDF = 10^{-4} for different modulation formats is shown in Table 1. From this table, it is clear that the DST-LFDMA system and the DFT-LFDMA system provide the same PAPR performance whereas the PAPR of the DST-IFDMA system is higher than that of the DFT-IFDMA system by about 1.5 dB and 1 dB for the QPSK and 16QAM, respectively.

8 Conclusion

In this paper, we propose a new transceiver scheme based on the DST for future wireless communications, namely DST-SC-FDMA. The system model of the DST-SC-FDMA system has been derived and its performance has been studied and compared with the existing systems. Simulation results have been shown that the DST-SC-FDMA system provides superior BER performance than the DFT-SC-FDMA and the OFDMA systems. It is also found that the BER performance of the DST-SC-FDMA system is the same as that of the DCT-SC-FDMA system. Results have also been demonstrated that the PAPR performance of the DST-LFDMA system is the same as that of the DFT-LFDMA system and better than that of the DCT-LFDMA and the OFDMA systems.

9 References

[1] 3rd Generation Partnership Project (3GPP) Technical Specification Group Radio Access Network; Physical Layer Aspects for Evolved Universal Terrestrial Radio Access (UTRA) (Release 7). 3GPP TR 25.814, V7.1.0, September 2006

[2] Myung H.G., Goodman D.J.: 'Single-carrier FDMA a new air interface for long term evolution' (John Wiley & Sons, Ltd., 2008)

[3] Myung H.G., Lim J., Goodman D.J.: 'Single-carrier FDMA for uplink wireless transmission', *IEEE Veh. Technol. Mag.*, 2006, **1**, (3), pp. 30–38

[4] Al-kamali F.S., Dessouky M.I., Sallam B.M., Shawki F., Al-Hanafy W., Abd El-Samie F.E.: 'Joint low-complexity equalization and carrier frequency offsets compensation scheme for MIMO SC-FDMA systems', *IEEE Trans. Wirel. Commun.*, 2012, **11**, (3), pp. 869–873

[5] Al-kamali F.S., Dessouky M.I., Sallam B.M., Abd El-Samie F.E., Shawki F.: 'A new single-carrier FDMA system based on the discrete

cosine transform'. ICCES'9 Conf., Cairo, Egypt, 14–16 December 2009, pp. 555–560

[6] Al-kamali F.S., Dessouky M.I., Sallam B.M., Shawki F., Abd El-Samie F.E.: 'Equalization and CFOs compensation for the SC-FDMA system with different basis functions'. *Wirel. Pers. Commun.*, 2012, **67**, pp. 113–138

[7] http://en.wikipedia.org/wiki/Discrete_sine_transform

[8] Al-kamali F.S., Dessouky M.I., Sallam B.M., Abd El-Samie F.E., Shawki F.: 'Transceiver scheme for SC-FDMA Implementing the wavelet transform and the PAPR reduction methods', *IET Commun.*, 2010, **4**, (1), pp. 69–79

[9] 3rd Generation Partnership Project, 3GPP TS 25.101 – Technical Specification Group Radio Access Network; User Equipment (UE) Radio Transmission and Reception (FDD) (Release 7), Section B.2.2, September 2007

4

Transparent settlement model between mobile network operator and mobile voice over Internet protocol operator

Luzango Pangani Mfupe[1,2], Mjumo Mzyece[1], Anish Mathew Kurien[1]

[1]Department of Electrical Engineering, FSATI, Tshwane University of Technology, Private Bag X680, Pretoria 0001, South Africa
[2]Meraka Institute, Council for Scientific and Industrial Research (CSIR), PO Box 395, Pretoria 0001, South Africa
E-mail: Lmfupe@csir.co.za

Abstract: Advances in technology have enabled network-less mobile voice over internet protocol operator (MVoIPO) to offer data services (i.e. voice, text and video) to mobile network operator's (MNO's) subscribers through an application enabled on subscriber's user equipment using MNO's packet-based cellular network infrastructure. However, this raises the problem of how to handle interconnection settlements between the two types of operators, particularly how to deal with users who now have the ability to make 'free' on-net MVoIP calls among themselves within the MNO's network. This study proposes a service level agreement-based transparent settlement model (TSM) to solve this problem. The model is based on concepts of achievement and reward, not violation and punishment. The TSM calculates the MVoIPO's throughput distribution by monitoring the variations of peaks and troughs at the edge of a network. This facilitates the determination of conformance and non-conformance levels to the pre-set throughput thresholds and, subsequently, the issuing of compensation to the MVoIPO by the MNO as a result of generating an economically acceptable volume of data traffic.

1 Introduction

Traditionally, in mobile networks, off-net traffic is considered to be circuit-switched traffic that originates from a mobile network operator (MNO) subscriber, terminated to other networks through a gateway (point of interconnection). The point of interconnection is essentially where call information is collected for the purpose of inter-operator settlements. On-net traffic is considered to be an internal form of circuit-switched traffic that flows between subscribers of the same MNO without passing through a point of interconnection [1]. However, in a mobile voice over internet protocol (MVoIP) context, an MVoIP operator (MVoIPO) such as Viber [2], does not own any network infrastructure but uses software applications to provide services within an MNO's network. In this particular context, the off-net traffic is considered to be packet-switched traffic which originates from an MVoIP user who is subscribed to a packet-based mobile network, terminated by the same mobile network, or other networks, to a non-MVoIP user through a point of interconnection.

In contrast, on-net, or peer-to-peer (P2P), traffic is considered to be packet-switched traffic which has originated from one MVoIP user who is subscribed to a mobile network, terminated at another MVoIP user subscribed in the same or a different mobile network and, vice versa, without passing through a point of interconnection. Additionally, two MVoIP users from the same MVoIPO can communicate in P2P mode regardless of which mobile network they are subscribed to. It is worth noting that the cost of an on-net mobile voice call is considered to be equal to the total sum of the call origination cost and the call termination cost [3]. However, the fact that on-net MVoIP traffic traverses to and from the application-enabled user equipment (UE) within an MNO's network, without necessarily passing through a point of interconnection, raises a fundamental problem of how the primary operator (PO) and the secondary operator (SO) (in this case the MNO and MVoIPO, respectively) can coexist with mutual economic benefits. This problem cannot easily be solved through traditional settlement mechanisms, and sometimes it is the source of conflicts between the two operators. A classic example is a case in the United Kingdom where the PO known as T-mobile decided to disable the internet telephony feature from its subscribers' handsets blocking them from accessing MVoIP services provided by an SO known as Truphone [4]. Throughout this paper we use the following terms interchangeably; on-net and P2P referring to MVoIP calls made and terminated within an MNO's network. MNO and MVoIPO, respectively, refer to PO and SO.

1.1 Establishment of on-net MVoIP calls within third generation partnership project (3GPP) universal mobile telecommunications system (UMTS) network

Fig. 1 shows how a P2P mobile VoIP call is established between two UEs subscribed to the MNO within a 3GPP UMTS network [5] [The UMTS network architecture is used for illustration purposes only. The model proposed in this paper can be applied in any third generation and beyond packet-based cellular network architecture(s).]. Both application-enabled UEs must also be registered with the MVoIPO via a session initiation protocol (SIP) [6]. This protocol is central to the concept of internet multimedia subsystem envisaged to effectively enable the unbundling of voice services in the data-based next generation mobile networks such as 3G UMTS and the 4G long-term evolution (LTE) [7]. SIP is often utilised to facilitate the end-to-end application layer interaction between the server and UE. Afterwards, either of the UEs can initiate a P2P call to the other by sending a call request (SIP signalling message) that may contain information such as codec type and session type via the proxy server. The SIP proxy server detached from the mobile network infrastructure (i.e. can be hosted in the cloud) routes the call request across the UMTS terrestrial radio access network (UTRAN), via the gateway general packet radio service (GPRS) support node (GGSN) and the Serving GPRS SN (SGSN), to the corresponding UE. Subsequently, the requested UE might send a reply (SIP signalling message) via the same route. Subsequently, an end-to-end data tunnel is established at the UMTS user-plane. Packet data carrying the actual voice conversation between the two calling parties (UE 'a' and UE 'b') will be tunnelled back and forth across the UTRAN, the GGSN and the SGSN.

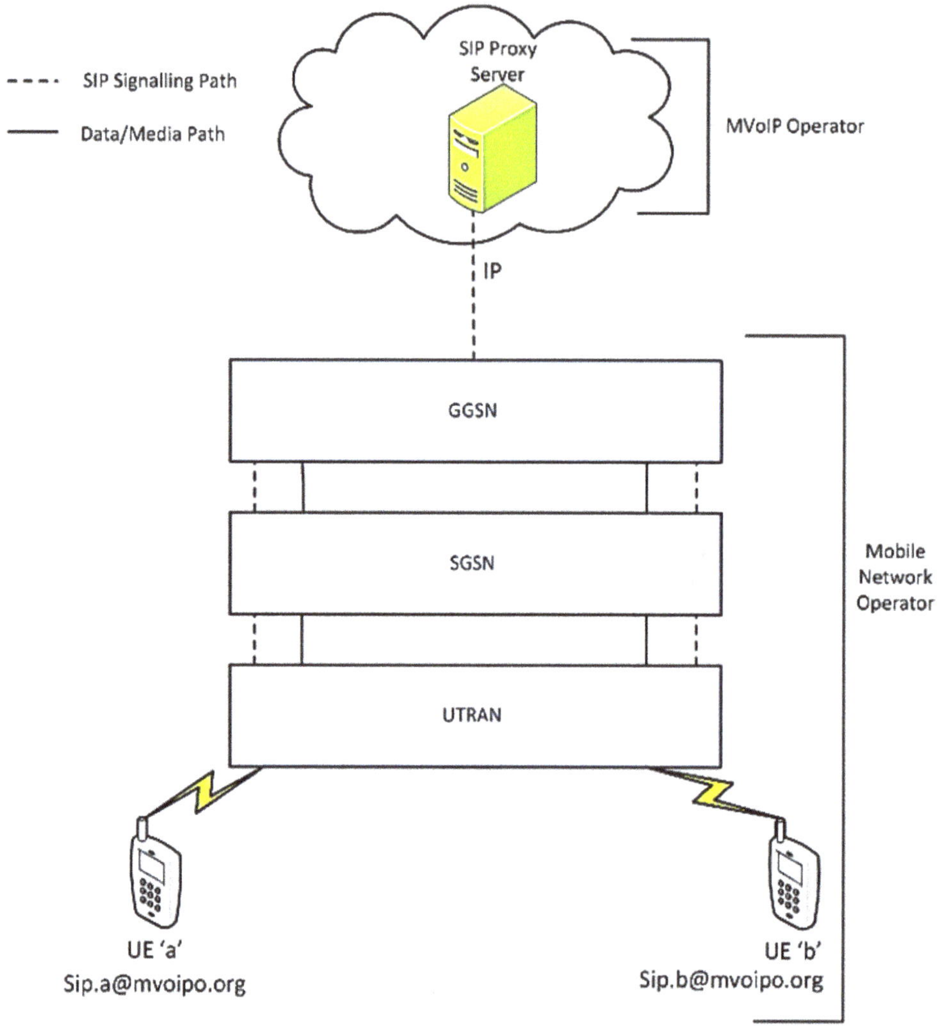

Fig. 1 *Illustration of on-net MVoIP call flow within UMTS network*

1.2 Contributions of this paper

Decoupling of transport and service has introduced new types of network-less operators at the 'edge' of the network (particularly in the application layer) that can offer best-effort complementary or competing services to the ones offered by the network owner [6]. A recent report [8] has indicated that mobile data traffic grew by 81% in 2013 and forecasted that this trend will surpass 15.9 Exabyte by 2018 which is 11-fold over 2013. Application-enabled MVoIP traffic contributed a significant amount of data in this growth [8]. This paradigm shift comes with profound challenges to the existing interconnection arrangements because the network owner has little or no visibility to the types of services offered by the network users [9]. It is noteworthy that most research on MVoIP tends to mainly focus on how a network user should be charged by the network owner for using network resources (i.e. methods for compositing, aggregating, correlating and description of charging records) or how to structure a bilateral cost sharing arrangement between network user and network owner. This is not directly relevant to how the on-net MVoIP traffic should be settled between the MNO and MVoIPO. Furthermore, we note in [10] that a network user faces consequences from the network owner, for over-utilisation of network resources (exceeding the pre-set service level agreement (SLA) thresholds), where it is rewarded for under-utilisation of network resources. To the best of our knowledge there is not yet a practical model that adequately considers the interconnection settlement mechanism that addresses the coexistence between MNO and network-less MVoIPO, particularly the key issue of how to handle P2P MVoIP calls originated and terminated within an MNO's network. We argue that a network-less operator offering services over internet protocol (IP) only, such as an MVoIPO, should be compensated by the network owner, the MNO, for generating an economically viable amount of extra data traffic. This is because this extra traffic is because of the joint efforts of both operators. More specifically, this paper makes two main contributions:

- Proposing a new inter-operator settlement scheme, the transparent settlement model (TSM), based on flexible compensation, that is, with multiple possible forms of monetary or non-monetary compensation paid by the network owner (the MNO) to the network user (the MVoIPO).
- Replacing the two central concepts of conventional SLA-based schemes, namely violation and punishment (i.e. users who violate the set SLA thresholds are punished by having their services restricted or removed), with two new concepts, achievement and reward. In this new paradigm, an MVoIPO should be compensated for generating an economically acceptable amount of data traffic (i.e. when an MVoIPO goes above the set SLA thresholds, it gets rewarded instead of getting punished). This is based on the fact that the MNO benefits from the internet data bundle charges paid by users of the MVoIP application.

The rest of this paper is arranged as follows. Section 2 highlights related work. Section 3 presents the TSM framework and design.

The simulation experiments, results and analysis are discussed in Section 4. Section 5 concludes this paper.

2 Related work

There is a significant body of literature and approaches dedicated to the settlements schemes of different types of operators in different layers of IP-based networks that cover the technical, economic and regulatory aspects of interconnection [10–12]. Reviews of pricing schemes and emerging charging methods for services offered by disparate IP-based operators are found in [13–16]. Techno-economic analysis of dynamic pricing framework for secondary users accessing LTE networks using dynamic spectrum access techniques [17]. Approaches and concepts that have a more direct implication to this paper include but are not limited to the following:

- 'Parameter-based static pricing schemes in dynamic environment [10, 16–18]' provide a mechanism to predefine usage prices with respect to the SLA contract thereby allowing each network user to be charged in a specific class of services. The most interesting method in this category is the cumulus pricing scheme (CPS); it is based on a flat-rate-type contract between the network user and the network owner such that the network user specifies the expected resource usage.

The SLA is monitored periodically by the feedback mechanism (cumulus points) in the pre-set time scales. During these pre-set times if the user (SO) violates the contract by exceeding the pre-set usage thresholds the network owner (PO) will take an action such as warning or contract re-negotiation. The CPS utilises IP traffic characteristics, such as the mean and standard deviation, as a way of solving the feasibility problem of internet tariffing encountered by internet service providers (ISPs) when pricing multi-service networks in the internet. The feasibility problem refers to the trade-off between customers and their interest in predictable and transparent tariffs against ISPs interest in achieving economic efficiency in operating the network and technical efficiency for the accounting operation that is characterised by predefined prices.

- 'The digital marketplace approach [19]' reduces the risk of conflicts among stakeholders in the wireless network business value chain. Broadly, the authors argue that the adoption of IP as a transport mechanism of choice might be a source of conflict among stakeholders. For example, a conflict may arise when a mobile user demands certain services which are prohibited by an MNO but can be made available by a third party provider through the application layer. Therefore the authors proposed an open market-based approach that would effectively allow MNOs and third party operators to coexist with mutual benefits.
- 'Real option pricing approaches [20]' deal with the issues of coexistence between the POs (MNOs) offering services in their networks and the network-less SOs such as MVoIPOs offering services only above IP over PO's networks. It is argued that for such coexistence to succeed, two central aspects must be studied: (i) types of differentiated services offered by the two operators and (ii) the objectives and risk behaviours of the two operators (i.e. the objectives of the two operators greatly differs: the MNO focuses on recovering its costs at the minimum risk possible through traditional pricing schemes. Meanwhile the MVoIPO focuses on attracting as many users as possible using innovative pricing schemes at competitive rates.).
- 'Packet voice interconnections through the IP exchange (IPX) [21]'. IPX is a predecessor of the GPRS roaming exchange. It is a global private network developed by the global system mobile (GSM) association (GSM-A). IPX is a service-aware system that supports end-to-end SLAs and the principles of cascading interconnection payments. The IPX supports three different types of interconnection models: bilateral transport only, bilateral service transit and multilateral hub service.

IPX paves the way for a global MNOs consolidation in response to the rising stakes of the new breed of third party application service providers such as mobile virtual network operators (MVNOs). IPX empowers MNOs with the means to resist interconnection with third party operators particularly when the services (e.g. rich communication services and applications) to be offered by the latter are a threat or do not conform to the existing billing arrangements. We therefore argue that this approach does not sufficiently address the settlement issue of a third party's on-net MVoIP calls within MNO's network.

3 TSM framework

3.1 Modelling and design

This section presents an overview of our proposed four-stage framework mechanism that enables the MNO and the MVoIPO to handle interconnection settlements, particularly the on-net traffic generated by MVoIPO users within the MNO's network. To function properly, the TSM framework is subdivided further into two distinct cycles: operational and contractual. In the operational cycle, consisting of stages 1, 2 and 3 of the framework, the MVoIPO's throughput is monitored and measured at the edge of MNOs network against the set SLA thresholds in real-time, and compensation is issued by the MNO to the MVoIPO. The operational cycle works on short-term timescales (typically minutes). In stage 4 of the framework, the economic merit of the SLA is re-evaluated in order to ascertain its viability on medium-to-long-term timescales (e.g. weeks, months).

3.1.1 TSM framework is based on the following underlying assumptions:

- *Transparency:* As the name suggests, it is in the best interest of both operators to disclose all technical and trade information relevant to the contract.
- *Thresholds:* The PO and the SO initially agree that the latter will eventually qualify for compensation after a consistent generation of uplink and downlink packet data throughput in the network owned by the former. The actual generated packet data throughput levels must consistently be equal to or exceed or equal a pre-set threshold of N bandwidth units per second where $N > 0$.
- *Conformance events:* Compensation will be dynamically issued based on the sum of conformance events (i.e. the moments of time when MVoIP throughput is equal to or above the set thresholds). The conformance events will be monitored and statistically collected at the GGSN to the nth percentile of time.
- *Type of compensation:* There are different possible types of compensation for the MVoIPO to pre-select from in addition to money, for example, prioritisation of MVoIP traffic during congested periods [17]. Fig. 2 shows the four-stage TSM framework.

The framework works as follows:

- *Stage 1:* The agreed MVoIP throughput thresholds (e.g. 'Y' ≥ 'X' bandwidth units) are set by both operators, followed by pre-selection of the type of compensation for the MVoIPO.
- *Stage 2:* The monitoring period starts during which the MVoIP throughput is actively measured and monitored at the edge of the MNO's core network.
- *Stage 3:* The MNO begins to issue compensation to the MVoIPO according to the usage patterns of the throughput (uplink and downlink) generated by users of the MVoIP application.
- *Stage 4:* Revisits the economic model before executing another transparent settlement cycle. The techno-economic model here refers to the game-theoretic decision-making modelling approach

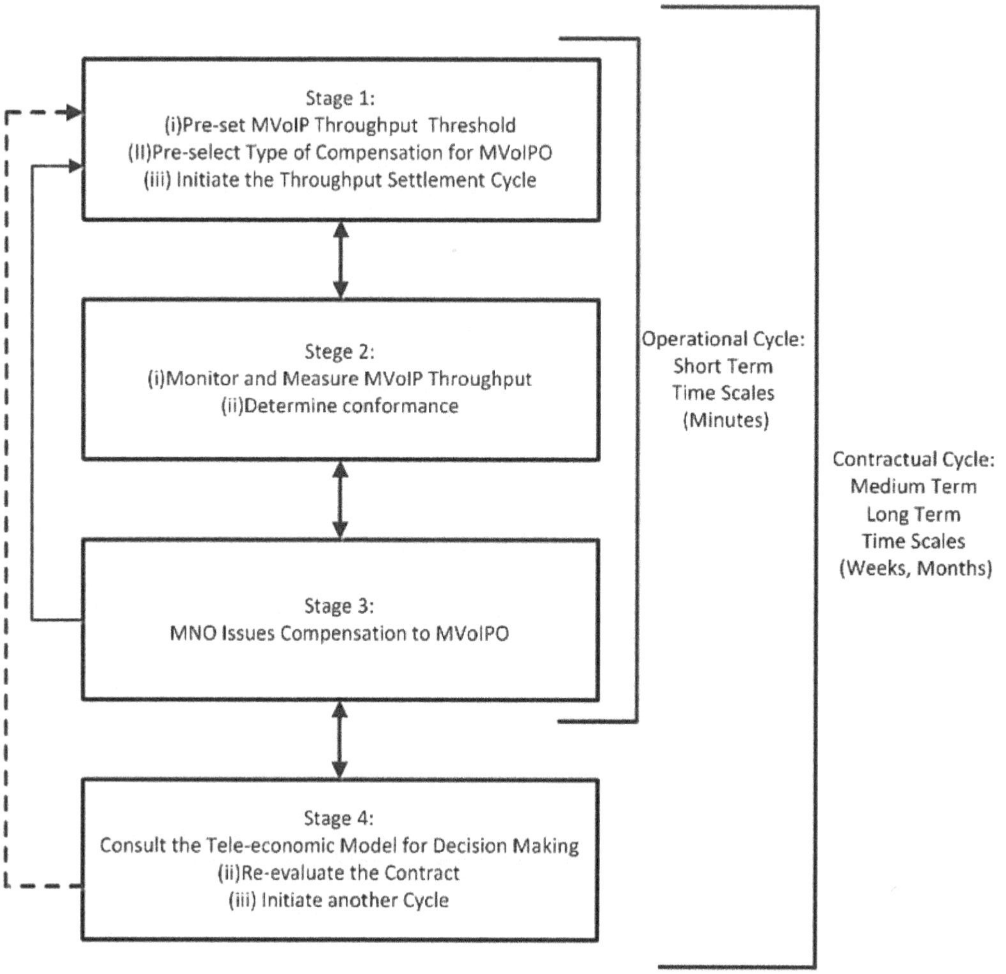

Fig. 2 *High-level view of the SLA-based TSM framework*

used to govern coexistence between the two types of operators [17, 22, 23]. This approach examines the decisions made by the MNO to invest in excess network capacity as well as the trade-off between the substitutionary effects and positive network externality effects that may be introduced by the MVoIPO.

- *Agreement re-evaluation:* Both operators will review the economic merits of TSM at the end of each contractual cycle on medium-to long-term timescales.

3.2 Monitoring and measuring the MVoIP throughput at the edge of MNO's CN in discrete-time

The discrete-time-based method used in this section allows the continuous fluctuations of the on-net MVoIP traffic to be measured in the predefined fixed-sized windows of time as a stationary process. In [24], it is suggested that when monitoring a continuous time process at a constant measuring distance, the continuous time intervals are transformed into a discrete-time distribution. This observation is true with respect to the Poisson process as it is transformed into a discrete-time process with the correlation between inter-arrival times of duration being zero.

Suppose that the MVoIP throughput measured at the MNO's GGSN is of pure chance traffic (PCT) I type (PCT type I refers to the traffic stream that has a Poisson arrival process and exponential call holding times). Furthermore, suppose that the MVoIP call arrivals are independent identically distributed. The density function of time intervals from each arrival to the first measurement is uniformly distributed and equal to $1/h$ (where h refers to the constant measuring distance in a given time period), whereas the probability of observing zero measurement during the call holding time is denoted by $p(0)$. If the exponential distribution holding time interval is $F(t) = 1 - e^{-\mu t}$ with a finite mean μ^{-1}, then according to [25] the following discrete distribution will be observed

$$p(0) = 1 - \frac{1}{\mu h}(1 - e^{-\mu h}) \tag{1}$$

$$p(k) = \frac{1}{\mu h}(1 - e^{-\mu h})^2 . e^{-(k-1)\mu h}, \quad k = 1, 2, \ldots \tag{2}$$

Therefore the jth derivatives of the probability generating function $Z(z)$ when the value of $z = 1$ is given by [24]

$$Z^{(j)}(1) = \frac{j!}{jh} \cdot \frac{1}{(e^{\mu h} - 1)^{j-1}} \tag{3}$$

In particular, (3) yields the mean m_1, the variance σ_j^2 and the form factor ε (the form factor refers to the measure of accuracy in the measurements) [25]

$$m_1 = Z^{(j)}(1) = \frac{1}{\mu h} \tag{4}$$

$$\sigma_j^2 = Z^{(2)}(1) + Z^{(1)}(1) - Z^{(1)}(1)^2 \tag{5}$$

$$\varepsilon = \left(\frac{\sigma_j^2}{m_1}\right)^2 = \mu h . \frac{e^{\mu h} + 1}{e^{\mu h} - 1} \geq 2 \tag{6}$$

However, the intensity rate and the variance of the items of call traffic that has a unit holding time ($\mu = 1, \; = A$) arriving inside the time interval T are yielded by [26]

$$\mu_j = A \tag{7}$$

$$\sigma_j^2 = \frac{A}{T}.\varepsilon \tag{8}$$

Therefore, by inserting the value of the form factor ε obtained in (6) in (8), we can express the variance as

$$\sigma_j^2 = \frac{2A}{T} \tag{9}$$

The related derivations of the variance of PCT I traffic type are found in [27–29].

Suppose that the conformance of MVoIP throughput generated by all hosts and received at the MNO's GGSN needs to be monitored (the conformance and non-conformance of MVoIP throughput is according to the SLA threshold assumptions described earlier). First, we divide the throughput monitoring time between $t=0$ to Z into h steps, such that each step has a size (Z/h) (e.g. monitoring time lengths in hours). Subsequently, we subdivide each step into small fixed-size throughput-monitoring windows (partitions) in minutes. Secondly, the total throughput at that particular partition needs to be calculated. We calculate the total throughput at the j monitoring window as follows

$$\mathrm{Te}_j = \frac{\sum_{t=(z/h)_j}^{(z/h)_{j+1}} ARU}{(Z/H)}, \quad j = 1, 2, 3, \ldots, h \tag{10}$$

where Te_j is the total amount of data transmitted in bits per second, A is the MVoIP application's transmitted data from the source to the destination per host, R is the number of repetitions per host per second and U is the number of hosts.

Thirdly, the average of the data measured within the window $(z/h)_j$, $j=1, 2, \ldots$ (where j is the monitoring time length in minutes) is determined. Assuming that each j consists of n observations (samples) such that $(z/h)_{j-n+1}, \ldots, (z/h)_j$, then we compute their mean by

$$\left(\frac{\bar{z}}{h}\right)_j = \frac{1}{n}\sum_{k=j-n+1}^{j}\left(\frac{z}{h}\right)_k \tag{11}$$

TSM Algorithm

Input: *{Pre-set throughput threshold}*

{Pre-set throughput monitoring window time lengths}

{Pre-select the type of compensation to secondary operator}

Output: *{Expected economic benefits to secondary operator and primary operator}*

Sequence:

1. *Transform the continuous Poissonian MVoIP traffic process into a discrete distribution and determine its theoretical mean and variance (Equations (1) to (9));*
2. *Partition the MVoIP traffic and determine the throughput within the partition (Equation (10));*
3. *Determine the mean value of the observations within the partition (Equation (11));*
4. *Determine the normalised variance of the mean of the observations inside the partition (Equation (12));*
5. *Construct the confidence interval for the mean throughput inside the partition (Equation (13));*
6. *Issue a conformance event to MVoIPO only if new throughput measurements are equal to or above the pre-set threshold (Equation (14));*
7. *Issue a pre-selected type of compensation to MVoIPO based on the accumulation of conformance events (Equation (15));*
8. *Revisit the decision-making model (economic) model in order to re-evaluate the contract before initiating another TSM cycle (this is performed on medium- to long-term timescales e.g., weeks, months).*

Fig. 3 *Algorithmic form of the TSM framework*

from which we derive the normalised variance of the mean as

$$\hat{\sigma}_j^2 = \frac{1}{\mathrm{Te}_j^2}\frac{1}{n-1}\sum_{k=j-n+1}^{j}\left(\left(\frac{z}{h}\right)_k - \left(\frac{\bar{z}}{h}\right)_j\right)^2 \quad (12)$$

That is, the normalised variance is an unbiased estimator of the original distribution such that $\hat{\sigma}_j^2$ has to be close to the theoretical variance derived in (9).

Fourthly, the nth percentile confidence interval estimates are constructed [30] to observe the MVoIP throughput conformance using the Student t-distribution [31] as follows

$$100(1-\alpha)\% = \left(\left(\frac{\bar{z}}{h}\right)_j \pm \left(\frac{t_{1-\alpha}}{2},\ n-1\right)\hat{\sigma}_j^2\right) \quad (13)$$

where $((t_{1-\alpha}/2),\ n-1)$ is the $(1-(\alpha/2))$ fractal of the Student t-distribution with $n-1$(degree of freedom), whereas $(1-\alpha)$ is the probability describing the level of the confidence and n is the number of observations. The values of α are obtained in the t-distribution table [32].

Fifthly, let conf_j denote the conformance event to be gained by the SO in the throughput-monitoring window (such events occur each time when the new throughput measurements found in the confidence interval are found to be within or above the pre-set SLA threshold). We calculate the conformance events by

$$\mathrm{conf}_j \to \left(\frac{z}{h}\right)_{j+1} \geq \mathrm{N}_j \quad (14)$$

where N_j is the pre-set SLA threshold.

Lastly, the accumulation of the conformance events results in the issuing of a pre-selected type of compensation to the SO by the PO such that

$$\varphi(\mathrm{comp}) = \sum_{j=1}^{n}\mathrm{conf}_j \quad (15)$$

where $\phi(\mathrm{comp})$ is the compensation with pre-selected type of choices (e.g. the prioritised MVoIP traffic) as discussed earlier.

In summary, the description of the proposed four-stage TSM framework presented in Fig. 2 can be presented in algorithmic form as shown in Fig. 3.

4 Simulation experiments

This section describes a simulation experiment for the logical interconnection between the MNO and the MVoIPO at the application layer. The simulation scenarios are constructed in OPNET Modeller version 15.0 [33] by re-using the expert service prediction module in order to analyse the MVoIP traffic growth and the SLA monitoring.

4.1 Simulation parameters

The following inputs to the simulation model have been used during different simulation runs. Table 1 introduces key parameters used in the simulation experiments.

Two simulation rounds, each consisting of five runs, were performed using five different seeds for random number generation (RNG). A traffic growth rate of 10% and a background traffic-scaling factor of 1.5 in best-effort mode were assumed. The UEs used a G.729 voice codec with a rate of 8 kbps and a constant packetisation interval of 20 ms (OFF – silent periods and ON – talk spurts). The call inter-arrival followed the Poisson process with exponential call holding times.

Furthermore, 30 UEs were used to model simultaneous users of the MVoIP application in a single cell with the assumption that non-MVoIP users were utilising the remaining cell capacity. In this way, the uplink and downlink throughput thresholds of ≥30 000 bps were set. It is important to note that the pre-set thresholds were only the expected MVoIPO throughput but not the actual throughput, which was expected to vary during the settlement cycle. Additionally, the time lengths of monitoring windows at the GGSN were set to 3 and 5 min.

Table 1 Voice application, profile, traffic growth and SLA attributes

Parameters	Values
Voice application and profile attributes	
application start time offset, s	Poisson (2)
application duration	end of profile (the application will end when the profile duration has expired)
application inter-repetition time, s	exponential (2)
application number of repetitions	unlimited
application repetition's pattern	serial (application start time is computed by adding the inter-repetition time to time at which the previous session completed)
profile repetition mode	serial
profile inter-repetition time, s	exponential (3600 s)
profile number of repetitions	unlimited
voice codec	G.729
call signalling protocol	SIP
application type of service	best-effort (0)
UE's P2P communication	enabled
traffic growth and SLA attributes	
simulation rounds	2 × 5 runs
traffic growth rate type	compound
growth rate	10%
background traffic-scaling factor	1.5
RNG	five seeds
number of UEs	30
MVoIP throughput threshold N_j (uplink, downlink)	≥30 000 bps
simulation duration	60 min
throughput-monitoring windows time lengths	3 min, 5 min
confidence interval (CI)	98%
monitoring location	GGSN

4.2 Simulation results

Tables 2 and 3 show the results of the MVoIP throughput (in the uplink and downlink directions) measured during the 3 and 5 min monitoring window time lengths.

Results recorded in Table 2 demonstrate that the MVoIP throughput monitored in the uplink direction conforms to the pre-set threshold (≥30 000 bps) for about 98th percentile of the simulation period

Table 2 Mean MVoIP uplink throughput (bps) received at the GGSN during the 3 and 5 min throughput-monitoring windows collected over the simulation period of 1 h

Monitoring window	3 min	5 min
seed 1	35 218 bps	31 907 bps
seed 2	47 027 bps	42 874 bps
seed 3	47 731 bps	44 860 bps
seed 4	47 874 bps	45 271 bps
seed 5	36 017 bps	34 047 bps
mean throughput	42 773 bps	39 792 bps
98% CIs (lower bound–upper bound)	35 900–49 646 bps	32 773–46 811 bps

Table 3 Mean MVoIP downlink throughputs (bps) received at the GGSN during the 3 and 5 min throughput-monitoring windows collected over the simulation period of 1 h

Monitoring window	3 min	5 min
seed 1	24 404 bps	21 664 bps
seed 2	38 718 bps	35 062 bps
seed 3	39 013 bps	35 973 bps
seed 4	39 010 bps	36 470 bps
seed 5-	24 768 bps	22 651 bps
mean throughput	33 183 bps	30 364 bps
98% confidence intervals (lower bound–upper bound)	24 148-33 880 bps	23 687-37 041 bps

of 1 h. Hence, the MVoIPO will receive full compensation from the MNO for the uplink throughput. A statistical graph in Fig. 4 illustrates conformance and non-conformance levels by the MVoIPO traffic at UMTS GTP (GPRS tunnelling protocol) level in the uplink direction received at the GGSN node when the throughput-monitoring window time length was set to 3 min. It is worth noting that the bars in the graph show the statistical average for each monitoring window.

In this scenario, the average value of conformance by the MVoIPO was 35 218 bps; that is, it was above the pre-set threshold (of ≥30 000 bps). However, in the first 360 s, the MVoIPO was below the threshold two times.

Results in Table 3 indicate that the MVoIP throughput monitored in the downlink direction partially conforms to the pre-set threshold of ≥30 000 bps for about 98th percentile of the simulation period of 1 h (i.e. the lower confidence interval bounds fall below the pre-set throughput thresholds, whereas the upper confidence interval bounds fall within the set throughput threshold). In other words, the MVoIPO will be receiving partial compensation from the MNO for the downlink throughput.

Fig. 5 contains a statistical graph illustrating levels of conformance and non-conformance levels by the MVoIPO traffic at UMTS GTP level in the downlink direction received at the GGSN node when the throughput-monitoring window time length was set to 3 min. In this scenario, the average value of non-conformance by the MVoIPO was 24 404 bps. It was below the pre-set throughput threshold (of ≥30 000 bps) 100% of the time, in which it was below the threshold 20 times.

4.3 Analysis

This section analyses the implications of the results on key factors affecting the successful implementation of the TSM.

Throughput threshold:

• It was observed that the pre-set threshold (≥30 000 bps) is roughly about 60% of the MVoIPO's peak throughput. This

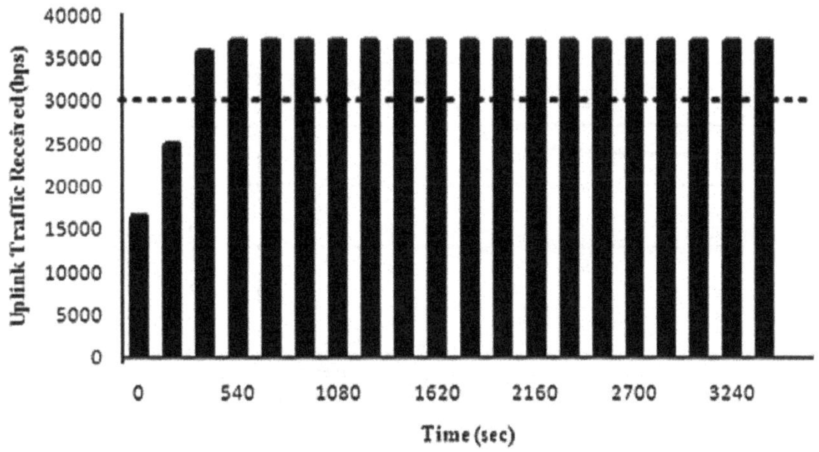

Fig. 4 *MVoIPO's uplink throughput received at the MNO's GGSN*
(Seed = 1; monitoring window time = 3 min; statistics collected over a period of 1 h)

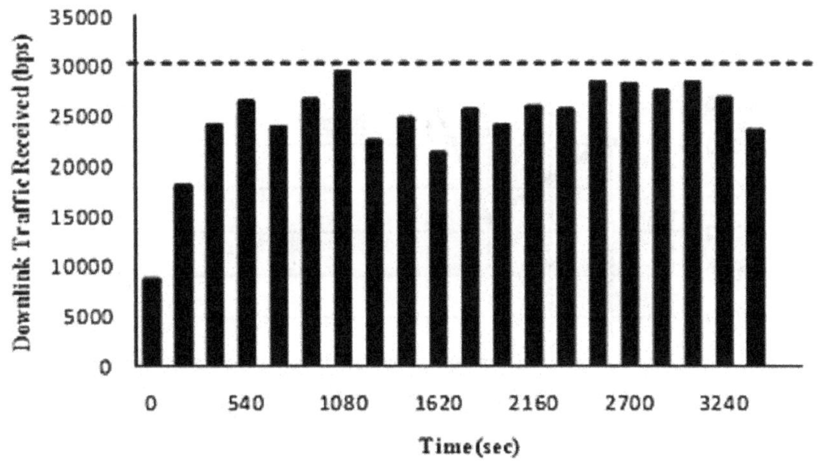

Fig. 5 *MVoIP downlink throughput received at the GGSN*
(Seed = 1; monitoring window time = 3 min; statistics collected over period of 1 h)

raises the question of how to select the optimal threshold values that will offer mutual benefits for both operators (i.e. both the MNO and the MVoIPO). On one hand, setting too large a value might cause the MVoIPO not to benefit from the MNO's incentives despite generating a substantial amount of data traffic.

On the other hand, setting too small a value might enable the MVoIPO to benefit from the MNO's incentives even if it does not generate an economically acceptable amount of data traffic. One possible solution for this crucial issue is to allow a grace period before the formal commencement of the SLA so that both operators can learn about each other's behaviours (i.e. the MNO learns about the MVoIPO's usage patterns, whereas the MVoIPO learns about the MNO's service availability and quality).

Monitoring window time length:

- The next notable factor is the choice of the right monitoring window time lengths. According to [25, 26], the arrival of telephony traffic at intervals of up to 15 min can be treated as a stationary Poisson process. Larger intervals are considered as non-stationary because of the possible intensity variations that may introduce difficulties in taking measurements.

In this paper, smaller throughput-monitoring window time lengths were used to avoid complications related to non-stationary data even though the number of monitored events that fell below the pre-set throughput threshold was roughly the same in the downlink direction for both the smaller 3 min window and the bigger 5 min window. It is worth observing, however, that an SLA contractual cycle may span several weeks or months, in which case the use of smaller throughput-monitoring window time lengths would result in an increased frequency of monitoring. However, according to [34], raw packet counting is a costly exercise; therefore it is proposed that using bigger throughput-monitoring window time lengths of up to 15 min would be more cost efficient.

Accuracy of the measurements:

- The accuracy of the measurements is largely attributable to the use of the variance of the MVoIP throughput distribution in the confidence interval calculations. The variance, being a second-order statistic, provides much needed granularity of measurements in each monitoring window. It is asserted in [28, 29] that more hidden characteristics of the load are revealed by the use of the second- and higher-order statistics than by the first-order statistics. Fig. 6 compares the sample mean and the variance of the received uplink MVoIP throughput with a 98% confidence interval.

In Fig. 6, the sample mean (in the first graph), a first-order statistic, does not reveal all the hidden behaviour (unevenness characteristics) of the uplink MVoIP throughput. By contrast, the variance (in the second graph), a second-order statistic, reveals both the peaks and the low points of the uplink MVoIP throughput,

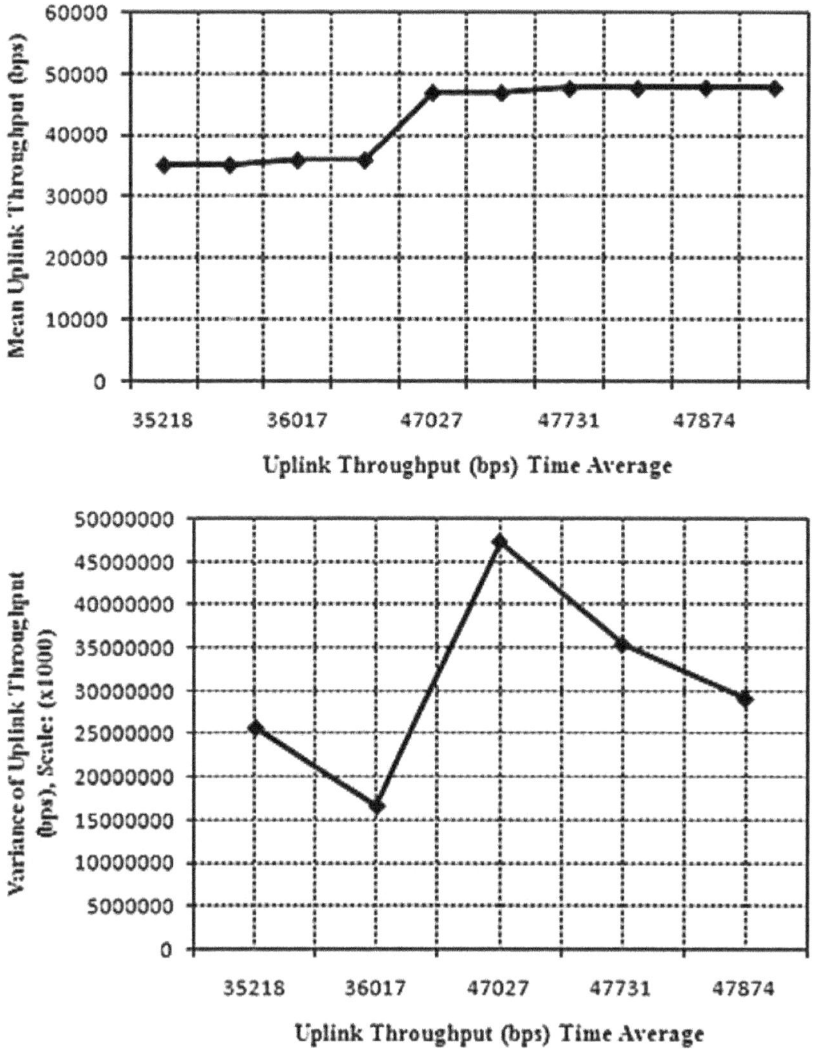

Fig. 6 *Mean (upper graph) and the variance (lower graph) at 98% confidence intervals for the aggregated uplink MVoIP throughput measured during the 3 min monitoring windows*

which is vitally important for the accuracy of the measurements used in the TSM.

Issuing of compensation to the SO:

- Another important factor relates to the issuing of compensation to the MVoIPO, particularly the exact amount of conformance events required to adequately and accurately determine this matter. We propose that this area should be left to the two operators to negotiate.

5 Conclusion

It is worth noting that despite the on-going explosion of data usage and the need for rollout of next generation high-speed packet-based cellular network infrastructure; many MNOs still find it difficult to gain maximum economic benefits from data traffic. In this paper, we have proposed a settlement model between network owner the MNO and network user the mobile voice internet operator (MVoIPO) for the on-net MVoIP data traffic within the MNO's network. More broadly, our model differs from well-known SLA-based schemes by replacing the two traditional concepts of violation and punishment (i.e. users who violate the set SLA thresholds are punished by having their services restricted or terminated), with two new concepts, achievement and reward. We argued that an MVoIPO should be compensated for generating an economically acceptable amount of data traffic. In other words, MVoIPO should still be incentivised instead of getting punished even when it occasionally goes below the set SLA thresholds. This is due to the fact that MNO will benefit from the internet access charges paid by users of the MVoIP application(s). Our approach can be instrumental in creating innovative business models for POs and SOs to effectively serve the future web 3.0 consumers. Additionally, this approach can aid in establishing 'radio frequency (RF) spectrum usage settlement models' between POs and SOs in the emerging opportunistic networks that are using dynamic spectrum access (DSA) techniques.

6 References

[1] Kanervisto J.: 'MVNO Pricing Structures in Finland (Publications of Transport and Communications No. 21/2005)', Helsinki, Ministry of Transport and Communications Finland, 2005

[2] Viber: 'Free calls, text and picture sharing with anyone, anywhere!'. Available at http://www.viber.com/, accessed September 2014

[3] Armstrong M., Wright J.: 'Mobile call termination', *Econ. J.*, 2009, **119**, (538), pp. 270–307

[4] Judge P.: 'T-mobile blocks rival's VoIP calls', Techworld. Available at http://www.news.techworld.com/mobile-wireless/9191/t-mobile-blocks-voip-rivals-calls/, accessed January 2012

[5] 3rd Generation Partnership Project (3GPP): 'Network Architecture (T1TRQ 3GPP 23.002-330)', Standards Committee T1 Telecommunications, version 3.3.0 edition, 2000

[6] Rosenberg J., Schulzrine H., Camarillo G., ET AL.: 'Session Initiation Protocol (SIP) - RFC 3261 (Request for Comments (RFC) No. 3261)', Fremont, Internet Engineering Task Force (IETF), 2002

[7] Curpen R., Ballan T., Sandu F., Costache C., Cerchez C.: 'Demonstrator for voice communication over LTE'. 2014 Tenth Int. Conf. on Communications (COMM), Bucharest, 29–31 May 2014, pp. 1–4

[8] Cisco: 'Visual Networking Index: Global Mobile Data Traffic Forecast Update, 2013–2018'. Available at http://www.cisco.com/c/en/us/solutions/collateral/service-provider/visual-networking-index-vni/white_paper_c11-520862.html, accessed September 2014

[9] Marcus S., Elixmann D.: 'The future of IP interconnection'. Final Report, European Commission, Brussels, EU-Commission, 2008

[10] Stiller B., Reichl P., Gerke J., Hasan H., Flury P.: 'Charging and accounting in high-speed networks', *Future Gener. Comput. Syst.*, 2003, **19**, (1), pp. 101–109

[11] Davoyan R., Effelsberg W.: 'Inter-carrier compensation between providers of different layers: advantages of transmission initiator determination'. Proc. Ninth Int. Conf. on Int. Federation for Information Processing (IFIP), TC6 Networking, Chennai, India, May 2010, pp. 373–381

[12] Davoyan R., Effelsberg W.: 'A new bilateral arrangement between interconnected providers'. Proc. Sixth Int. Workshop on Internet Charging and QoS Technologies (ICQT): Network Economics for Next Generation Networks, Aachen, Germany, May 2009, pp. 85–96

[13] Noam E.M.: 'Interconnection practices', in Cave M., Majumdar S., Vogeslang I. (Eds.): 'Handbook of telecommunications economics' (Elsevier, Amsterdam, Netherlands, 2000), vol. 1, pp. 387–409

[14] Gizelis C., Vergados D.D.: 'A survey of pricing schemes in wireless networks', *IEEE Commun. Surv. Tutor.*, 2011, **13**, pp. 126–145

[15] Blade V., Ghosal D.: 'Pricing approaches in the design of next-generation wireless networks: a review and a unified proposal', *IEEE Commun. Surv.*, 2007, **9**, (2), pp. 88–101

[16] Faria F., Nogueira J.M.: 'Exploring context based charging for composite services'. Proc. Int. Conf. Intelligence in Next Generation Networks (ICIN), Berlin, Germany, October 2010, pp. 1–6

[17] Dixit S., Periyalwar S., Yanikomeroglu H.: 'Secondary user access in LTE architecture based on a base-station-centric framework with dynamic pricing', *IEEE Trans. Veh. Technol.*, 2013, **62**, (1), pp. 284–296

[18] Wang J., Leung V.C.M.: 'Integrating predictable planning and unpredictable dynamics in network utility maximization: an improved quota-based market model'. Proc. Int. IEEE Symp. on Signal Processing an Information Technology, Vancouver, BC, Canada, August 2006, pp. 478–483

[19] Bush J., Irvine J., Dunlop J.: 'Digital marketplace for tussle in the next generation wireless networks'. Proc. 70th IEEE Int. Conf. on Vehicular Technology (VTC), Anchorage, Alaska, USA, September 2009, pp. 1–5

[20] D'Acquisto G., Cassara P., Alcuri L.: 'American options based service pricing for virtual operators'. Proc. IEEE Int. Workshops on Network Operations and Management Symp. (NOMS), Salvador, Bahia, Brazil, April 2008, pp. 197–184

[21] GSM Association: 'Packet Voice Interconnection Service Schedule to AA.80', 2010

[22] Mfupe L., Mzyece M., Kurien A.: 'A tele-economic approach for MNOs and software based M-VoIP operators to co-exist using game theory'. Proc. Int. Conf. Southern African Telecommunication Networks and Applications (SATNAC), Ezwilini, Swaziland, September 2009

[23] Ac'imovic-Raspopovi'c V., Kosti'c-Ljubisavljevi'c A., Mladenovi'c S.: 'Competitive pricing using Cournot game in the next generation networks'. Proc. Int. Conf. Telecommunications in Modern Satellite Cable and Broadcasting Services (TELKIS), Nis, Serbia, October 2011, pp. 297–300

[24] Iversen V.B.: 'Teletraffic engineering and network planning' (Technical University of Denmark, Lyngby, Denmark, 2010)

[25] Iversen V.B.: 'On the accuracy in measurements of time intervals and traffic intensities with applications to teletraffic and simulation'. PhD thesis, IMSOR, Technical University of Denmark, Lyngby, Denmark, 1976

[26] Palm C.: 'Accuracy of measurements in determining traffic volumes by the scanning method', *Tek. Medd. från K. Telegrafstyrelsen*, 1941, **7**, (9), pp. 97–115

[27] Glabowski M.: 'Modelling systems with multiservice overflow erlang and engset traffic streams', *Int. J. Adv. Telecommun.*, 2008, **1**, (1), pp. 14–26

[28] Riordan J.: 'Appendix 1: derivation of moments of overflow traffic', in Wilkinson R.: 'Theories of toll traffic engineering in the USA', *Bell Syst. Tech. J. (BSTJ)*, 1956, **40**, (4), pp. 507–514

[29] Mandjes M., Saniee I., Stolyar A.: 'Load characterization and anomaly detection for voice over IP traffic', *IEEE Trans. Neural Netw.*, 2005, **16**, (5), pp. 1019–1026

[30] Jain R.: 'The art of computer systems performance analysis: techniques for experimental designs, measurement, simulation, and modelling' (Wiley-Interscience, 1991, 1st edn.)

[31] Student: 'The probable error of a mean', *Biometrika*, 1908, **6**, (1), pp. 1–25

[32] Statsoft: 'Electronic Statistics Textbook'. Available at http://www.statsoft.com/textbook/distribution-tables/#t, accessed November 2011

[33] OPNET Technologies: 'OPNET Modeller 15.0'. Available at http://www.opnet.com, accessed November 2011

[34] Cushnie J., Hutchison D., Oliver H.: 'Evolution of charging and billing models for GSM and future mobile internet services (External Publication Report No. HPL-IRI-2000-4)'. Available at http://www.hpl.hp.com/techreports/2000/HPL-IRI-2000-4.pdf, accessed December 2011

5

Kinematics, dynamics and control design of 4WIS4WID mobile robots

Ming-Han Lee, Tzuu-Hseng S. Li

aiRobots Laboratory, Department of Electrical Engineering, National Cheng Kung University, 1 University Road, Tainan, Taiwan 70101
E-mail: thsli@mail.ncku.edu.tw

Abstract: Kinematic and dynamic modelling and corresponding control design of a four-wheel-independent steering and four-wheel-independent driving (4WIS4WID) mobile robot are presented in this study. Different from the differential or car-like mobile robot, the 4WIS4WID mobile robot is controlled by four steering and four driving motors, so the control scheme should possess the ability to integrate and manipulate the four independent wheels. A trajectory tracking control scheme is developed for the 4WIS4WID mobile robot, where both non-linear kinematic control and dynamic sliding-mode control are designed. All of the stabilities of the kinematic and dynamic control laws are proved by Lyapunov stability analysis. Finally, the feasibility and validity of the proposed trajectory tracking control scheme are confirmed through computer simulations.

1 Introduction

Many structures of vehicles or mobile robots have been presented, such as two-wheel-steering (2WS), four-wheel-steering (4WS), differential wheel, skid-steer and omni-directional drive. In recent years, because of the rapid expansion of electric motors and their high manoeuvrability, vehicles or mobile robots based on the four-wheel-independent steering and four-wheel-independent driving (4WIS4WID) structure have been presented and researched in many areas such as transportation, agriculture and service industry.

The 2WS or car-like structure [1–4] is like that of a traditional car, but in this structure only the front wheels can steer and they are coupled by mechanical linkages. The 4WS [5, 6] can steer the front wheels and rear wheels, but the wheels are coupled with mechanical linkages so that whether they are 2WS or 4WS, the steering angles are limited and are restricted by their mechanical structures. This limitation means that the vehicles or robots of 2WS or 4WS cannot implement some motions, for example, lateral moving.

The differential wheeled mobile robot [7–11] has two parallel driving wheels on the right side and left side of the robot and one or two caster wheels to balance the robot. The two driving wheels drive independently, and this type of robot uses the velocity difference between the two wheels to turn the direction and move. Another similar type is skid-steer, which uses the same way of driving, but different from the differential wheeled mobile robot, it may have four or more driving wheels [12, 13] or be equipped with tracks [14] without castor wheels such as skid-steer loaders or backhoes.

The omni-directional mobile robots [15–18] are equipped with omni wheels so that they can move in any direction instantly without turning the direction of the wheels or changing the robot orientation. Although this type has high manoeuvrability, it still has some disadvantages. The first disadvantage is that all the wheels do not roll in the direction of the robot movement, so the efficiency is low. The second disadvantage is omni-wheel itself contains many wheels/rollers which lead greater resistance to rotate and imply greater loss of energy.

There are some vehicles and wheeled mobile robots, Nissan Pivo 2, Nissan Pivo 3, Toyota Fine-T, Toyota Fine-X, humanoid robot Justin [19], OK-1 [20], Care-O-bot 3 [21] and others [22–24] which adopt the 4WIS4WID structure. The 4WIS4WID mobile robot is equipped with four wheels, but the four wheels are not coupled by mechanical linkages. Each wheel is steered independently using a motor, which is equipped between the body and each wheel of the mobile robot. On the basis of the structure of the 4WIS4WID mobile robot, the manoeuvrability is more diversified so it can implement some motions such as lateral moving or zero-turn radius. The 4WIS4WID mobile robot can achieve the same actions as car-like, 4WS and skid-steer mobile robots. Moreover, the 4WIS4WID mobile robot can also implement the same motions as the omni-directional mobile robot. However, the 4WIS4WID mobile robot has to turn the orientation of wheels so it cannot instantly move in any direction like the omni-directional mobile robot.

The first purpose of this paper is to present the kinematic and dynamic models, which can be utilised to design controllers to deal with control issues for 4WIS4WID mobile robot. The 4WIS4WID mobile robot has high manoeuvrability; the control problem is more complex because it has to manipulate four-wheel motors and four steering motors to move smoothly. Reference [25] presents an omni-directional steer-by-wire system to control the steering angle of four wheels, where the instantaneous centre of rotation (ICR) is considered. When robot moves straight, the ICR is undefined or infinite. This disadvantage means that the control law has to be distinguished into three conditions, turning right, moving straight and turning left; then it has to switch them. The kinematic control method proposed in [26] is also based on the ICR, where one extra control mode has to be induced when the desired rotational velocity of mobile robot is zero. Furthermore, the velocity control scheme developed in [26] also needs to be divided into three control modes based on three conditions. In [27], Selekwa and Nistler proposed a kinematic control law to control the steering angle and wheel velocity of four wheels, where the virtual rear and front steering angles should be given first. This phenomenon limits its application. Moreover, if the virtual rear or front steering angle is zero, the wheel steering angle cannot be calculated. If the steering angle of any wheel is zero, the wheel velocity cannot be calculated or is infinite.

To design a better controller for the 4WIS4WID mobile robot, we present a kinematic model that includes eleven state variables, and a dynamic model, which is developed by the Lagrangian equation. On the basis of the proposed models, we design a controller that combines kinematic control with dynamic control to implement the trajectory tracking control problem. Through the non-linear kinematic control law, the velocity and steering angle of every wheel can be determined appropriately. In dynamic control law, this paper proposes a trajectory tracking control scheme based on sliding-mode control (SMC) to reduce the tracking error with system uncertainties and external disturbances. The SMC and its extensions have been applied to many kinds of mobile robots [5, 28–30].

The main contributions of this paper are as follows: (i) both kinematic and dynamic models of the 4WIS4WID mobile robot are developed. They can help researchers to design and develop other controllers and/or applications for the 4WIS4WID mobile robot. (ii) both non-linear kinematic control and SMC-based dynamic control schemes for the trajectory tracking problem of 4WIS4WID mobile robot are presented and the stabilities are confirmed by the Lyapunov stability theory. Moreover, the comparison with the proportional–integral (PI) dynamic control is given. (iii) both the model and control system are not examined based on the concept of ICR, so the control algorithm is not necessarily divided into some conditions addressed in the literature [25, 26].

The organisation of this paper is as follows: in Section 2, both the kinematic and dynamic models of the 4WIS4WID mobile robot are addressed. Section 3 describes the trajectory tracking problem for the 4WIS4WID mobile robot. Furthermore, the kinematic and dynamic controllers are presented. The trajectory tracking simulation results based on the proposed control scheme proposed are illustrated in Section 4. Finally, Section 5 gives conclusions.

2 System description

This section presents the kinematic and dynamic models of the 4WIS4WID mobile robot. One can find that the concept of ICR is not necessary to setup these models.

2.1 Kinematic model

A simplified 4WIS4WID mobile robot system with four wheels, without considering the motions of pitch, roll and heave, is shown in Fig. 1 and the configuration is depicted in Fig. 2. The robot is driven by four-wheel motors and through another four motors to turn these four-wheel motors. x, y and θ are the postures of the robot in the world coordinate frame; w_i denotes the wheel i; δ_i denotes the steering angle of wheel i; a and b are, respectively, the length between the centroid of the robot and each wheel. Before developing the model of 4WIS4WID mobile robot, three configurations for the robot are assumed as below.

Assumption 1: The position of the four wheels in the robot coordinate frame is predefined as $(x^r_{w1}, y^r_{w1}) = (a, b)$, $(x^r_{w2}, y^r_{w2}) = (-a, b)$, $(x^r_{w3}, y^r_{w3}) = (-a, -b)$ and $(x^r_{w4}, y^r_{w4}) = (a, -b)$. The (x^r_{wi}, y^r_{wi}) denotes the coordinates of wheel i in the robot coordinate frame. v and v_i are, respectively, the linearity velocity of the robot and wheel i, where $v = \sqrt{v_x^2 + v_y^2}$ and $v_i = \sqrt{v_{xi}^2 + v_{yi}^2}$. ω is the angular velocity of the robot's body. ω_i is the steering velocity of wheel i.

Assumption 2: The positions of centre of mass and centroid of the 4WIS4WID mobile robot are the same.

Assumption 3: Both the radius and mass of all wheels are assumed to be the same.

Under no slipping condition, the relationships of the velocity between each wheel and the body of robot are obtained by the

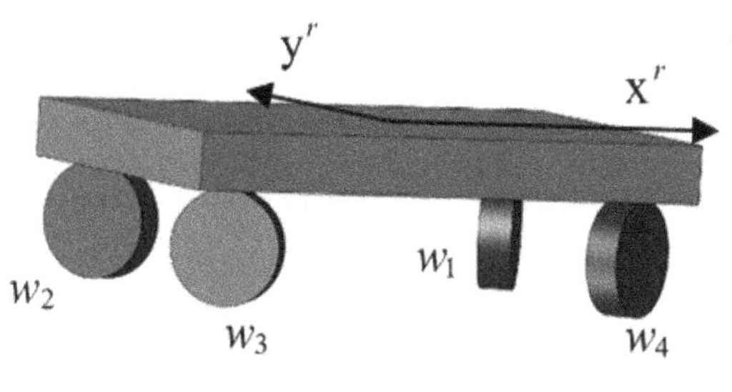

Fig. 1 *4WIS4WID robot system*

rigid body constraint for the robot which are represented as

$$v_{xi} = v_i \cos(\delta_i) = v_x - y^r_{wi}\omega \tag{1}$$

$$v_{yi} = v_i \sin(\delta_i) = v_y + x^r_{wi}\omega \tag{2}$$

By substituting the parameters of the four wheels into (1) and (2), the relationship of the four wheels and body of the robot can be represented as

$$\boldsymbol{P}\begin{bmatrix} v_x \\ v_y \\ \omega \end{bmatrix} = \boldsymbol{X}\begin{bmatrix} v_1 \\ v_2 \\ v_3 \\ v_4 \end{bmatrix} \tag{3}$$

where
$$\boldsymbol{P} = \begin{bmatrix} 1 & 0 & -b \\ 0 & 1 & a \\ 1 & 0 & -b \\ 0 & 1 & -a \\ 1 & 0 & b \\ 0 & 1 & -a \\ 1 & 0 & b \\ 0 & 1 & a \end{bmatrix}, \boldsymbol{X} = \begin{bmatrix} c(\delta_1) & 0 & 0 & 0 \\ s(\delta_1) & 0 & 0 & 0 \\ 0 & c(\delta_2) & 0 & 0 \\ 0 & s(\delta_2) & 0 & 0 \\ 0 & 0 & c(\delta_3) & 0 \\ 0 & 0 & s(\delta_3) & 0 \\ 0 & 0 & 0 & c(\delta_4) \\ 0 & 0 & 0 & s(\delta_4) \end{bmatrix},$$

and where $c(\cdot)$ and $s(\cdot)$ denote the trigonometric functions $\cos(\cdot)$ and $\sin(\cdot)$, respectively. Moreover, the pseudo-inverse matrix of $\boldsymbol{P}$ is as follows

$$\boldsymbol{P}^+ = \begin{bmatrix} 1/4 & 0 & 1/4 & 0 & 1/4 & 0 & 1/4 & 0 \\ 0 & 1/4 & 0 & 1/4 & 0 & 1/4 & 0 & 1/4 \\ -\dfrac{b}{K} & \dfrac{a}{K} & -\dfrac{b}{K} & -\dfrac{a}{K} & \dfrac{b}{K} & -\dfrac{a}{K} & \dfrac{b}{K} & \dfrac{a}{K} \end{bmatrix}$$

where $K = 4a^2 + 4b^2$. Premultiply (3) by $\boldsymbol{P}^+$, one can obtain the relationship of the four wheels and robot's body as follows

$$\begin{bmatrix} v_x \\ v_y \\ \omega \end{bmatrix} = \begin{bmatrix} \dfrac{c(\delta_1)}{4} & \dfrac{c(\delta_2)}{4} & \dfrac{c(\delta_3)}{4} & \dfrac{c(\delta_4)}{4} \\ \dfrac{s(\delta_1)}{4} & \dfrac{s(\delta_2)}{4} & \dfrac{s(\delta_3)}{4} & \dfrac{s(\delta_4)}{4} \\ W_1 & W_2 & W_3 & W_4 \end{bmatrix}\begin{bmatrix} v_1 \\ v_2 \\ v_3 \\ v_4 \end{bmatrix} \tag{4}$$

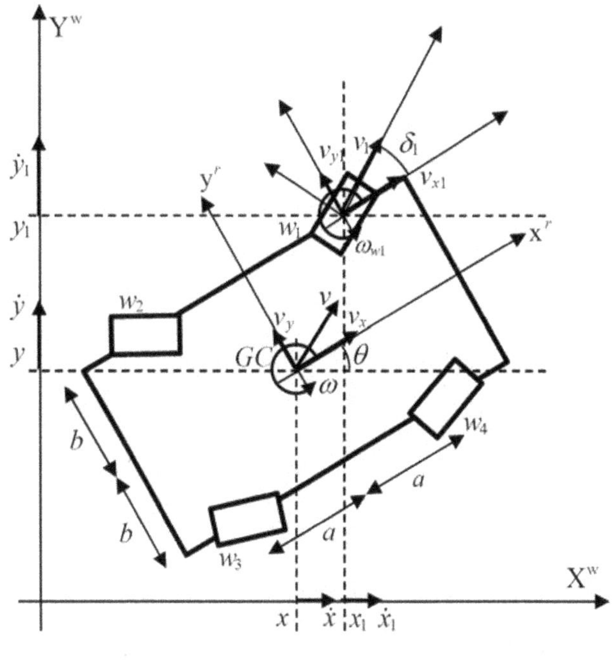

Fig. 2 *Configuration of 4WIS4WID robot system in world coordinate frame*

where $W_i = \left(-y^r_{wi}c(\delta_i) + x^r_{wi}s(\delta_i)\right)/\left(4(x^r_{wi})^2 + 4(y^r_{wi})^2\right)$. Note that $\boldsymbol{P}^{+}\boldsymbol{P} = \boldsymbol{I}_3$, where $\boldsymbol{I}_3 \in \Re^{3\times 3}$ is the identity matrix.

Combining the three states of the robot body, wheel rotating angle and wheel steering angle, there are 11 state variables to represent the position and posture of the mobile robot

$$\boldsymbol{q} = \begin{bmatrix} x & y & \theta & \varphi_1 & \varphi_2 & \varphi_3 & \varphi_4 & \delta_1 & \delta_2 & \delta_3 & \delta_4 \end{bmatrix}^{\mathrm{T}} \quad (5)$$

After some calculations, one can obtain the kinematic model of 4WIS4WID mobile robot as follows

$$\dot{\boldsymbol{q}} = \begin{bmatrix} \dot{x} \\ \dot{y} \\ \dot{\theta} \\ \dot{\varphi}_1 \\ \dot{\varphi}_2 \\ \dot{\varphi}_3 \\ \dot{\varphi}_4 \\ \dot{\delta}_1 \\ \dot{\delta}_2 \\ \dot{\delta}_3 \\ \dot{\delta}_4 \end{bmatrix} = \begin{bmatrix} \frac{c_1}{4} & \frac{c_2}{4} & \frac{c_3}{4} & \frac{c_4}{4} & 0 & 0 & 0 & 0 \\ \frac{s_1}{4} & \frac{s_2}{4} & \frac{s_3}{4} & \frac{s_4}{4} & 0 & 0 & 0 & 0 \\ W_1 & W_2 & W_3 & W_4 & 0 & 0 & 0 & 0 \\ r^{-1} & 0 & 0 & 0 & 0 & 0 & 0 & 0 \\ 0 & r^{-1} & 0 & 0 & 0 & 0 & 0 & 0 \\ 0 & 0 & r^{-1} & 0 & 0 & 0 & 0 & 0 \\ 0 & 0 & 0 & r^{-1} & 0 & 0 & 0 & 0 \\ 0 & 0 & 0 & 0 & 1 & 0 & 0 & 0 \\ 0 & 0 & 0 & 0 & 0 & 1 & 0 & 0 \\ 0 & 0 & 0 & 0 & 0 & 0 & 1 & 0 \\ 0 & 0 & 0 & 0 & 0 & 0 & 0 & 1 \end{bmatrix} \begin{bmatrix} v_1 \\ v_2 \\ v_3 \\ v_4 \\ \omega_1 \\ \omega_2 \\ \omega_3 \\ \omega_4 \end{bmatrix} = \boldsymbol{J}\boldsymbol{v} \quad (6)$$

where $\dot{\boldsymbol{q}}$ is defined as the velocity vector in the world coordinate frame; φ_i denotes the angular of wheel i; c_i and s_i are represented as $\cos(\delta_i+\theta)$ and $\sin(\delta_i+\theta)$, respectively.

2.2 Dynamic model

To derive the dynamic model, the kinetic energy of the mobile robot is described as

$$\begin{aligned} E &= E_x + E_y + E_\theta + E_\varphi + E_\delta \\ &= \frac{1}{2}m_{\mathrm{b}}v_x^2 + \frac{1}{2}m_{\mathrm{b}}v_x^2 + \frac{1}{2}I_\theta\dot{\theta}^2 \\ &\quad + \frac{1}{2}\sum_{i=1}^{4}\left(m_{\mathrm{w}}v_i^2 + I_\varphi\dot{\varphi}_i^2\right) + \frac{1}{2}\sum_{i=1}^{4}I_\delta\dot{\delta}^2 \\ &= \frac{1}{2}m_{\mathrm{b}}v_x^2 + \frac{1}{2}m_{\mathrm{b}}v_x^2 + \frac{1}{2}I_\theta\dot{\theta}^2 \\ &\quad + \frac{1}{2}\sum_{i=1}^{4}\left(m_{\mathrm{w}}\left(v_{ix}^2 + v_{iy}^2\right) + I_\varphi\dot{\varphi}_i^2\right) + \frac{1}{2}\sum_{i=1}^{4}I_\delta\dot{\delta}^2 \\ &= \frac{1}{2}m_{\mathrm{b}}v_x^2 + \frac{1}{2}m_{\mathrm{b}}v_x^2 + \frac{1}{2}I_\theta\dot{\theta}^2 \\ &\quad + \frac{m_{\mathrm{w}}}{2}\sum_{i=1}^{4}\left(4v_x^2 + 4v_y^2 + 4\left(a_i^2 + b_i^2\right)\dot{\theta}^2\right) \\ &\quad + \frac{I_\varphi}{2}\sum_{i=1}^{4}\dot{\varphi}_i^2 + \frac{I_\delta}{2}\sum_{i=1}^{4}\dot{\delta}^2 \\ &= \frac{1}{2}mv_x^2 + \frac{1}{2}mv_y^2 + \frac{1}{2}I\dot{\theta}^2 + \frac{I_\varphi}{2}\sum_{i=1}^{4}\dot{\varphi}_i^2 + \frac{I_\delta}{2}\sum_{i=1}^{4}\dot{\delta}^2 \end{aligned} \quad (7)$$

where m_{b} and m_{w} are, respectively, the weight of the robot's body and each wheel; I_θ, I_φ and I_δ are the moment of inertia about the rotating of the robot's body, the rolling of the wheel and the steering of the wheel, respectively; $m = m_{\mathrm{b}} + 4m_{\mathrm{w}}$ and $I = I_\theta + 4m_{\mathrm{w}}(a^2+b^2)$.

Owing to $v_x^2 + v_y^2 = \dot{x}^2 + \dot{y}^2$, (7) can be rewritten as

$$E = \frac{1}{2}m\dot{x}^2 + \frac{1}{2}m\dot{y}^2 + \frac{1}{2}I\dot{\theta}^2 + \frac{I_\varphi}{2}\sum_{i=1}^{4}\dot{\varphi}_i^2 + \frac{I_\delta}{2}\sum_{i=1}^{4}\dot{\delta}^2 \quad (8)$$

Without considering the disturbances, the Lagrange formula is used to derive the dynamic equation

$$\frac{\mathrm{d}}{\mathrm{d}t}\left(\frac{\partial E}{\partial \dot{\boldsymbol{q}}_j}\right) - \frac{\partial E}{\partial \boldsymbol{q}_j} + \boldsymbol{F}(\dot{\boldsymbol{q}}) + \boldsymbol{\tau}_{\mathrm{d}} = \boldsymbol{N}(q)\boldsymbol{\tau}, \quad (j = 1, 2, 3, \ldots, 11) \quad (9)$$

where $\boldsymbol{\tau}$ is the torque input vector, $\boldsymbol{F}(\dot{\boldsymbol{q}})$ denotes the surface friction forces and $\boldsymbol{\tau}_{\mathrm{d}}$ is the bounded disturbances. After doing some calculations, the dynamic model is obtained and represented as follows

$$\boldsymbol{M}\ddot{\boldsymbol{q}} + \boldsymbol{V}_m\dot{\boldsymbol{q}} + \boldsymbol{F} + \boldsymbol{\tau}_{\mathrm{d}} = \boldsymbol{N}\boldsymbol{\tau} \quad (10)$$

The metrics in (10) are represented as

$$\boldsymbol{M} = \mathrm{diag}\{m, m, I, I_\varphi, I_\varphi, I_\varphi, I_\varphi, I_\delta, I_\delta, I_\delta, I_\delta\}$$

$$\boldsymbol{V}_m = 0$$

$$\boldsymbol{N} = \begin{bmatrix} \frac{c_1}{r} & \frac{c_2}{r} & \frac{c_3}{r} & \frac{c_4}{r} & 0 & 0 & 0 & 0 \\ \frac{s_1}{r} & \frac{s_2}{r} & \frac{s_3}{r} & \frac{s_4}{r} & 0 & 0 & 0 & 0 \\ T_1 & T_2 & T_3 & T_4 & 0 & 0 & 0 & 0 \\ 1 & 0 & 0 & 0 & 0 & 0 & 0 & 0 \\ 0 & 1 & 0 & 0 & 0 & 0 & 0 & 0 \\ 0 & 0 & 1 & 0 & 0 & 0 & 0 & 0 \\ 0 & 0 & 0 & 1 & 0 & 0 & 0 & 0 \\ 0 & 0 & 0 & 0 & 1 & 0 & 0 & 0 \\ 0 & 0 & 0 & 0 & 0 & 1 & 0 & 0 \\ 0 & 0 & 0 & 0 & 0 & 0 & 1 & 0 \\ 0 & 0 & 0 & 0 & 0 & 0 & 0 & 1 \end{bmatrix}$$

$$\boldsymbol{F} = \begin{bmatrix} f_{\mathrm{d}\varphi_1} & f_{\mathrm{d}\varphi_2} & f_{\mathrm{d}\varphi_3} & f_{\mathrm{d}\varphi_4} & f_{\mathrm{d}\delta_1} & f_{\mathrm{d}\delta_2} & f_{\mathrm{d}\delta_3} & f_{\mathrm{d}\delta_4} \end{bmatrix}^{\mathrm{T}}$$

$$\boldsymbol{\tau}_{\mathrm{d}} = \begin{bmatrix} \tau_{\mathrm{d}\varphi_1} & \tau_{\mathrm{d}\varphi_2} & \tau_{\mathrm{d}\varphi_3} & \tau_{\mathrm{d}\varphi_4} & \tau_{\mathrm{d}\delta_1} & \tau_{\mathrm{d}\delta_2} & \tau_{\mathrm{d}\delta_3} & \tau_{\mathrm{d}\delta_4} \end{bmatrix}^{\mathrm{T}}$$

and

$$\boldsymbol{\tau} = \begin{bmatrix} \tau_{\varphi_1} & \tau_{\varphi_2} & \tau_{\varphi_3} & \tau_{\varphi_4} & \tau_{\delta_1} & \tau_{\delta_2} & \tau_{\delta_3} & \tau_{\delta_4} \end{bmatrix}^{\mathrm{T}}$$

where diag is the abbreviation of diagonal matrix; $T_i = r^{-1}[x_i s(\delta_i) - y_i c(\delta_i)]$; $f_{\mathrm{d}\varphi_i}$ and $f_{\mathrm{d}\delta_i}$ are the friction forces affected on the rolling and steering wheel i, respectively; $\tau_{\mathrm{d}\varphi_i}$ and $\tau_{\mathrm{d}\delta_i}$ are the bounded disturbances affected on the rolling and steering wheel i, respectively; and τ_{φ_i} and τ_{δ_i} are the torques to roll and steer wheel i, respectively. Differentiating (6) and substituting into (10), then multiplying the result by $\boldsymbol{J}^{\mathrm{T}}$, (10) is transferred into another form as follows

$$\bar{\boldsymbol{M}}(\boldsymbol{q})\dot{\boldsymbol{v}} + \bar{\boldsymbol{V}}_m(\boldsymbol{q})\boldsymbol{v} + \bar{\boldsymbol{F}} + \bar{\boldsymbol{\tau}}_{\mathrm{d}} = \bar{\boldsymbol{N}}(\boldsymbol{q})\boldsymbol{\tau} \quad (11)$$

where $\bar{\boldsymbol{M}} = \boldsymbol{J}^{\mathrm{T}}\boldsymbol{M}\boldsymbol{J} \in \Re^{8\times 8}$, $\bar{\boldsymbol{V}}_m = \boldsymbol{J}^{\mathrm{T}}\boldsymbol{M}\dot{\boldsymbol{J}} \in \Re^{8\times 8}$, $\bar{\boldsymbol{F}} = \boldsymbol{J}^{\mathrm{T}}\boldsymbol{F} \in \Re^{8\times 1}$, $\bar{\boldsymbol{\tau}}_{\mathrm{d}} = \boldsymbol{J}^{\mathrm{T}}\boldsymbol{\tau}_{\mathrm{d}} \in \Re^{8\times 1}$ and $\bar{\boldsymbol{N}} = \boldsymbol{J}^{\mathrm{T}}\boldsymbol{N} \in \Re^{8\times 8}$.

The variables of $\bar{\boldsymbol{M}}$, $\bar{\boldsymbol{V}}_m$ and $\bar{\boldsymbol{N}}$ are

$$\bar{\boldsymbol{M}} = \begin{bmatrix} \bar{M}_{11} & \bar{M}_{21} & \bar{M}_{31} & \bar{M}_{41} & 0 & 0 & 0 & 0 \\ \bar{M}_{21} & \bar{M}_{22} & \bar{M}_{32} & \bar{M}_{42} & 0 & 0 & 0 & 0 \\ \bar{M}_{31} & \bar{M}_{32} & \bar{M}_{33} & \bar{M}_{43} & 0 & 0 & 0 & 0 \\ \bar{M}_{41} & \bar{M}_{42} & \bar{M}_{43} & \bar{M}_{44} & 0 & 0 & 0 & 0 \\ 0 & 0 & 0 & 0 & I_\delta & 0 & 0 & 0 \\ 0 & 0 & 0 & 0 & 0 & I_\delta & 0 & 0 \\ 0 & 0 & 0 & 0 & 0 & 0 & I_\delta & 0 \\ 0 & 0 & 0 & 0 & 0 & 0 & 0 & I_\delta \end{bmatrix}$$

$$\bar{M}_{ii} = IW_i^2 + \frac{m}{16}\left(c_i^2 + s_i^2\right) + \frac{I_\varphi}{r^2}, \quad i = 1,\ 2,\ 3,\ 4$$

$$\bar{M}_{ij} = IW_jW_i + \frac{m}{16}\left(c_jc_i + s_js_i\right)$$

$$i = 2,\ 3,\ 4;\ j = 1,\ 2,\ 3,\ 4;\ i \neq j$$

$$\bar{\boldsymbol{V}}_m = \begin{bmatrix} \bar{V}_{11} & \bar{V}_{21} & \bar{V}_{31} & \bar{V}_{41} & 0 & 0 & 0 & 0 \\ \bar{V}_{21} & \bar{V}_{22} & \bar{V}_{32} & \bar{V}_{42} & 0 & 0 & 0 & 0 \\ \bar{V}_{31} & \bar{V}_{23} & \bar{V}_{33} & \bar{V}_{43} & 0 & 0 & 0 & 0 \\ \bar{V}_{32} & \bar{V}_{24} & \bar{V}_{34} & \bar{V}_{44} & 0 & 0 & 0 & 0 \\ 0 & 0 & 0 & 0 & 0 & 0 & 0 & 0 \\ 0 & 0 & 0 & 0 & 0 & 0 & 0 & 0 \\ 0 & 0 & 0 & 0 & 0 & 0 & 0 & 0 \\ 0 & 0 & 0 & 0 & 0 & 0 & 0 & 0 \end{bmatrix}$$

$$\bar{V}_{ij} = IW_j\dot{W}_i + m\left(c_j\dot{c}_i + s_j\dot{s}_i\right)/16$$

$$i = 1, 2, 3, 4; j = 1, 2, 3, 4$$

$$\bar{\boldsymbol{N}} = \begin{bmatrix} \bar{N}_{11} & \bar{N}_{21} & \bar{N}_{31} & \bar{N}_{41} & 0 & 0 & 0 & 0 \\ \bar{N}_{21} & \bar{N}_{22} & \bar{N}_{32} & \bar{N}_{42} & 0 & 0 & 0 & 0 \\ \bar{N}_{31} & \bar{N}_{23} & \bar{N}_{33} & \bar{N}_{43} & 0 & 0 & 0 & 0 \\ \bar{N}_{32} & \bar{N}_{24} & \bar{N}_{34} & \bar{N}_{44} & 0 & 0 & 0 & 0 \\ 0 & 0 & 0 & 0 & 1 & 0 & 0 & 0 \\ 0 & 0 & 0 & 0 & 0 & 1 & 0 & 0 \\ 0 & 0 & 0 & 0 & 0 & 0 & 1 & 0 \\ 0 & 0 & 0 & 0 & 0 & 0 & 0 & 1 \end{bmatrix}$$

$$\bar{N}_{ii} = \frac{W_i(y_ic(\delta_i) + x_is(\delta_i))}{r} + \frac{s_i^2 + c_i^2}{4r} + \frac{1}{r}, \quad i = 1,\ 2,\ 3,\ 4$$

and

$$\bar{N}_{ij} = \frac{W_i(y_jc(\delta_j) + x_js(\delta_j))}{r} + \frac{s_is_j + c_ic_j}{4r}$$
$$i = 1,\ 2,\ 3,\ 4;\ j = 1,\ 2,\ 3,\ 4;\ i \neq j$$

Property 1: $\bar{\boldsymbol{M}}$ is a symmetric positive-definite matrix.

Property 2: $\dot{\bar{\boldsymbol{M}}} - 2\bar{\boldsymbol{V}}_m$ is the skew symmetric.

Premultiply (11) by $\bar{\boldsymbol{M}}^{-1}$ and regarding the friction forces $\bar{\boldsymbol{F}}$ and disturbances torques $\bar{\boldsymbol{\tau}}_{\mathrm{d}}$ as the system uncertainties and disturbances, then the dynamic equation of the 4WIS4WID can be described as

$$\begin{aligned}\dot{\boldsymbol{v}}(t) &= -\bar{\boldsymbol{M}}^{-1}\bar{\boldsymbol{V}}_m\boldsymbol{v}(t) + \bar{\boldsymbol{M}}^{-1}\bar{\boldsymbol{N}}\boldsymbol{\tau}(t) - \bar{\boldsymbol{M}}^{-1}\bar{\boldsymbol{F}} - \bar{\boldsymbol{M}}^{-1}\bar{\boldsymbol{\tau}}_{\mathrm{d}} \\ &\equiv -\boldsymbol{A}\boldsymbol{v}(t) + \boldsymbol{B}\boldsymbol{\tau}(t) + \boldsymbol{d}(t)\end{aligned} \tag{12}$$

where $\boldsymbol{A} = \bar{\boldsymbol{M}}^{-1}\bar{\boldsymbol{V}}_m$, $\boldsymbol{B} = \bar{\boldsymbol{M}}^{-1}\bar{\boldsymbol{N}}$ and $\boldsymbol{d} = -\bar{\boldsymbol{M}}^{-1}\bar{\boldsymbol{F}} - \bar{\boldsymbol{M}}^{-1}\bar{\boldsymbol{\tau}}_{\mathrm{d}}$.

3 Trajectory tracking control design

In this section, a trajectory tracking problem of the 4WIS4WID mobile robot is discussed and both non-linear kinematic and dynamic SMC controllers for the 4WIS4WID mobile robot are designed.

3.1 Problem statement

Suppose that a reference robot $\boldsymbol{q}_{\mathrm{d}} = [x_{\mathrm{d}}(t) \quad y_{\mathrm{d}}(t) \quad \theta_{\mathrm{d}}(t)]^{\mathrm{T}}$ moves along a time-varying trajectory. The control objective is to let the robot track $\dot{\boldsymbol{q}}_{\mathrm{d}}$ and the tracking error between the reference robot and the actual robot approach zero. On the basis of this objective, the main task is to design a control law for the 4WIS4WID mobile robot to ensure the actual robot is able to track the reference one.

3.2 Kinematic controller design

Before designing the controllers, the vector of the tracking error is defined as follows

$$\begin{aligned}\boldsymbol{q}_{\mathrm{e}} &= \begin{bmatrix} x_{\mathrm{e}} \\ y_{\mathrm{e}} \\ \theta_{\mathrm{e}} \end{bmatrix} = \boldsymbol{R}(\theta)\cdot\tilde{\boldsymbol{q}}_{\mathrm{e}} \\ &= \begin{bmatrix} \cos\theta & \sin\theta & 0 \\ -\sin\theta & \cos\theta & 0 \\ 0 & 0 & 1 \end{bmatrix}\begin{bmatrix} x_{\mathrm{d}} - x \\ y_{\mathrm{d}} - y \\ \theta_{\mathrm{d}} - \theta \end{bmatrix}\end{aligned} \tag{13}$$

and the tracking error dynamic is denoted as

$$\dot{\boldsymbol{q}}_{\mathrm{e}} = \dot{\boldsymbol{q}}_{\mathrm{d}} - \dot{\boldsymbol{q}} = \begin{bmatrix} \dot{x}_{\mathrm{e}} & \dot{y}_{\mathrm{e}} & \dot{\theta}_{\mathrm{e}} \end{bmatrix}^{\mathrm{T}} \tag{14}$$

where $\boldsymbol{R}(\theta)$ is the rotation matrix. After some computing, $\dot{\boldsymbol{q}}_{\mathrm{e}}$ can be derived as follows

$$\begin{aligned}C\dot{\boldsymbol{q}}_{\mathrm{e}} &= \begin{bmatrix} y_{\mathrm{e}}\dot{\theta} + v_{x\mathrm{d}} - v_x \\ -x_{\mathrm{e}}\dot{\theta} + v_{y\mathrm{d}} - v_y \\ \omega_{\mathrm{d}} - \omega \end{bmatrix} \\ &= \begin{bmatrix} y_{\mathrm{e}}\dot{\theta} + \frac{1}{4}\sum_{i=1}^{4}(v_{i\mathrm{d}}\cos\delta_{i\mathrm{d}}) - \frac{1}{4}\sum_{i=1}^{4}(v_i\cos\delta_i) \\ -x_{\mathrm{e}}\dot{\theta} + \frac{1}{4}\sum_{i=1}^{4}(v_{i\mathrm{d}}\sin\delta_{i\mathrm{d}}) - \frac{1}{4}\sum_{i=1}^{4}(v_i\sin\delta_i) \\ \sum_{i=1}^{4}(v_{i\mathrm{d}}W_{i\mathrm{d}}) - \sum_{i=1}^{4}(v_iW_i) \end{bmatrix}\end{aligned} \tag{15}$$

where $v_{i\mathrm{d}} = \sqrt{v_{xi\mathrm{d}}^2 + v_{yi\mathrm{d}}^2}$, $\delta_{i\mathrm{d}} = \arctan2(v_{yi\mathrm{d}}, v_{xi\mathrm{d}})$, $W_{i\mathrm{d}} = \left(-y_{wi}^{r}c(\delta_{i\mathrm{d}}) + x_{wi}^{r}s(\delta_{i\mathrm{d}})\right) / \left(4\left(x_{wi}^{r}\right)^2 + 4\left(y_{wi}^{r}\right)^2\right)$, $v_{xi\mathrm{d}} = v_{x\mathrm{d}} - y_{wi}^{r}\omega_{\mathrm{d}}$ and $v_{yi\mathrm{d}} = v_{y\mathrm{d}} + x_{wi}^{r}\omega_{\mathrm{d}}$.

The kinematic tracking control law for the 4WIS4WID mobile robot is derived as follows. Define the Lyapunov function candidate as

$$V_k = \frac{1}{2}\left(x_{\mathrm{e}}^2 + y_{\mathrm{e}}^2 + \theta_{\mathrm{e}}^2\right) \geq 0 \tag{16}$$

Taking the time derivative of V_k implies

$$\begin{aligned}\dot{V}_k &= x_e\dot{x}_e + y_e\dot{y}_e + \theta_e\dot{\theta}_e \\ &= x_e\left[\frac{1}{4}\sum_{i=1}^{4}(v_{id}\cos\delta_{id}) - \frac{1}{4}\sum_{i=1}^{4}(v_i\cos\delta_i)\right] \\ &\quad + y_e\left[\frac{1}{4}\sum_{i=1}^{4}(v_{id}\sin\delta_{id}) - \frac{1}{4}\sum_{i=1}^{4}(v_i\sin\delta_i)\right] \\ &\quad + \theta_e\left[\sum_{i=1}^{4}(v_{id}W_{id}) - \sum_{i=1}^{4}(v_iW_i)\right]\end{aligned} \tag{17}$$

Let the control laws be defined as

$$v_{ic} = z_1\sqrt{a_i^2 + b_i^2} \tag{18}$$

$$\delta_{ic} = z_2 + \arctan 2(b_i,\ a_i) \tag{19}$$

where $a_i = v_{id}\cos\delta_{id} + k_x x_e - k_\theta y_{wi}\theta_e$ and $b_i = v_{id}\sin\delta_{id} + k_y y_e + k_\theta x_{wi}\theta_e$; k_x, k_y and k_θ are all positive constants; both z_1 and z_2 are defined as

$$(z_1,\ z_2) = \begin{cases}(+1,\ 0), & |\delta_{id} - \delta_i| \le \pi/2 \\ (-1,\ \pi), & \text{otherwise}\end{cases} \tag{20}$$

By substituting both (18) and (19) into (17), one can obtain the results as follows.

Although $(z_1, z_2) = (+1, 0)$, then

$$\begin{aligned}&\sum_{i=1}^{4}(v_{id}\cos\delta_{id} - v_i\cos\delta_i) \\ &= \sum_{i=1}^{4}\left(v_{id}\cos\delta_{id} - \sqrt{a_i^2 + b_i^2}\,\frac{a_i}{\sqrt{a_i^2 + b_i^2}}\right) \\ &= \sum_{i=1}^{4}(-k_x x_e + k_\theta y_{wi}\theta_e) = -4k_x x_e\end{aligned} \tag{21}$$

$$\begin{aligned}&\sum_{i=1}^{4}(v_{id}\sin\delta_{id} - v_i\sin\delta_i) \\ &= \sum_{i=1}^{4}\left(v_{id}\sin\delta_{id} - \sqrt{a_i^2 + b_i^2}\cdot\frac{b_i}{\sqrt{a_i^2 + b_i^2}}\right) \\ &= \sum_{i=1}^{4}(-k_y y_e - k_\theta x_{wi}\theta_e) = -4k_y y_e\end{aligned} \tag{22}$$

$$\begin{aligned}&\sum_{i=1}^{4}(v_{id}W_{id}) - \sum_{i=1}^{4}(v_iW_i) \\ &= \frac{1}{4}\sum_{i=1}^{4}\left(-\frac{y_{wi}^r(-k_x x_e + k_\theta y_{wi}^r\theta_e)}{(x_{wi}^r)^2 + (y_{wi}^r)^2} + \frac{x_{wi}^r\left(-k_y y_e - k_\theta x_{wi}^r\theta_e\right)}{(x_{wi}^r)^2 + (y_{wi}^r)^2}\right) \\ &= \frac{1}{4}\sum_{i=1}^{4}\left(-\frac{k_\theta\theta_e(y_{wi}^r)^2 + k_\theta\theta_e(x_{wi}^r)^2}{(x_{wi}^r)^2 + (y_{wi}^r)^2}\right) \\ &= -k_\theta\theta_e\end{aligned} \tag{23}$$

Although $(z_1, z_2) = (-1, \pi)$, then

$$\begin{aligned}&\sum_{i=1}^{4}(v_{id}\cos\delta_{id} - v_i\cos\delta_i) \\ &= \sum_{i=1}^{4}\left[v_{id}\cos\delta_{id} - \left(-\sqrt{a_i^2 + b_i^2}\cdot - \frac{a_i}{\sqrt{a_i^2 + b_i^2}}\right)\right] \\ &= \sum_{i=1}^{4}(-k_x x_e + k_\theta y_{wi}\theta_e) = -4k_x x_e\end{aligned} \tag{24}$$

$$\begin{aligned}&\sum_{i=1}^{4}(v_{id}\sin\delta_{id} - v_i\sin\delta_i) \\ &= \sum_{i=1}^{4}\left[v_{id}\sin\delta_{id} - \left(-\sqrt{a_i^2 + b_i^2}\cdot - \frac{b_i}{\sqrt{a_i^2 + b_i^2}}\right)\right] \\ &= \sum_{i=1}^{4}(-k_y y_d - k_\theta x_{wi}\theta_e) = -4k_y y_e\end{aligned} \tag{25}$$

and

$$\begin{aligned}&\sum_{i=1}^{4}(v_{id}W_{id}) - \sum_{i=1}^{4}(v_iW_i) \\ &= \frac{1}{4}\sum_{i=1}^{4}\left(-\frac{y_{wi}^r(-k_x x_e + k_\theta y_{wi}^r\theta_e)}{(x_{wi}^r)^2 + (y_{wi}^r)^2} + \frac{x_{wi}^r\left(-k_y y_e - k_\theta x_{wi}^r\theta_e\right)}{(x_{wi}^r)^2 + (y_{wi}^r)^2}\right) \\ &= -k_\theta\theta_e\end{aligned} \tag{26}$$

According to the above results, one can find (17) becomes

$$\dot{V}_k = -4k_x x_e^2 - 4k_y y_e^2 - k_\theta\theta_e^2 \tag{27}$$

Note that (19) is the control signal of steering angle. To control the steering angle, the control input of the steering angular velocity of wheel i for the 4WIS4WID mobile robot is defined as

$$\dot{\delta}_i = \dot{\delta}_{ic} + k_{\delta_i}\delta_{ie} \tag{28}$$

where k_{δ_i} is a positive constant of wheel i.

Define the Lyapunov function candidate for the four steering angles of 4WIS4WID mobile robot as

$$V_\delta = \frac{1}{2}\sum_{i=1}^{4}\delta_{ie}^2 \ge 0 \tag{29}$$

where $\delta_{ie} = \delta_{ic} - \delta_i$. Taking the time derivative of V_δ and considering (28) imply

$$\dot{V}_\delta = \sum_{i=1}^{4}\delta_{ie}\dot{\delta}_{ie} = \sum_{i=1}^{4}\delta_{ie}\left(\dot{\delta}_{ic} - \dot{\delta}_i\right) \tag{30}$$

By choosing appropriate gains of k_x, k_y, k_θ and k_{δ_i}, $\dot{V}_k$ and $\dot{V}_\delta$ can be guaranteed less than zero for all $t \ge 0$ while $\boldsymbol{q}_e \ne 0$ and $\delta_{ie} \ne 0$. Therefore, using the kinematic control laws, (18), (19) and (28) to control the 4WIS4WID mobile robot, the tracking error (13) is asymptomatically stable.

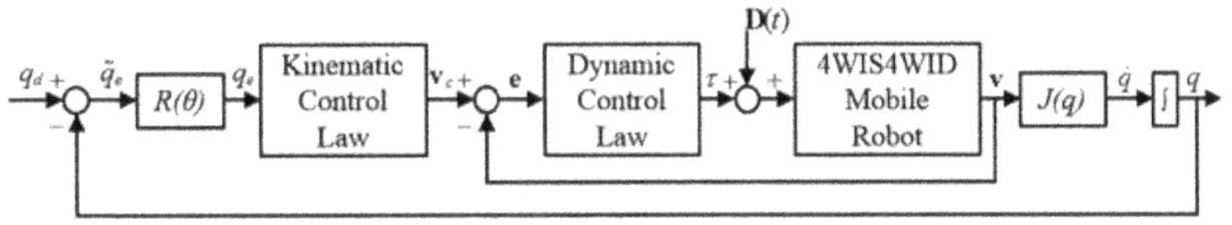

Fig. 3 *Trajectory tracking control scheme of the 4WIS4WID mobile robot*

3.3 Dynamic controller design

In this section, a dynamic tracking controller, SMC, will be designed for the 4WIS4WID mobile robot. Define the velocity error as

$$e = \begin{bmatrix} e_{\varphi_1} \\ e_{\varphi_2} \\ e_{\varphi_3} \\ e_{\varphi_4} \\ e_{\delta_1} \\ e_{\delta_2} \\ e_{\delta_3} \\ e_{\delta_4} \end{bmatrix} = \begin{bmatrix} v_{1c} \\ v_{2c} \\ v_{3c} \\ v_{4c} \\ \dot{\delta}_{1c} \\ \dot{\delta}_{2c} \\ \dot{\delta}_{3c} \\ \dot{\delta}_{4c} \end{bmatrix} - \begin{bmatrix} v_1 \\ v_2 \\ v_3 \\ v_4 \\ \dot{\delta}_1 \\ \dot{\delta}_2 \\ \dot{\delta}_3 \\ \dot{\delta}_4 \end{bmatrix} = v_c(t) - v(t) \tag{31}$$

and by choosing a PI-like sliding surface as

$$\begin{aligned} S(t) &= \begin{bmatrix} s_1 & s_2 & s_3 & s_4 & s_5 & s_6 & s_7 & s_8 \end{bmatrix}^{\mathrm{T}} \\ &= e + \boldsymbol{\lambda} \int_0^t e(\tau)\, d\tau \end{aligned} \tag{32}$$

where $\boldsymbol{\lambda} \in \Re^{8\times 8}$ is a diagonal positive-definite matrix and $\boldsymbol{\lambda}$ is selected to stabilise the sliding surface. Although the sliding surface $S(t)$ approaches zero, the tracking errors will also converge to zero. Therefore, the objective is to design a controller to let the system errors reach on the sliding surface. To derivate the equivalent control laws $\tau_{eq}(t)$, both the uncertainties and disturbances are assumed to be zero. The time derivative of (32) is

$$\dot{S}(t) = \dot{e} + \boldsymbol{\lambda} e = \dot{v}_c(t) + A\dot{v}(t) - B\tau(t) + \boldsymbol{\lambda} e(t) \tag{33}$$

Given $\dot{S}(t) = 0$, then the equivalent control law $\tau_{eq}(t)$ can be obtained as follows

$$\tau_{eq}(t) = B^{-1}(\dot{v}_c(t) + Av(t) + \boldsymbol{\lambda} e(t)) \tag{34}$$

Once the $\tau_{eq}(t)$ has been determined, one should derive the auxiliary discontinuous control law $\tau_r(t)$ to overcome the uncertainties and disturbances of the mobile robot. Suppose that A and B have the bounded uncertain parameters and the bounded disturbances $d(t)$ are existing, (12) is rewritten as

$$\dot{v}(t) \equiv -\hat{A}v(t) + \hat{B}\tau(t) + d(t) \tag{35}$$

where $\hat{A} = A + \Delta A$ and $\hat{B} = B + \Delta B$. $\Delta A \in \Re^{8\times 8}$ and $\Delta B \in \Re^{8\times 8}$ are the bounded system uncertainties which are generated by perturbations of system parameters. By considering the uncertainties and disturbances, (33) is rewritten as

$$\begin{aligned} \dot{S}(t) &= \dot{v}_c(t) + \hat{A}v(t) - \hat{B}\tau(t) + d(t) + \boldsymbol{\lambda} e(t) \\ &= \dot{v}_c(t) + (A + \Delta A)v(t) - (B + \Delta B)\tau(t) \\ &\quad + d(t) + \boldsymbol{\lambda} e(t) \\ &= \dot{v}_c(t) + \hat{A}v(t) - \hat{B}\tau(t) + \boldsymbol{\lambda} e(t) + (\Delta A U(t) \\ &\quad - \Delta B\tau(t) + d(t)) \\ &= \dot{v}_c(t) + Av(t) - B\tau(t) + \boldsymbol{\lambda} e(t) + D(t) \end{aligned} \tag{36}$$

where $D(t) = \Delta A U(t) - \Delta B\tau(t) + d(t) \in \Re^{8\times 1}$, and the element D_i of $D(t)$ is defined as $|D_i| \le \bar{D}_i$, where $\bar{D}_i$ is a bound positive constant of D_i.

Therefore, in order to ensure the system is stable, an auxiliary discontinuous control law is added and defined as

$$\tau_r(t) = B^{-1} K \mathbf{Sat}(S) \tag{37}$$

where $K \in \Re^{8\times 8}$ is the switch gain matrix concerned with upper bounds of uncertainties and K is positive-definite and diagonal

$$\mathbf{Sat}(S,\ \varepsilon) = \begin{bmatrix} \mathrm{sat}(s_1,\ \varepsilon) \\ \mathrm{sat}(s_2,\ \varepsilon) \\ \mathrm{sat}(s_3,\ \varepsilon) \\ \mathrm{sat}(s_4,\ \varepsilon) \\ \mathrm{sat}(s_5,\ \varepsilon) \\ \mathrm{sat}(s_6,\ \varepsilon) \\ \mathrm{sat}(s_7,\ \varepsilon) \\ \mathrm{sat}(s_8,\ \varepsilon) \end{bmatrix}$$

ε is a small positive constant and sat is the saturation function which is adopted to suppress the chattering behaviour. Sat is represented as

$$\mathrm{sat}(s_i,\ \varepsilon) = \begin{cases} \mathrm{sign}(s_i), & |s_i| > \varepsilon \\ s_i/\varepsilon, & |s_i| \le 0 \end{cases}$$

Table 1 Parameters used in simulations

Description	Symbol	Value	Unit
coordinate of wheel 1	$(x^r_{w1},\ y^r_{w1})$	(0.2, 0.25)	m
coordinate of wheel 2	$(x^r_{w2},\ y^r_{w2})$	(−0.2, 0.25)	m
coordinate of wheel 3	$(x^r_{w3},\ y^r_{w3})$	(−0.2, −0.25)	m
coordinate of wheel 4	$(x^r_{w4},\ y^r_{w4})$	(0.2, −0.25)	m
wheel radius	r	0.05	m
robot's body mass	m_b	26	kg
vehicle inertia moment	I_z	1.5	kgm^2
wheel mass	m_w	1	kg
wheel rolling torque limitation	τ_{φ_i}	5	Nm
wheel rolling inertia moment	I_φ	0.003	kgm^2
wheel steering torque limitation	τ_{δ_i}	1	Nm
wheel steering inertia moment	I_δ	0.005	kgm^2
limit of steering angle	δ_i	$(-\pi,\ \pi)$	rad

Table 2 Control parameters

Description	Symbol	Value
gains of (18) and (19)	k_x, k_y, k_θ	4, 4, 3
gains of (28)	k_{δ_i}	5
integral gains of (32)	λ_i	5
gains of (37)	k_i	10

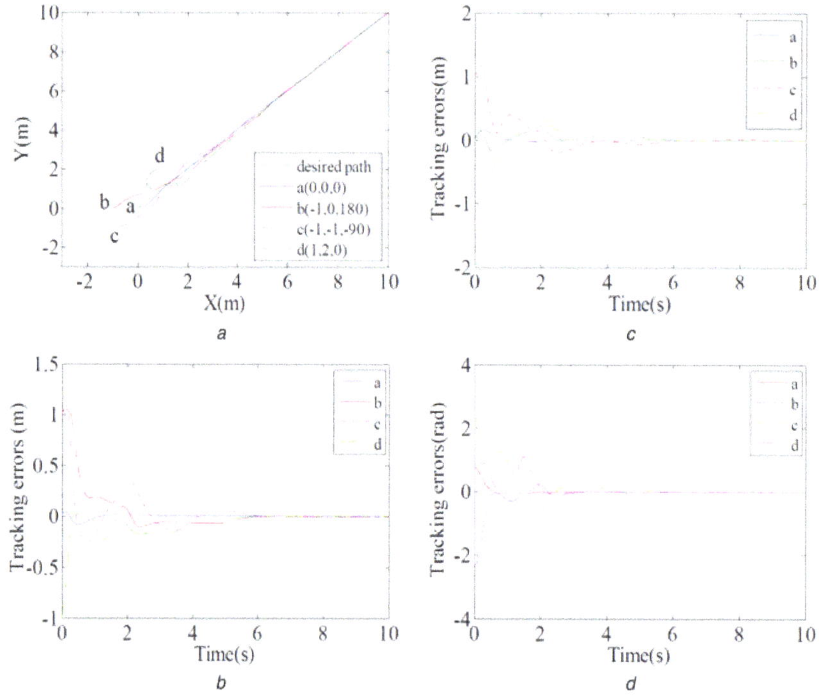

Fig. 4 *Simulation results of the straight trajectory of the four initial positions*
a x–y plot of the 4WIS4WID mobile robot
b–*d* Tracking errors of x_e, y_e and θ_e

The element k_{ii} in $\boldsymbol{K}$ is designed as

$$k_i = \bar{D}_i + \eta_i, \quad i = 1, 2, 3, \ldots, 8 \tag{38}$$

where $\eta_i > 0$.

Combining $\boldsymbol{\tau}_{eq}(t)$ and $\boldsymbol{\tau}_r(t)$, the consequential SMC control law for the mobile robot is

$$\begin{aligned} \boldsymbol{\tau}(t) &= \boldsymbol{\tau}_{eq}(t) + \boldsymbol{\tau}_r(t) \\ &= \boldsymbol{B}^{-1}\big(\dot{\boldsymbol{v}}_c(t) + \boldsymbol{A}\boldsymbol{v}(t) + \boldsymbol{\lambda}\boldsymbol{e}(t) + \boldsymbol{K} \cdot \mathbf{Sat}(\boldsymbol{S})\big) \end{aligned} \tag{39}$$

To show that the dynamic control law is stable for the tracking control, the Lyapunov function candidate is defined as

$$V_{SMC} = \frac{1}{2}\boldsymbol{S}^{T}(t)\boldsymbol{S}(t) \geq 0 \tag{40}$$

By taking the time derivative of V_{SMC}

$$\begin{aligned} \dot{V}_{SMC} &= \boldsymbol{S}(t)^{T}\dot{\boldsymbol{S}}(t) \\ &= \boldsymbol{S}(t)^{T}(\dot{\boldsymbol{v}}_c(t) + \boldsymbol{A}\boldsymbol{v}(t) \\ &\quad - \boldsymbol{B}\boldsymbol{\tau}(t) + \boldsymbol{\lambda}\boldsymbol{e}(t) + \boldsymbol{D}(t)) \\ &= \boldsymbol{S}(t)^{T}(-\boldsymbol{K} \cdot \mathbf{Sat}(\boldsymbol{S}) \\ &\quad + \boldsymbol{D}(t)) \leq \sum_{i=1}^{8}(-\eta_i|s_i(t)|) \end{aligned} \tag{41}$$

Thus, V_{SMC} is bounded for all time despite the uncertainties existed. The influence created by the uncertainties and disturbances could be reduced. According to the results of (27), (30) and (41), using the control laws (18), (19), (28) and (39), one can find that the 4WIS4WID mobile robot will converge exponentially to the reference trajectory. All of the dynamic errors satisfy $\boldsymbol{q}_e \to 0$ and $\boldsymbol{e} \to 0$ as $t \to \infty$.

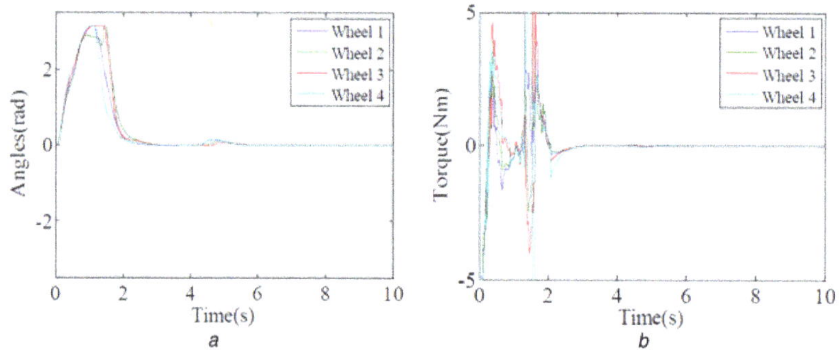

Fig. 5 *Steering angle and rolling torque of the four wheels of the initial position d*
a Steering angle
b Rolling torque

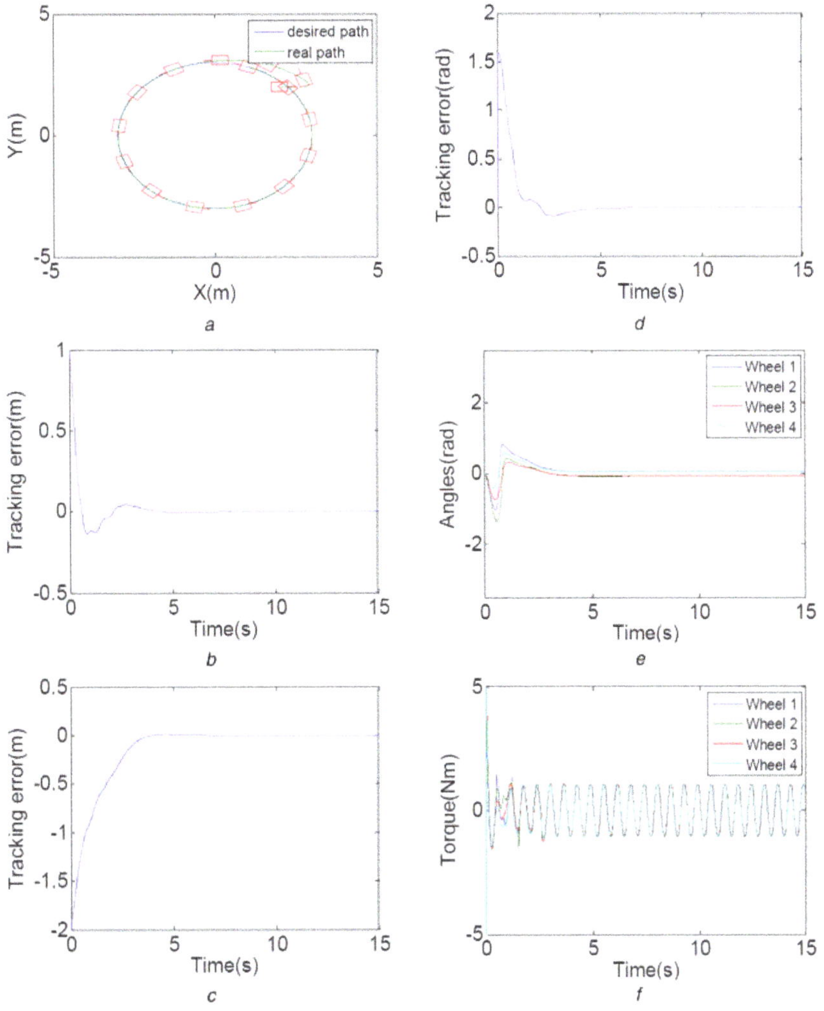

Fig. 6 *Simulation results of Case (I)*
a x–y plot of the 4WIS4WID mobile robot
b–d Tracking errors of x_e, y_e and θ_e
e Steering angle of four wheels
f Rolling torques of four wheels

The whole control scheme is shown in Fig. 3, where the inner and outer loops are used to control the dynamics and kinematics of 4WIS4WID mobile robot, respectively. The kinematic control law is implemented by equations (18), (19) and (28) and the dynamic control law is realised by (39).

4 Simulations

In this section, three computer simulations are performed to demonstrate the effectiveness and feasibility of the proposed control schemes for the 4WIS4WID mobile robot. The following simulations will define the trajectory and the initial position (X, Y, θ) of the robot. All the parameters of the mobile robot are listed in Table 1. Note that the limit of the steering angle in Table 1 means the steering motors cannot turn the wheel infinitely in the same direction because the wires used to transmit signals and power to the wheels might be twisted to breaking. They can only turn clockwise or counterclockwise in a half circle. Table 2 lists all the control parameters. In Section 4.1, a straight trajectory is used to test the feasibility of the control scheme. After the test in Section 4.2, two curvilinear trajectories are selected to demonstrate the performances using the presented control scheme.

4.1 Test by a straight trajectory

The straight trajectory utilised to test the control scheme is generated as follows

$$\begin{aligned} X_d(t) &= t \\ Y_d(t) &= t \\ \theta_d(t) &= \arctan 2\left(\dot{Y}_d(t),\, \dot{X}_d(t)\right) \end{aligned}$$

Four initial positions are selected to track the trajectory, which are $a(X, Y, \theta) = (0, 0, 0)$, $b(X, Y, \theta) = (-1, 0, 180°)$, $c(X, Y, \theta) = (-1, -1, -90°)$ and $d(X, Y, \theta) = (1, 2, 0)$. Fig. 4*a* shows the x–y plot of the trajectory tracking result of the four initial positions. The tracking errors of the four initial positions of the straight trajectory are shown in Figs. 4*b–d*. Figs. 5*a* and *b* represent the four-wheel steering angles and torques of the initial position $(X, Y, \theta) = (10, 3, 90°)$, respectively. One can find that as time increases, the tracking errors of this system are converging to zero. The simulation results demonstrate that the 4WIS4WID mobile robot can be stably driven using the control scheme.

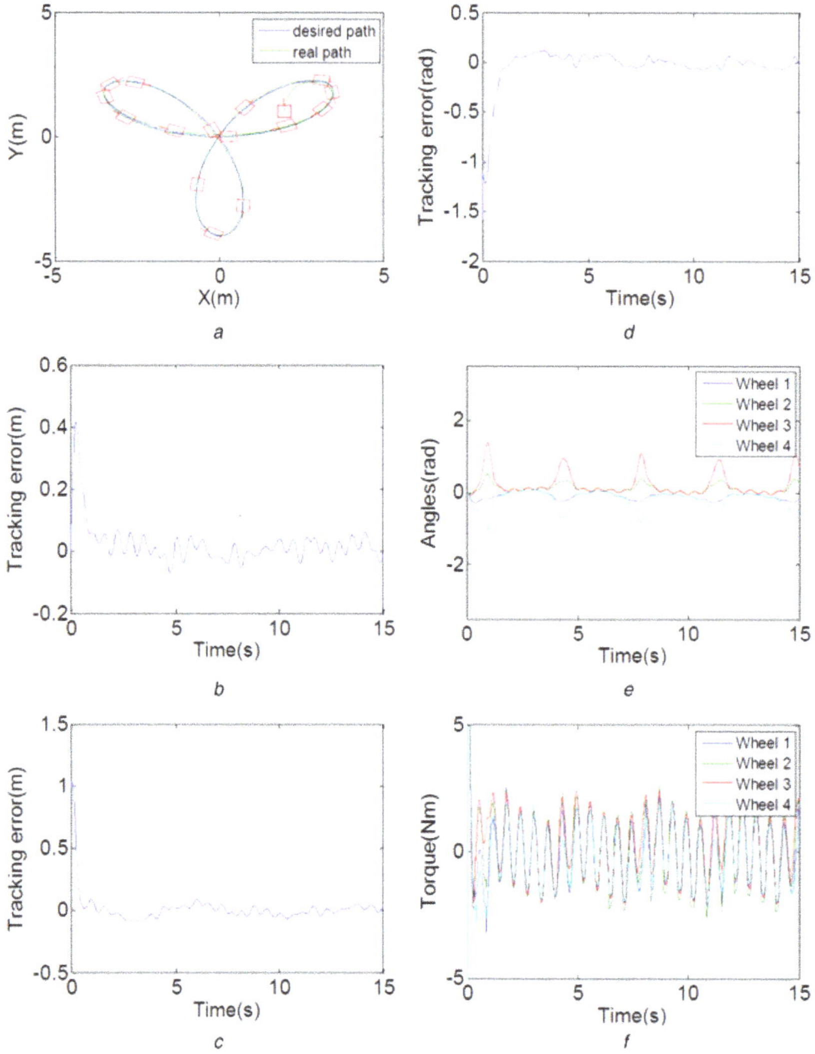

Fig. 7 *Simulation results of Case (II)*
a *x–y* plot of the 4WIS4WID mobile robot
b–d Tracking errors of x_e, y_e and θ_e
e Steering angle of four wheels
f Rolling torques of four wheels

4.2 Curvilinear trajectory simulations

In this section, two trajectories are utilised to demonstrate the results and performances of the proposed control schemes.

For Case (I), a circle trajectory is generated by

$$X_d(t) = 3\cos(0.5t)$$
$$Y_d(t) = 3\sin(0.5t)$$
$$\theta_d(t) = \arctan 2\big(\dot{Y}_d(t),\ \dot{X}_d(t)\big)$$

If the initial position of the robot is at $(X, Y, \theta) = (2, 2, 0)$ and the uncertainties and disturbances are given and bounded by

$$\boldsymbol{D}(t) = 5\sin(10t)\begin{bmatrix}1 & 1 & 1 & 1 & 1 & 1 & 1 & 1\end{bmatrix}^{\mathrm{T}}$$

Simulation results of tracking trajectory control are shown in Fig. 6, where tracking errors, four-wheel steering angles and torques are given.

For Case (II), a trajectory is generated by

$$X_d(t) = 2\sin(1.2t) + 2\cos(0.6t)$$
$$Y_d(t) = 2\sin(0.6t) + 2\cos(1.2t)$$
$$\theta_d(t) = \arctan 2\big(\dot{Y}_d(t),\ \dot{X}_d(t)\big)$$

The initial position of the robot is at $(X, Y, \theta) = (2, 1, 90°)$. The random uncertainties and disturbances given by the following equation are added

$$\boldsymbol{D}(t) = 5\sin(10t)\begin{bmatrix}1 & 2 & 2 & 1 & 1 & 2 & 2 & 1\end{bmatrix}^{\mathrm{T}}$$

The simulation results are shown in Fig. 7, which includes the tracking trajectory, tracking errors, four-wheel steering angles and torques.

Both simulation results demonstrate that the proposed trajectory tracking control scheme can make the 4WIS4WID mobile robot track the desired trajectories successfully and stably although the uncertainties and disturbances exist.

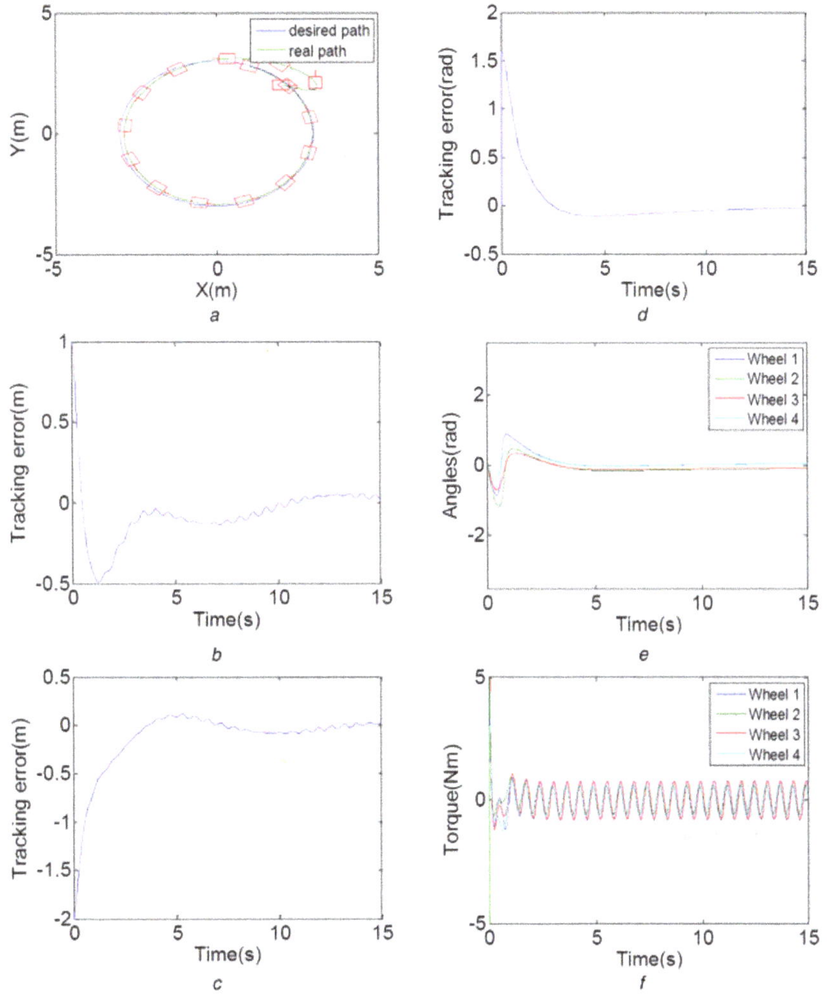

Fig. 8 *Simulation results of Case (I) by the PI dynamic controller*
a *x–y* plot of the 4WIS4WID mobile robot
b–d Tracking errors of x_e, y_e and θ_e
e Steering angle of four wheels
f Rolling torques of four wheels

4.3 Comparative study

The purpose of this section is to demonstrate the superiority of the proposed controller, where the kinematic controller is unchanged and the SMC dynamic controller is replaced by the PI controller. The control law (39) is rewritten as $\boldsymbol{\tau}(t) = \boldsymbol{K}_{\mathrm{P}}\boldsymbol{e} + \boldsymbol{K}_{\mathrm{I}}\int_0^t \boldsymbol{e}(\tau)\,\mathrm{d}\tau$, where $\boldsymbol{K}_{\mathrm{P}} = \mathrm{diag}\{5, 5, 5, 5, 2, 2, 2, 2\}$ and $\boldsymbol{K}_{\mathrm{I}} = \mathrm{diag}\{0.5, 0.5, 0.5, 0.5, 0.5, 0.5, 0.5, 0.5\}$. For the Case (I), the simulation results of the PI dynamic controller are depicted in Fig. 8. By comparing Fig. 6 with Fig. 8, one can find that the convergent speed of the tracking errors in Fig. 8 is slower than that in Fig. 6. Furthermore, the tracking errors of y_e and θ_e have some tremors in Fig. 8 and these phenomena are influenced by $\boldsymbol{D}(t)$. It shows that using the SMC is better than PI control to suppress the uncertainties and disturbances in this research.

5 Conclusions

This paper has proposed new kinematic and the dynamic models for 4WIS4WID mobile robot. These two models can be used to examine some new control schemes and/or other applications for the 4WIS4WID mobile robot. On the basis of these models, the dynamic trajectory tracking control for the 4WIS4WID mobile robot has been presented. The control scheme does not include the parameters of the ICR, so the discontinuous condition when the 4WIS4WID mobile robot moves straight does not encounter. To obtain a better performance for the trajectory tracking control, the proposed kinematic control scheme is derived by the non-linear control and the dynamic control scheme is established by the SMC control technique. Once all of the controller parameters are assigned correctly, the control law can reduce the trajectory tracking errors and suppress the affection of the system uncertainties and external disturbances efficiently. Using the Lyapunov stability theory, the stabilities of the kinematic and dynamic controllers have been proved. The simulation results demonstrate that whether the desired path is a straight line or curvilinear path, the 4WIS4WID mobile robot can track the path successfully. In comparison with the PI dynamic control, the SMC control is indeed effective in suppressing the influences of uncertainties and disturbances for the trajectory tracking control of 4WIS4WID mobile robot.

6 Acknowledgments

This work supported in part by the Ministry of Science and Technology of the Republic of China under grant NSC101-2221-E-006-193-MY3, in part by the Ministry of Education, and the aim for the Top University Project to the National Cheng Kung University (NCKU) are greatly appreciated.

7 References

[1] Yeh Y.C., Li T.H.S., Chen C.Y.: 'Adaptive fuzzy sliding-mode control of dynamic model based car-like mobile robot', *Int. J. Fuzzy Syst.*, 2009, **11**, (4), pp. 272–286

[2] Egerstedt M., Hu X., Stotsky A.: 'Control of mobile platforms using a virtual vehicle approach', *IEEE Trans. Autom. Control*, 2001, **46**, (11), pp. 1777–1782

[3] Hwang C.L., Chang N.W.: 'Fuzzy decentralized sliding-mode control of a car-like mobile robot in distributed sensor-network spaces', *IEEE Trans. Fuzzy Syst.*, 2008, **16**, (1), pp. 97–109

[4] Kumar U., Sukavanam N.: 'Backstepping based trajectory tracking control of a four wheeled mobile robot', *Int. J. Adv. Robot. Syst.*, 2008, **5**, (4), pp. 403–410

[5] Nishihara T.O., Kumamoto H.: 'Automatic path-tracking controller of a four-wheel steering vehicle', *Veh. Syst. Dyn.*, 2009, **47**, (10), pp. 1205–1227

[6] Song Y.D., Chen H.N., Li D.Y.: 'Virtual-point-based fault-tolerant lateral and longitudinal control of 4 W-steering vehicles', *IEEE Trans. Intell. Transp. Syst.*, 2011, **12**, (4), pp. 1343–1351

[7] Ruiz U., Murrieta-Cid R., Marroquin J.L.: 'Time-optimal motion strategies for capturing an omnidirectional evader using a differential drive robot', *IEEE Trans. Robot.*, 2013, **29**, (5), pp. 1180–1196

[8] Ailon A., Zohar I.: 'Control strategies for driving a group of nonholonomic kinematic mobile robots in formation along a time-parameterized path', *IEEE/ASME Trans. Mechatronics*, 2012, **17**, (2), pp. 326–336

[9] Zhang Q., Lapierre L., Xiang X.: 'Distributed control of coordinated path tracking for networked nonholonomic mobile vehicles', *IEEE Trans. Ind. Inform.*, 2013, **9**, (1), pp. 472–484

[10] Park B.S., Yoo S.J., Park J.B., Choi Y.H.: 'A bioinspired neurodynamics-based approach to tracking control of mobile robots', *IEEE Trans. Ind. Electron.*, 2012, **59**, (8), pp. 3211–3220

[11] Censi A., Franchi A., Marchionni L., Oriolo G.: 'Simultaneous calibration of odometry and sensor parameters for mobile robots', *IEEE Trans. Robot.*, 2013, **29**, (2), pp. 475–492

[12] Yi J., Wang H., Zhang J., Song D., Jayasuriya S., Liu J.: 'Kinematic modeling and analysis of skid-steered mobile robots with applications to low-cost inertial-measurement-unit-based motion estimation', *IEEE Trans. Robot.*, 2009, **25**, (5), pp. 1087–1097

[13] Yi J., Song D., Zhang J., Goodwin Z.: 'Adaptive trajectory tracking control of skid-steered mobile robots'. Proc. IEEE Int. Conf. on Robotics and Automation, 2007, pp. 2605–2610

[14] Morales J., Martinez J.L., Mandow A., Garcia-Cerezo A., Pedraza S.: 'Power consumption modeling of skid-steer tracked mobile robots on rigid terrain', *IEEE Trans. Robot.*, 2009, **25**, (5), pp. 1098–1108

[15] Kim K.B., Kim B.K.: 'Minimum-time trajectory for three-wheeled omnidirectional mobile robots following a bounded-curvature path with a referenced heading profile', *IEEE Trans. Robot.*, 2011, **27**, (4), pp. 800–808

[16] Indiveri G.: 'Swedish wheeled omnidirectional mobile robots: kinematics analysis and control', *IEEE Trans. Robot.*, 2009, **25**, (1), pp. 164–171

[17] Lins Barreto S.J.C., Scolari Conceicao A.G., Dorea C.E.T., Martinez L., de Pieri E.R.: 'Design and implementation of model-predictive control with friction compensation on an omnidirectional mobile robot', *IEEE/ASME Trans. Mechatronics*, 2014, **19**, (2), pp. 467–476

[18] Huang H.C.: 'SoPC-based parallel ACO algorithm and its application to optimal motion controller design for intelligent omnidirectional mobile robots', *IEEE Trans. Ind. Inform.*, 2013, **9**, (4), pp. 1828–1835

[19] Dietrich A., Wimbock T., Albu-Schaffer A., Hirzinger G.: 'Reactive whole-body control: dynamic mobile manipulation using a large number of actuated degrees of freedom', *IEEE Robot. Autom. Mag.*, 2012, **19**, (2), pp. 20–33

[20] Qian H., Lam T.L., Li W., Xia C., Xu Y.: 'System and design of an omni-directional vehicle'. Proc. IEEE Int. Conf. on Robotics and Biomimetics, Bangkok, Thailand, 2009, pp. 389–394

[21] Connette C.P., Pott A., Hägele M., Verl A.: 'Control of an pseudo-omnidirectional, non-holonomic, mobile robot based on an ICM representation in spherical coordinates'. Proc. IEEE Int. Conf. on Decision and Control, Cancun, Mexico, 2008, pp. 4976–4983

[22] Clavien L., Lauria M., Michaud F.: 'Instantaneous centre of rotation estimation of an omnidirectional mobile robot'. Proc. IEEE Int. Conf. on Robotics and Automation, Anchorage, AK, USA, 2010, pp. 5435–5440

[23] Ploeg J., van der Knaap A.C.M., Verburg D.J.: 'ATS/AGV-design, implementation and evaluation of a high performance AGV'. Proc. IEEE Intelligent Vehicles Symp., 2002, **1**, pp. 127–134

[24] Lin C.J., Hsiao S.M., Wang Y.H., Yeh C.H., Huang C.F., Li T.H.S.: 'Design and implementation of a 4WS4WD mobile robot and its control applications'. Proc. IEEE Int. Conf. on Systems Science and Engineering, Budapest, Hungary, 2013, pp. 235–240

[25] Lam T.L., Qian H., Xu Y.: 'Omnidirectional steering interface and control for a four-wheel independent steering vehicle', *IEEE/ASME Trans. Mechatronics*, 2010, **15**, (3), pp. 329–338

[26] Jiang S.Y., Song K.T.: 'Differential flatness-based motion control of a steer-and-drive omnidirectional mobile robot'. Proc. IEEE Int. Conf. on Mechatronics and Automation 2013, Takamatsu, Japan, August 2013, pp. 1167–1172

[27] Selekwa M.F., Nistler J.R.: 'Path tracking control of four wheel independently steered ground robotic vehicles'. Proc. IEEE Conf. on Decision and Control and European Control Conf. (CDC-ECC), Orlando, FL, USA, December 2011, pp. 6355–6360

[28] Hwang C.L., Wu H.M.: 'Trajectory tracking of a mobile robot with frictions and uncertainties using hierarchical sliding-mode under-actuated control', *IET Control Theory Appl.*, 2013, **7**, (7), pp. 952–965

[29] Jean J.H., Lian F.L.: 'Robust visual servo control of a mobile robot for object tracking using shape parameters', *IEEE Trans. Control Syst. Technol.*, 2012, **20**, (6), pp. 1461–1472

[30] Rubagotti M., Della Vedova M.L., Ferrara A.: 'Time-optimal sliding-mode control of a mobile robot in a dynamic environment', *IET Control Theory Appl.*, 2011, **5**, (16), pp. 1916–1924

Analogue photonic link design charts for microwave engineering applications

Vincent J. Urick

Naval Research Laboratory, Washington, DC, USA
E-mail: vincent.urick@nrl.navy.mil

Abstract: A set of unique design charts for intensity-modulation direct-detection microwave photonics links is presented. The charts facilitate link design and analysis, clearly demonstrating performance trade-offs in terms of standard microwave performance metrics.

1 Introduction

The field of microwave photonics [1] has vast utility with applications ranging from links [2] to advanced signal processing [3, 4]. Despite its wide applicability, the radio-frequency (RF) performance trade-offs of such links are not easily accessible from the literature. This paper provides a straightforward, yet powerful, analysis resulting in design charts that can be utilised in many engineering disciplines.

2 Governing equations

The analysis is focused on an intensity-modulation direct-detection analogue photonic link employing an external Mach–Zehnder modulator (MZM) as shown in Fig. 1. In this case it is assumed, as is typical, that half of the generated photocurrent is lost in an impedance matching circuit at the photodiode output. The behaviour of such a link is well understood and can be cast in terms of the RF gain factor (g), noise factor (F), third-order spurious-free dynamic range (SFDR) and 1 dB compression dynamic range (CDR) [1]

$$g = I_{dc}^2 \pi^2 R_i R_o / (4V_\pi^2) \tag{1}$$

$$F = \frac{4V_\pi^2 N_o}{I_{dc}^2 \pi^2 R_i R_o k_B T_s} = 1 + \frac{V_\pi^2}{\pi^2 R_i}\left(\frac{4}{I_{dc}^2 R_o} + \frac{2e}{I_{dc} k_B T_s} + \frac{\text{RIN}}{k_B T_s}\right) \tag{2}$$

$$\text{SFDR} = \left[I_{dc}^2 R_o / (N_o B)\right]^{2/3} \tag{3}$$

$$\text{CDR}_{1\,\text{dB}} = 0.4516 \cdot I_{dc}^2 R_o / (4 N_o B) \tag{4}$$

respectively. In the above equations, I_{dc} is the average photocurrent, V_π is the MZM half-wave voltage, R_i and R_o are the input and output resistances, N_o is the output noise power spectral density, k_B is Boltzmann's constant, $T_s = 290$ K is the standard noise temperature, e is the electronic charge constant, B is the RF bandwidth and RIN is the relative intensity noise. This last term captures optically generated noise in excess of shot noise. Finally, (3) captures only the distortion associated with the ideal MZM transfer function – other lower-level sources requiring additional analysis have been treated in [5].

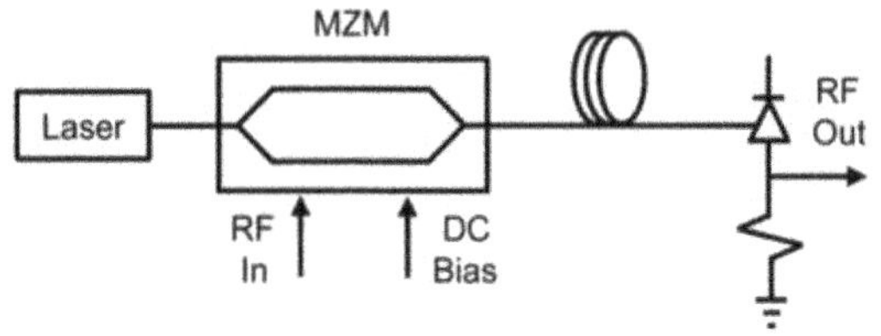

Fig. 1 *Direct-detection link employing an MZM*

Following the technique first proposed by Bucholtz *et al.* [6], the SFDR can be written as a function of F by multiplying (2) and (3)

$$\text{SFDR}(F) = \left(\frac{1}{F}\right)^{2/3}\left(\frac{4V_\pi^2}{\pi^2 R_i k_B T_s B}\right)^{2/3} \tag{5}$$

In similar fashion, (2) and (4) can be multiplied to yield

$$\text{CDR}_{1\,\text{dB}}(F) = \frac{1}{F}\frac{0.4516\, V_\pi^2}{\pi^2 R_i k_B T_s B} \tag{6}$$

As demonstrated in the following section, (5) and (6) capture the major trade-offs of the link in Fig. 1.

3 Design charts

Plots generated from (5) and (6) are shown in Figs. 2 and 3, respectively. In our previous work [6], only the fundamental limits on (5) were examined resulting in a plot similar to Fig. 2*a*. Here, the important effects of RIN are taken into account and analysis of (6) is also included. Although RF gain contours were included in [6], they are excluded here as the gain is easily determined by (1).

Figs. 2 and 3 are plotted for $R_i = R_o = 50\ \Omega$ and $B = 1$ Hz. The range of values for V_π and I_{dc} were chosen to coincide with the present and foreseeable state of the art. For example, the best performance in the microwave is $V_\pi \simeq 1$ V and $I_{dc} \simeq 100$ mA, whereas that in the millimetre wave is $V_\pi > 10$ V and $I_{dc} \simeq 10$ mA. Future high-power microwave photodetectors may achieve 1 A and maybe even 10 A. Photodetectors above 100 GHz may be limited to 1 mA or less average current. Sub-volt V_π levels with wide bandwidth will hopefully emerge; $V_\pi > 16$ V has limited utility. The RIN values were chosen judiciously in terms of laser performance, although other sources can add RIN such as optical amplifiers or propagation effects. An RIN level of −175 dBc/Hz is in line with the best reported results; many semiconductor lasers exhibit relaxation oscillations with peak RIN = −155 dBc/Hz.

For a fixed R_i and B, V_π contours are generated directly by (5) and (6). The I_{dc} curves are obtained from (2) plotted as a function of V_π with a fixed R_o and RIN. The intersection of any pair of I_{dc} and V_π contours uniquely determines the values of SFDR/$\text{CDR}_{1\,\text{dB}}$ and F. The detrimental effects of increasing RIN can be seen in the figures. For example, there is minimal benefit in going beyond $I_{dc} = 10$ mA when RIN = −155 dBc/Hz unless higher gain is desired. An important trade-off shown by this analysis is that between F and dynamic range. This is typically understood in regard to electronic components but sometimes lost in the photonics community – the trades are clearly shown here. The slope of the V_π contours by themselves reinforces the common misperception that $V_\pi \to 0$ is optimal.

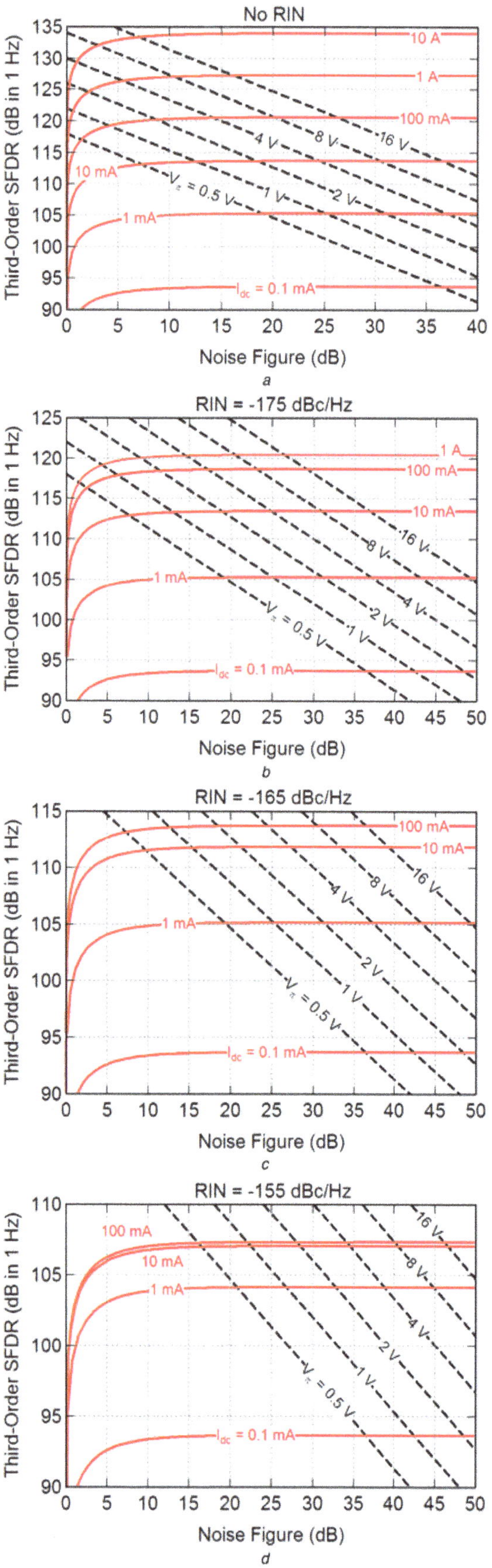

Fig. 2 *Design space for SFDR(F) as given by (5). The intersection of the DC photocurrent and MZM half-wave voltage contours indicate the allowable values*
a no relative intensity noise (RIN)
b –175 dBc/Hz RIN
c –165 dBc/Hz RIN
d –155 dBc/Hz RIN

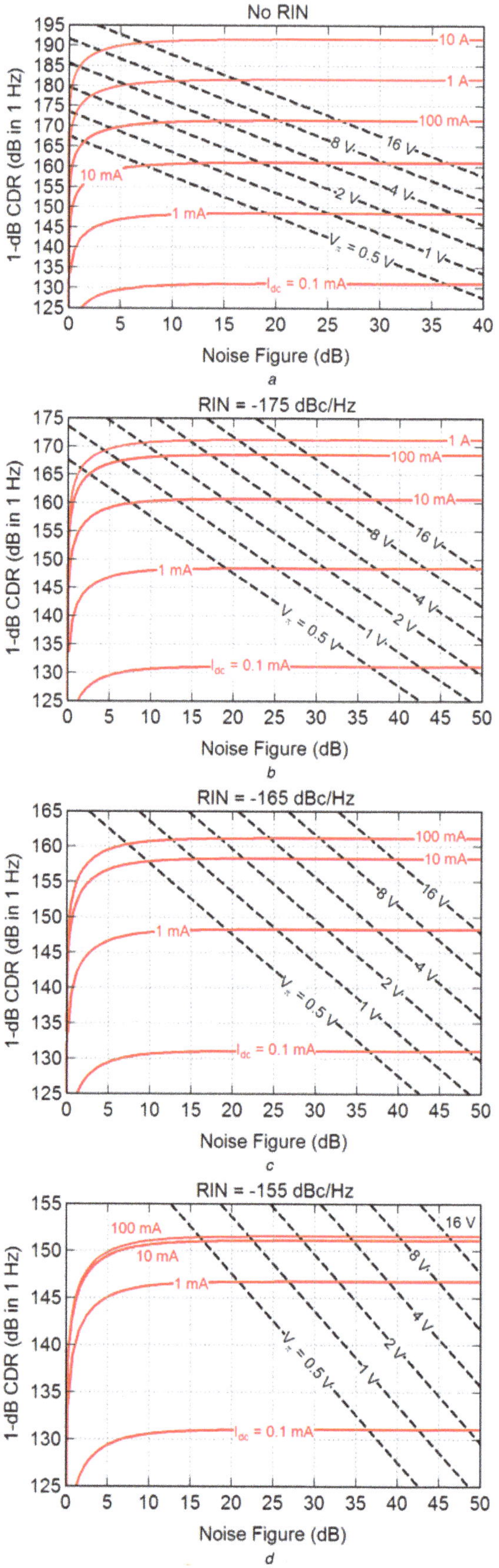

Fig. 3 *Design space for $CDR_{1\ dB}(F)$ as given by (6). The intersections of the DC photo current and MZM half-wave voltage contours indicate the allowable values*
a no relative intensity noise (RIN)
b –175 dBc/Hz RIN
c –165 dBc/Hz RIN
d –155 dBc/Hz RIN

However, the I_{dc} curves in the low-noise-figure limit demonstrate that a larger V_π can sometimes afford more dynamic range at the cost of increased F.

4 Summary and conclusion

Two simple equations have been presented that can be plotted to determine the trade space for an analogue photonic link employing intensity modulation via an MZM. The graphs presented provide a straightforward means to determine the bounds on performance for such a link in terms of component metrics. Although the results here concentrate on intensity-modulated links, the methodology can be adopted for the analysis of phase- or polarisation-modulated links as well. This paper can serve as a reference to guide analogue photonic link design and integration into electronic systems.

5 References

[1] Urick V.J., McKinney J.D., Williams K.J.: 'Fundamentals of microwave photonics' (Wiley, Hoboken, NJ, 2015)

[2] Cox III C.H., Ackerman E.I.: 'Recent advances in high-frequency (> 10 GHz) microwave photonic links', in Kaminow I.P., Li T., Willner A.E. (Eds.): 'Optical fiber telecommunications VIB' (Academic, 2013)

[3] Capmany J., More J., Gasulla I., Sanco J., Lloet J., Sales S.: 'Microwave photonic signal processing', *J. Lightwave Technol.*, 2013, **31**, (4), pp. 571–586

[4] Minasian R.A., Chan E.H.W., Yi X.: 'Microwave photonic signal processing', *Opt. Express*, 2013, **21**, (19), pp. 22918–22936

[5] Urick V.J., Diehl J.F., Draa M.N., McKinney J.D., Williams K.J.: 'Wideband analog photonic links: some performance limits and considerations for multi-octave implementations', *Proc. SPIE*, 2012, **8259**, pp. 1–14, doi: 825904-1-14

[6] Bucholtz F., Urick V.J., Godinez M., Williams K.J.: 'Graphical approach for evaluating performance limitations in externally modulated analog photonic links', *IEEE Trans. Microw. Theory Tech.*, 2008, **56**, (1), pp. 242–247

7

Assistive technology for relieving communication lumber between hearing/speech impaired and hearing people

Rini Akmeliawati[1], Donald Bailey[1,2], Sara Bilal[1], Serge Demidenko[2,3], Nuwan Gamage[4], Shujjat Khan[2], Ye Chow Kuang[4], Melanie Ooi[4], Gourab Sen Gupta[2]

[1]Faculty (Kulliyyah) of Engineering, International Islamic University Malaysia, Jl, Gombak 53100, Kuala Lumpur, Malaysia
[2]School of Engineering and Advanced Technology, Massey University, New Zealand, Private Bag 11222, Palmerston North 4442, New Zealand
[3]Centre of Technology, RMIT University Vietnam, 702 Nguyen Van Linh Blvd, Ho Chi Minh City, HCMC, Vietnam
[4]School of Engineering, Monash University Malaysia, Jl Lagoon Selatan, 46150, Selangor Darul Ehsan, Malaysia
E-mail: serge.demidenko@rmit.edu.vn

Abstract: This study proposes an automatic sign language translator, which is developed as assistive technology to help the hearing/speech impaired communities to communicate with the rest of the world. The system architecture, which includes feature extraction and recognition stages is described in detail. The signs are classified into two types: static and dynamic. Various types of sign features are presented and analysed. Recognition stage considers the hidden Markov model and segmentation signature. Real-time implementation of the system with the use of Windows7 and LINUX Fedora 16 operating systems with VMware workstation is presented in detail. The system has been successfully tested on Malaysian sign language.

1 Overview

'Sign language' (SL) is a highly structured non-verbal language utilising both manual and non-manual communications. Manual communication consists of movements and orientation of hand/arm conveying symbolic meaning, whereas non-manual communication involves mainly facial expression (as shown in Fig. 1*a*), head movement, body posture and orientation, which help in augmenting the meaning of the manual signs. Furthermore SL consists of static signs, which are mainly the alphabets (shown in Fig. 1*b*), and dynamic signs, which involve some motions as shown in Fig. 1*c*.

SLs have a systematic and complex structure of grammar that consists of isolated signs (words) and continuous signs (sentences) differing from one country to another.

Everyday communication with the hearing population poses a major challenge to those with hearing loss. For many people who were either born with hearing impairment or became impaired later in their lives, SL is used as their main language. Spoken languages such as English, Malay and others are often learnt only as a second language. As a result, their reading and writing skills are often below average as they mostly opt to converse in SL. Although some can read, many others fail in cases where reading is needed.

Common current options for alternative communication modes include cochlear implants, writing and interpreters. Cochlear implants are small and complex electronic devices that can help to provide a sense of sound to a person who is profoundly hearing/speech impaired or severely hard-of-hearing. The use of a cochlear implant requires both a surgical procedure and significant therapy to learn or re-learn the sense of hearing. Not everyone acquires learning at the same level with this device and the device is relatively expensive. Handwriting is another alternative for communicating with hearing/speech impaired people. However, most of the hearing/speech impaired people cannot communicate well through written language because they use SL as their preferable language for communicating. Interpreters are commonly used within the hearing/speech impaired community, but interpreters can charge high hourly rates and be awkward in situations where privacy is of high concern, such as at a doctor or lawyer's office. In addition, the number of interpreters is very limited particularly in developing countries like Malaysia [1].

'Automatic SL translator' (ASLT) is an automated system using advanced technology to translate a particular SL into a readable language such as English, Malay, Chinese etc. The existing ASLT systems generally use the following:

i. *DataGlove* or *CyberGlove* [2]: A specially built electronic glove worn by a signer. The glove has built-in sensors, which detect and transmit information on the hand posture as illustrated in Fig. 2.

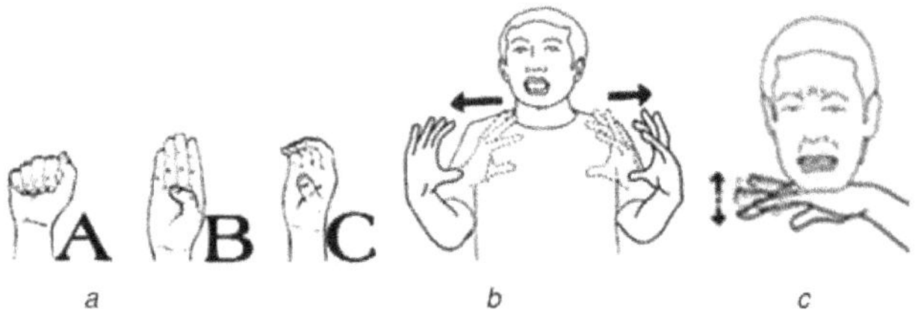

Fig. 1 *Overview*
a Alphabet letters '*A*', '*B*' and '*C*'
b Dynamic sign using two hands 'big'
c Facial expression in sign 'dirty' [1]

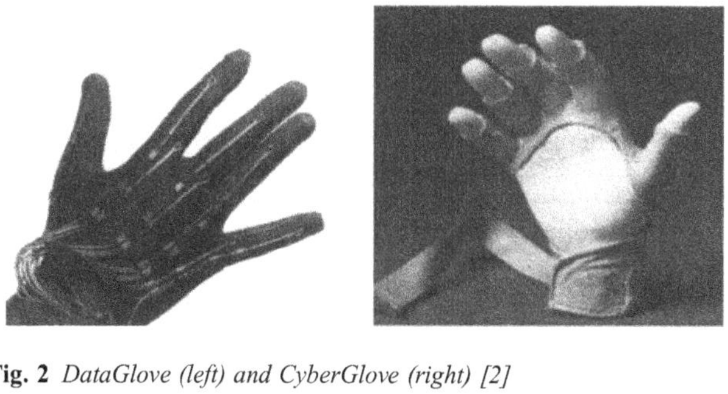

Fig. 2 *DataGlove (left) and CyberGlove (right) [2]*

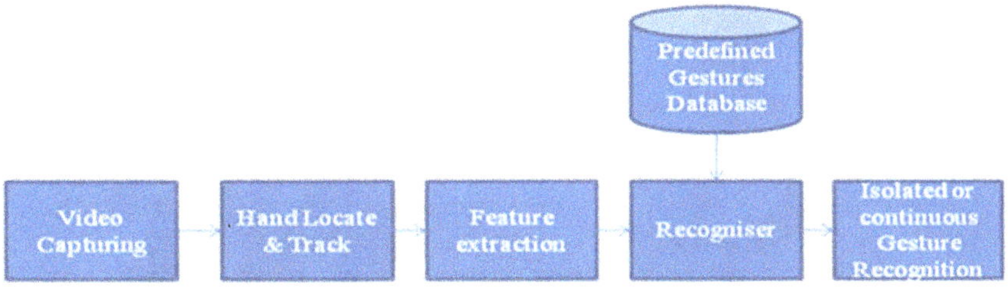

Fig. 3 *Structure of a complete ASLT*

ii. *Vision-based approaches:* where a camera is used to capture images of a person who is signing by using either a coloured glove or just bare hands. The major advantage of this approach compared with the application of the DataGlove or CyberGlove is the flexibility. It can be developed to include non-manual signs such as recognition of facial expressions and head movements as well as lip-reading. In addition, the position of the signer's hand with respect to other body parts could be identified by using the vision system. Owing to the above-mentioned advantages, this paper focuses on the vision-based ASLT.

This paper is organised as follows. Section 2 presents the system architecture, starting from the video image acquisition, hand location and tracking and feature extraction. Section 3 describes the recognition stage in which 'hidden Markov model' (HMM) and segmentation signature are detailed. Section 4 discusses the real-time implementation of the system. Section 5 discusses the performance of the system. Finally, this paper is concluded in Section 6.

2 System architecture

In general, the structure of vision-based automatic SL consists of four crucial stages; database collection, blob detection and tracking, feature extraction and the recognition stage.

The structure is shown in Fig. 3.

2.1 SL database collection

Most SLs databases are available only for educational and learning purposes. The SL database is required to build algorithms, expressing the nature of signs and covering the possibility of signing. Recent researches in SL recognition are conducted for SLs used in different countries.

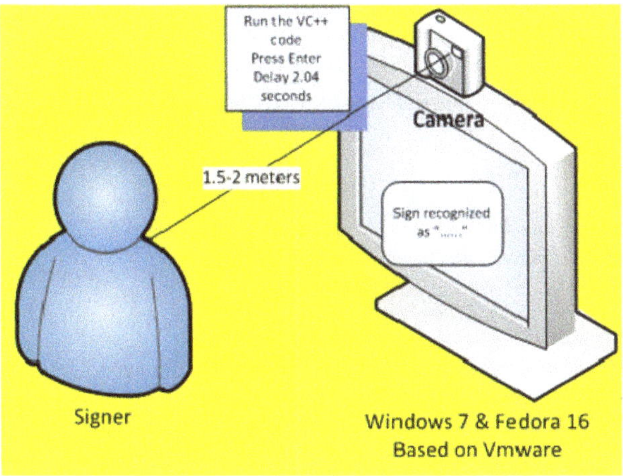

Fig. 4 *Sign language database collection system set-up*

The 'Malaysian SL' (MSL) database was collected by recording video of signers. The signer has to stand about 1.5–2 m in front of the camera, where his/her upper body is visible in the scene as shown in Fig. 4. Once the signer starts signing, the recording process is initiated. The video is then stored into the computer for further processing.

2.2 Blob detection and tracking

The second stage of the ASLT includes finding the face and two hand blobs in each video frame. There are several existing methods to accomplish these tasks such as skin colour approaches [3], motion constraint approaches [4] and the static background inclusion approaches [5]. However, it is often not reliable to model a skin colour where high variations of skin colours and different lighting conditions are present. Limitations arise from the fact that human skin can be defined in various colour spaces after applying colour normalisation. Therefore the model has to accept a wide range of colours, making it more susceptible to the noise. On the other hand, the motion systems assume that the hand is the fastest moving object in the image frame. This is not the case when the hand gesture does not carry a fast motion or the head gesture is stronger than the hand, that is, the head gesture is more active than that of the hand. In addition, the background inclusion approaches assume a static background. These approaches have many limitations in terms of reliability. Therefore those methods have drawbacks making them to be not a good choice for the SL recognition developed system. Therefore a novel hybrid system combining the appearance-based method and skin colour segmentation has been introduced:

(a) *Blob detection with Haar-like features:* Initialising the system with a face or specific hand shape will provide an adequate search region for skin pixels. 'Artificial neural network'-based methods, 'support vector machines' (SVMs) and other kernel methods have been used for face or hand posture detection. However, most of these algorithms use raw pixel values as features. This makes such systems sensitive to noise and changes in illumination. Instead, other approaches such as Haar-like features proposed by Viola and Jones [6] have been used in this paper to detect the face or hand as an initial stage. Meanwhile, the system has been initiated by detecting the face or hand region. The dimension of the colour space is not a big concern because the range of the skin pixels within the detected area has been obtained. This reduces the required memory space and processing time. To perform the skin detection from an image, the image needs to be converted from red–green–blue (RGB) to YC_bC_r colour space after the face or hand has been detected. Then, in order to find skin pixels that fall within the same colour space range of the detected hand or face, 10×10 pixels box from the centre of the face is extracted as shown in Fig. 5*c*. Finally, a range of skin pixels is specified based on the detected face, the distribution of skin pixels values is highlighted using C_b and C_r components and ($C_b - C_r$ or $C_r - C_b$) as an additional threshold while the luminance Y component is discarded. The segmented image is shown in Fig. 5*d*.

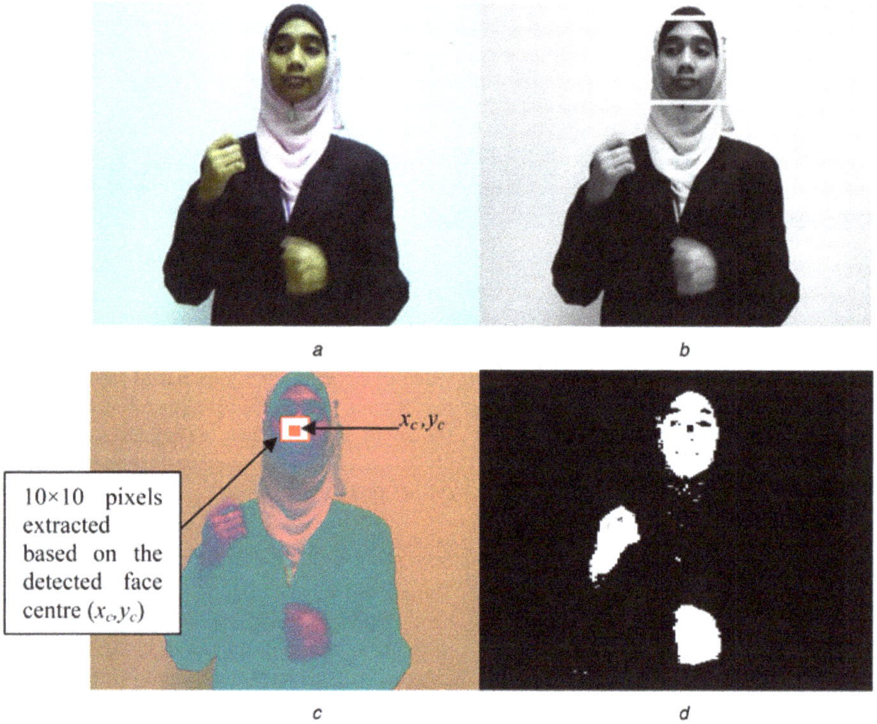

Fig. 5 *Blob detection with Haar-like features*
a Original RGB image
b Detected face on grey-scale image
c YC_bC_r image
d Segmented image based on C_b, C_r and $C_r - C_b$ threshold [7]

(b) *Blob labelling:* The detected and extracted face and two hand blobs must be identified and labelled. In this paper, the method developed in [8] that simultaneously labels contours and components in binary images is used. This method labels the blobs and identifies each blob based on certain criteria such as the size and perimeter. However in the developed system, the blob labelling has been achieved by applying the y-axis labelling approach as shown in Fig. 6, rather than using the size and the perimeter as mentioned in [8].

2.3 Feature extraction

Performance of any SL recognition system significantly depends on obtaining efficient features to represent pattern characteristics. There is no algorithm, which shows how to select the representation or to choose the features. The selection of features depends on the application [9]. There are many different methods to represent two-dimensional (2D) images such as boundary, topological, shape grammar and description of similarity. Features should be chosen so that they are intensive to noise-like pattern variations and the number of features should be small for easy computation. The hand posture shape feature, motion trajectory feature and hand position with respect to other human upper body (HUB) parts play an important role within the preparation stage of the gesture before SL recognition stage. There are static and dynamic signs in SL as mentioned earlier. The static signs usually come under the alphabet signs. This research has conducted tests on static signs as well as dynamic ones.

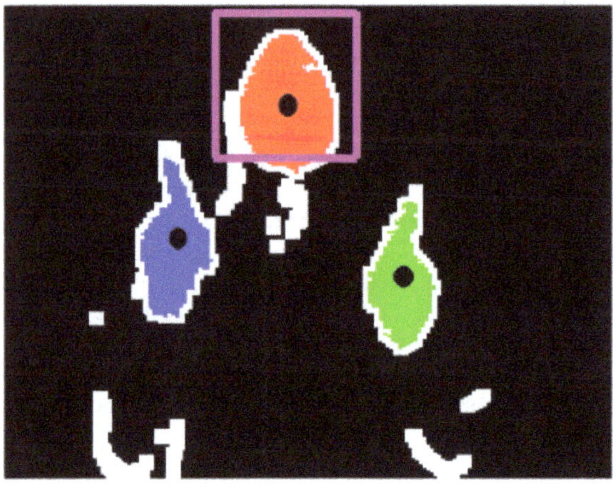

Fig. 6 *Blob labelling using different colours*

2.3.1 1D profile for finger detection in static signs: The boundary of a hand is extracted by subtracting the filled image from the eroded one. An important issue is to determine the internal profile of a detected hand boundary, which emphasises on the detection of fingers. This is done by allocating the centre of the target (x_c, y_c).

Then the radial distance $d_i = \sqrt{(x_i - x_c)^2 + (y_i - y_c)^2}$ from the centroid (x_c, y_c) to each array point (x_i, y_i) is computed. After smoothing d_i for noise reduction using Savitzky–Golay smoothing filters, the local maxima in the 1D curve has been determined. The number of the peaks in the vector above the centre range determines the fingers, whereas the approximate distance between the peaks identifies the finger which is opened. This procedure is shown in Fig. 7.

2.3.2 High-level features for dynamic signs: This research investigated three types of features, which are: 'geometric' (shape), 'motion' and 'location'.

Geometric features: Contours or edges are features, which can be used in any model-based technique as well as in non-model ones. The aim is to obtain similar values of features for similar hand

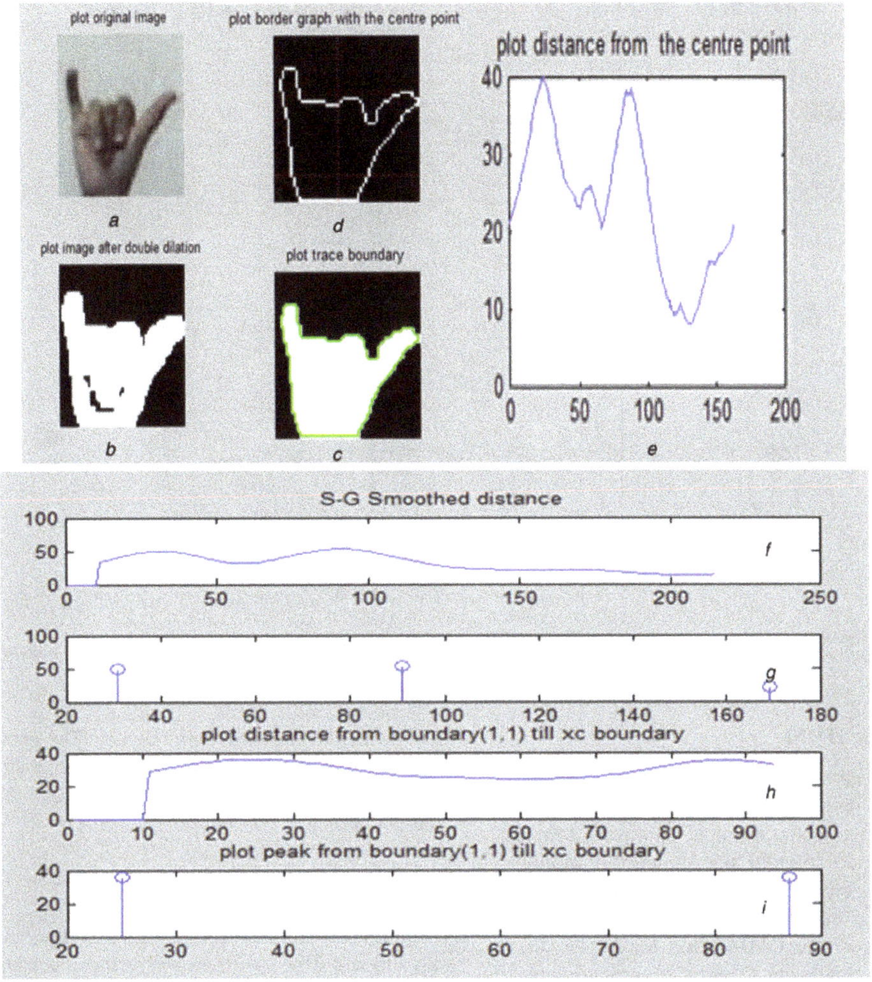

Fig. 7 *1D profile for finger detection in static signs*
a Detected hand
b Double dilation
c Filled image
d Boundary
e 1D profile
f Smoothed profile
g Local maxima
h and *i* Upper hand profile and maxima from its centre [10]

shapes and distant values for different shapes. This is done by choosing seven features (blob area, semi-major and semi-minor axes, orientation angle, compactness, roundness and eccentricity). These features are moment-based features from the grey-scale image. The definition of moments can be found in [11].

Motion features: To capture the dynamic characteristics of the hand posture, sequence of frames must be available with the hand motion. The main issue is to track the hand over the video frames and extract the centre point (x_c, y_c) from the hand posture over a sequence of successive frames. Following that, eight trajectory-based features are extracted from the detected blob. These eight features are stored in a feature vector. An n-dimensional feature vector represents numerical features of an object. These features are the centre point of the hand (x_c, y_c), difference between the consecutive centre points, average sum of x_c and y_c, velocity and angle.

Location features: In SL recognition systems, it is very important to find the hand location with respect to the head, the shoulder and the chest as this carries a lot of meaning. The HUB parts can be used as reference for static and dynamic signs. Various existing methods using geometric modelling, boosted classifiers and SVM have been introduced for HUB detection. However, real-time applications such as automatic SL recognition systems require a fast method for HUB part detection and tracking. Therefore, in this paper, a fast and robust search algorithm for HUB parts based on the study [12] has been introduced. It assumes that all body parts can be measured with respect to a head size. Initialising the system with finding a face provides an adequate search region for other HUB parts. The proposed system used Haar-like features and 'AdaBoost' algorithm for face region detection. Following that, other HUB regions are found based on a human figure adjusted for artists based on accurate 8-head size as shown in Fig. 8.

Tracking HUB region: Pose detection (or initialisation) is typically performed on the initial video frame followed by pose tracking where the pose parameters obtained from the current frame are used as the starting value for the subsequent video frame. Many methods exist to track pose using motion histories and optical flow technique. Other methods, such as continuously adaptive mean shift (CAMSHIFT) were designed for face and coloured object tracking using probability distribution [13]. CAMSHIFT is very sensitive to the change of the face colour over time. Therefore the tracking in this paper is based on the face centre point which is obtained from the skin binary image. This method of tracking performs better than the CAMSHIFT tracking method as shown in Figs. 9 and 10.

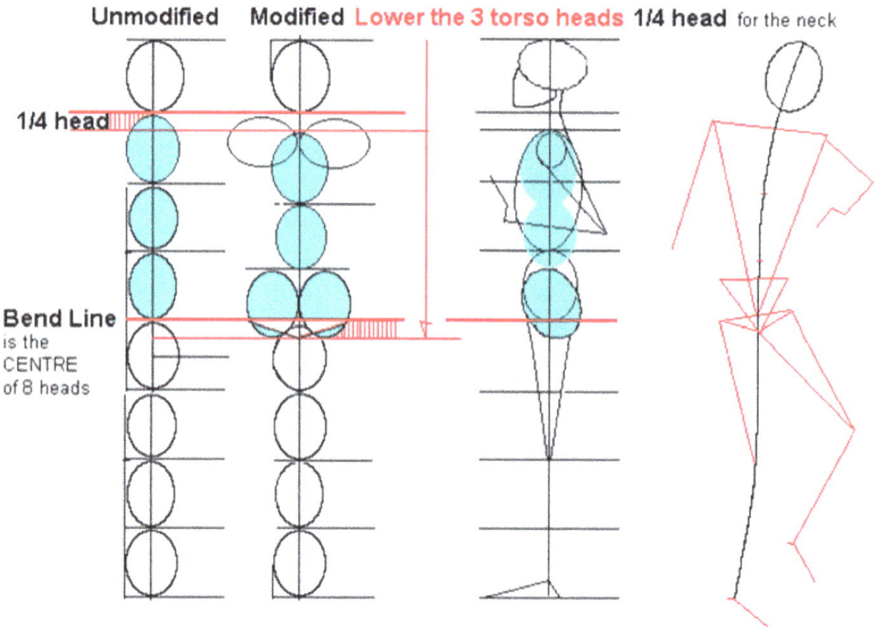

Fig. 8 *Accurate 8-head-high adult male human figure [12]*

3 Recognition of SL using HMM

In this research HMM has been chosen for training and testing the MSL translator system based on its ability to distinguish transitions between signs. It is sufficient to model a simple sign using a three state HMM. Meanwhile for more complex signs in terms of motion and sign duration more states are necessary [15]. Therefore in this paper a four state HMM have been used to model isolated signs, with no skip state. Training of the HMM-based recogniser was done using tools developed employing C and C++ programming languages under 'LINUX' operation system.

3.1 Isolated and continuous sign training, testing and recognition

Signs have been collected with at least ten repetitions for each sign. Then, signs passed through blob detection to feature extraction stages. After extraction of motion and geometric features from isolated and continuous signs, these features are concatenated in one text file. The most crucial issue during training and recognising isolated and continuous signs is the HMM topology, grammar structure and master label file (MLF) for words and sentence structure.

3.2 Grammar for isolated and continuous signs

3.2.1 Grammar for isolated signs: An example of the grammar file structure, which contains five isolated words 'car', 'chicken', 'dog', 'noodles' and 'school' is as follows

$$\$\text{gesture} = \text{car}|\text{chicken}|\text{dog}|\text{noodles}|\text{school}(\$\text{gesture})$$

3.2.2 Grammar for continuous signs: The grammar for sentences was chosen after analysing different grammar structures. Many different types of grammar were tested and the below sentence grammar structure was chosen

$$\text{pronoun/noun} + \text{verb} + \text{adjective} + \text{noun}$$

The grammar with both 'Start' sign and 'End' sign was tested. The start and end delay times are very crucial because it shows the beginning and ending of the sentence. In the developed system, it is not required for a signer to sign under certain restrictions, such as fixing a start sign. Therefore the start and end of the sign is taken as a random observation.

3.3 Recognition accuracy of isolated signs

The isolated signs comprised 42 signs from MSL database, which were collected randomly within the categories of pronoun, noun, verbs and adjective. The evaluation of different isolated sign combinations has been done by running different experiments in order to analyse the effect of feature vectors on the recognition accuracy.

3.3.1 Effect of features: example of four signs: Four signs 'green', 'yellow', 'water' and 'rice' have been chosen based on the similarities between each pair ('green' and 'yellow', 'water' and 'rice') to run the second set of experiments. Signs 'green' and 'yellow' are very similar to each other as illustrated in Figs. 11*a* and *b*. It is difficult even for human eyes to differentiate between these two signs.

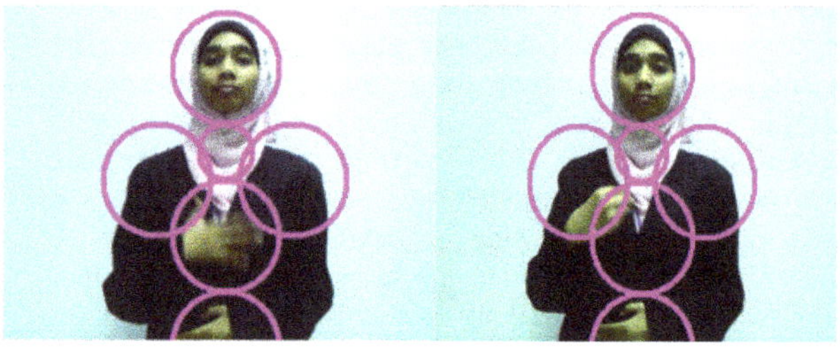

Fig. 9 *Tracking of HUB within two frames of sign 'white' using the proposed method [14]*

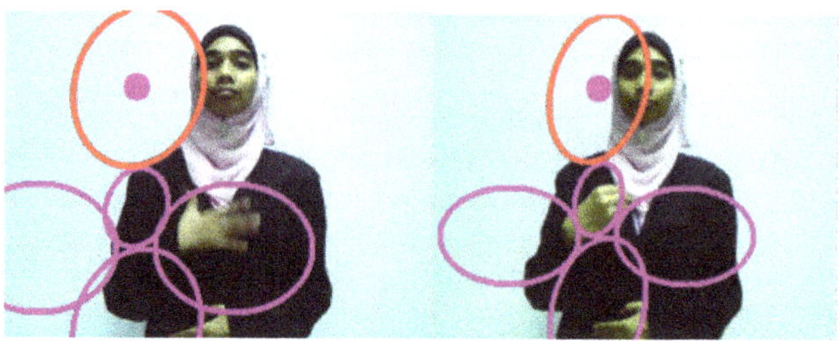

Fig. 10 *Tracking of HUB within two frames of sign 'white' using CAMSHIFT [13]*

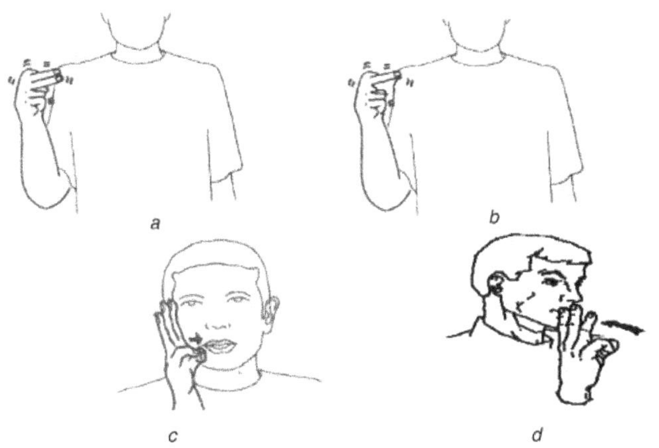

Fig. 11 *Signs*
a 'Green'
b 'Yellow'
c 'Rice'
d 'Water' from MSL database [1]

From Fig. 12, it is clear that the confusion mainly happened between signs that have similarities such as 'green' and 'yellow' because the motion trajectory is relatively the same. In such cases, one can estimate that geometric features could play a role in the accuracy enhancement. In fact, Fig. 13 shows that the existence of geometric features could drop the system accuracy. These could be verified as follows:

i. Signs 'water' and 'rice' have better accuracies in Fig. 13 than the confusion that happened between the two signs in Fig. 12. This is because of nature of the two signs; that is, slow dynamic motion (see Fig. 11).
ii. Signs 'green' and 'yellow' are very similar even for human vision (see Figs. 11*a* and *b*). Therefore none of the features could enhance the system accuracy in that case.

3.3.2 Effect of feature vectors: combining one- and two-hand signs: Another two sets of experiments have been launched while combining signs that use two hands and signs that use only one hand. The set of experiments employing eight feature vectors from one- and two-hand signs have been used. Although using eight feature vectors for one- and two-hand signs separately, the system accuracy was 75.9883 and 85.61%, respectively. However, if the two sets (for one- and two-hand signs) are combined (11 signs in total) the accuracy was 83.63%.

3.4 Recognition accuracy for continuous signs

Sentences with the grammatical structure 'noun/pronoun, verb, adjective, noun' were chosen for the recognition. In total 20 nouns, 5

```
------------------------ Overall Results ---------------
SENT: %Correct=70.00 [H=14, S=6, N=20]
WORD: %Corr=70.00, Acc=70.00 [H=14, D=0, S=6, I=0, N=20]
------------------------ Confusion Matrix --------------
       g   r   w   y
       r   i   a   e
       e   c   t   l
       e   e   e   l
       n       r   o  Del [ %c / %e]
gree   1   0   0   1    0 [50.0/5.0]
rice   0   6   0   0    0
wate   0   3   4   0    0 [57.1/15.0]
yell   2   0   0   3    0 [60.0/10.0]
Ins    0   0   0   0
```

Fig. 12 *Training and test results for 'yellow', 'green', 'water' and 'rice' signs (motion features)*

```
------------------------ Overall Results ---------------
SENT: %Correct=58.33 [H=7, S=5, N=12]
WORD: %Corr=58.33, Acc=58.33 [H=7, D=0, S=5, I=0, N=12]
------------------------ Confusion Matrix --------------
       g   r   w   y
       r   i   a   e
       e   c   t   l
       e   e   e   l
       n       r   o  Del [ %c / %e]
gree   1   0   0   0    0
rice   1   2   0   1    0 [50.0/16.7]
wate   0   0   1   0    0
yell   2   1   0   3    0 [50.0/25.0]
Ins    0   0   0   0
========================================================
```

Fig. 13 *Training and test results for 'yellow', 'green', 'water' and 'rice' signs (motion and geometric features)*

pronouns, 5 adjectives and 7 verbs with a total lexicon of 37 words were employed.

In general, the duration of each sentence was between 2 and 3 s. Two 'native' signers signed the sentences in a natural way as they communicate with each other in the 'Malaysian Federation of the Deaf' society. Therefore no specific pauses between the signs within the sentences took place. The signed sentences from each signer were divided into two sets. The system was trained using 90% of the collected sentences and 10% of them were used for validating the system. HMM with four states without any states skipped was developed for training the sentences.

Sets of experiments were utilised using two signers. 'Signer 1' signed 172 sentences and 'Signer 2' signed 202 sentences. The experiments have been repeated 50 times with different datasets for training and testing from the 172 and 202 sentences, separately. First, eight motion features have been used to test the system accuracy. The recognition accuracies of sentences were 55.02 and 55.52%. Then the geometric and motion features were combined together to train and test the system. Using 172 and 202 sentences signed by two signers, the achieved recognition accuracies were 39.39 and 42%, respectively.

Published studies suggest that the recognition performance of the isolated gesture approaches over a natural discourse is deteriorated by the dynamics of an uninterrupted communication [16–23]. These dynamic constraints are subjective and related to the language experience as well as word choice skills of the signer. Just like a native speaker, a native signer not only signs fluently, but also utilises a broader vocabulary as compared with inexperienced ones (non-natives). Similarly, sometimes an experienced person may exploit the parallel nature of SL by conveying multiple ideas employing compound gestures through the process called modulation [24]. Another challenging aspect of continuous recognition is so-called co-articulation, which is a process of smoothly connecting two lexically different glosses [25]. Since the joining glosses are mostly distinct in their lexical components, their co-articulation may cause significant transformation in a parameter space disturbing their spatial, morphological or temporal components. For example, referring to frames in Figs. 14*b*–*d* we can observe the preparatory transitions after the end of a noun 'I' to another sign 'Go' in Figs. 14*a* and *e*, respectively. These lexically insignificant transitions in shape and location are hard to distinguish as invalid signs and consequently they may be mistranslated. Without employing any proper segmentation or synchronisation, severe context modification is eminent because the intermittent sign in Fig. 14*b* can be recognised as a verb 'to cost' or noun 'money'.

A simple solution used by most of the existing systems is to introduce a synchronisation mechanism through a variety of *ad hoc* measures like pseudo pauses [26–28]. They are like the silent period between two words in a speech. These pauses provide the synchronisation, which makes the isolated gesture recognition system work

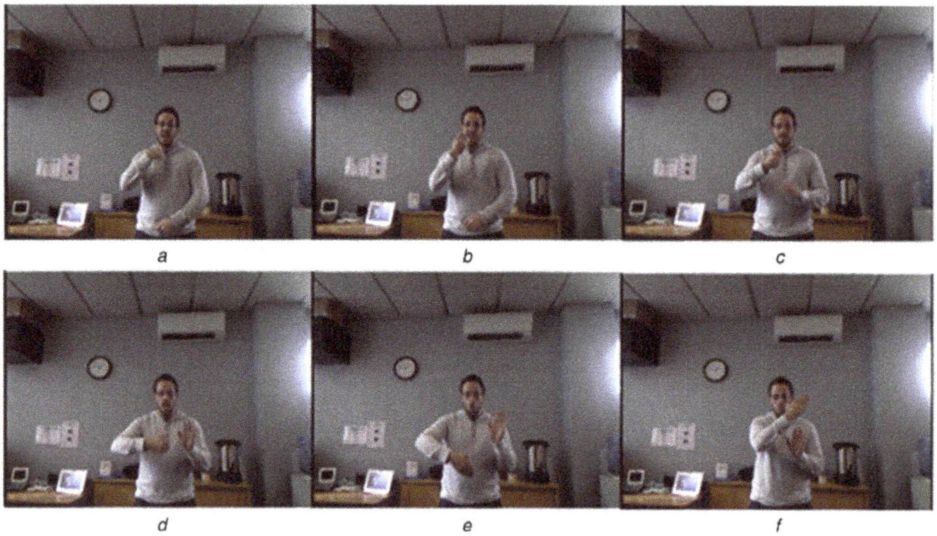

Fig. 14 *Recognition accuracy for continuous signs*
a and *e* Noun 'I' and verb 'Go' shown in frames
b–d Connected through Co-articulation frames

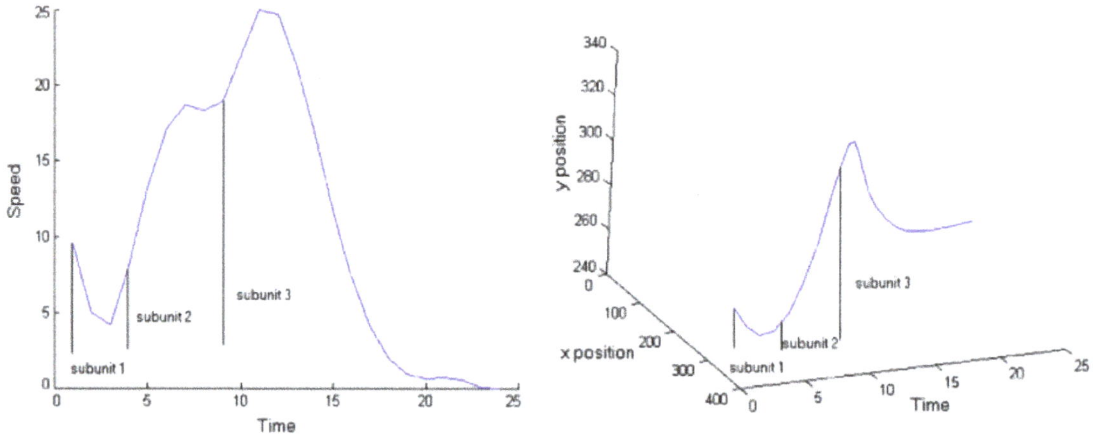

Fig. 15 *Subunit segmentation through directional variation [32]*

on a continuous discourse. The inter-sign pauses are explicitly inserted in many ways either by the exaggerated hold of every sign on its completion, bringing articulators back to a specific neutral position or by taking them out of the field of view. In another approach, they are explicitly triggered by some external means like signer must press a button/paddle after each sign by his/her toe. Obviously, these *ad hoc* segmentation measures simplify the recognition process by turning a continuous discourse into a co-articulation-free sequence of disjoint postures. Nevertheless, for a practical recognition system, these schemes are not too attractive because they disturb the natural prosody of a signer.

HMM-based approaches assume the co-articulation as a temporal variation. In such methods, each gesture in the vocabulary is modelled by an HMM [29, 30], which is a generative model based on likelihood and priors learned during the training phase. These approaches are reported to be robust and able to normalise any temporal inconsistencies. However, they require a huge amount of training data to obtain a system fully trained on a medium-sized vocabulary [16, 29]. These techniques are very useful in continuous speech recognition applications because of the phonological decomposition of a word into its basic units called phonemes. Now instead of training an HMM for each word, it is trained on its phonemes which numbers about 50–60, far lesser than the entire vocabulary. Unfortunately because of unavailability of any valid subunit of a sign or the very large amount of assumed subunits called 'cherems' (about 2500–3000), the HMM-based approaches have limited use. Alternatively, in the deterministic approaches also called 'direct segmentation methods', all valid signs in a

Fig. 16 *Verb sign 'to ask' is gesticulated by a single movement shown by arrows*
a Gesture for '*ask*'
b '*Asking*'

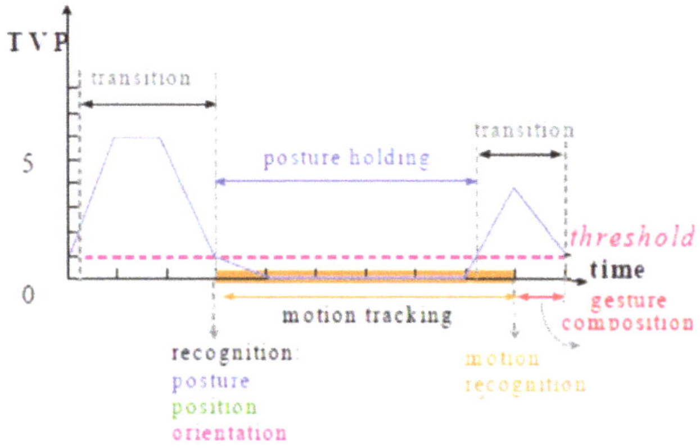

Fig. 17 *Time-varying parameter-based segmentation [33]*

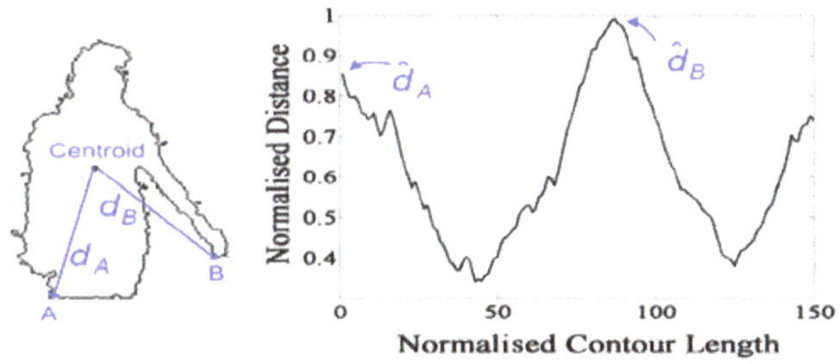

Fig. 18 *Motion signature formation [23]*

continuous sentence are extracted by detecting the sign boundaries with the help of different spatiotemporal features. Research [31] proposed a subunit extraction system based on the deterministic motion features like articulator's velocity. As shown in Fig. 15, it hypothesised the significant directional variation of an articulator as the main segmentation features of the subunits (cherems) of a sign.

Movement component of a gesture is the most significant part of a continuous discourse, which is considered for the segmentation. A majority of the existing models utilise the movement trajectories (2D or 3D) and their temporal derivatives (velocity and acceleration) as their segmentation features. Inspired from the pause-based

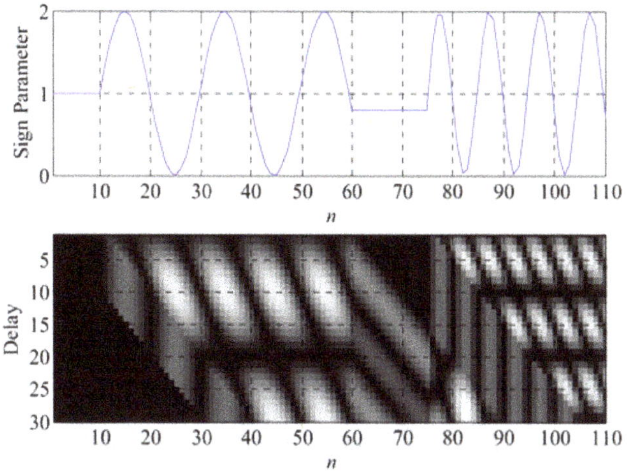

Fig. 20 *DAD signature of time-varying sign parameters*

speech segmentation, these schemes mainly rely on the energy of a continuous signal like local minima or the discontinuities.

The word segmentation of a natural SL discourse by a native signer results in high false positive rate because of unclear 'pauses' in the hand movement. The accuracy of most existing approaches deteriorates without imposing artificial pauses or exaggeration in the normal signing. Apart from the motion information of a gesture, there are few other unaddressed spatiotemporal cues to detect the sign boundaries. Some of these include a sudden change in articulator's direction, change in non-manual signs and short-termed repetitions. Sign repetition is frequently used in natural signing for a variety of signs like interrogative, explanative and indicative gestures. More importantly, the temporal references of a gesture are modified through the repetitions. For example, a verb sign 'to ask' is gesticulated by a single movement shown by arrows in Fig. 16*a*, but the repetition of the same sign (Fig. 16*b*) turns it into a present participle form 'asking'. Obviously, the sign repetition becomes a clear indication of the gesture boundary.

Research [33] proposes a deterministic segmentation scheme in which a word boundary is decided by observing the state of its time-varying parameters. As shown in Fig. 17, if the number of time-varying parameters drops below a specific threshold, the articulator is considered to be in a quasi-stationary state. Therefore the corresponding frame is taken as the end of the previous sign and all

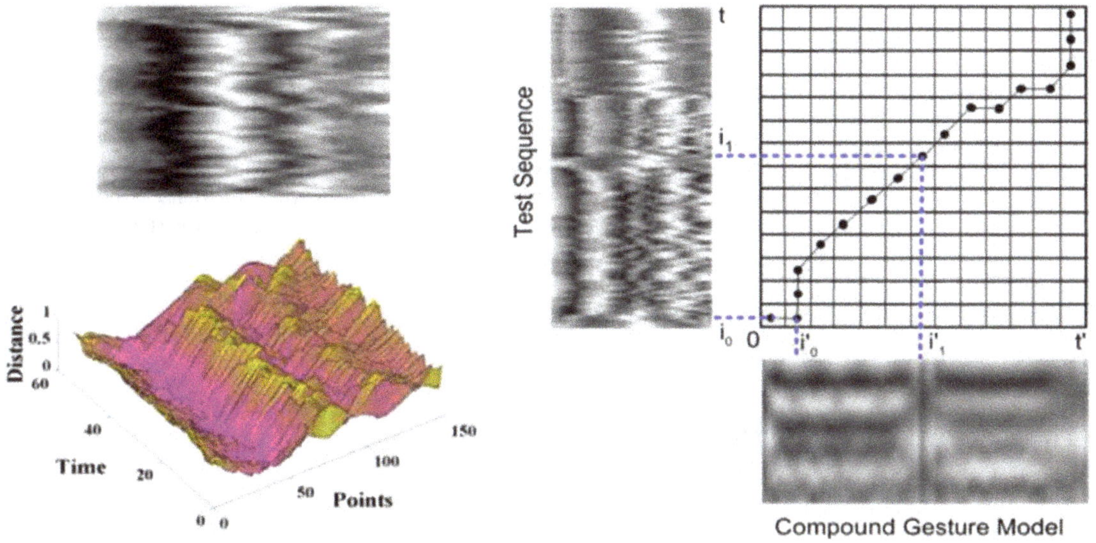

Fig. 19 *Left: signature modelling (2D and 3D); right: matching using dynamic time warping [23]*

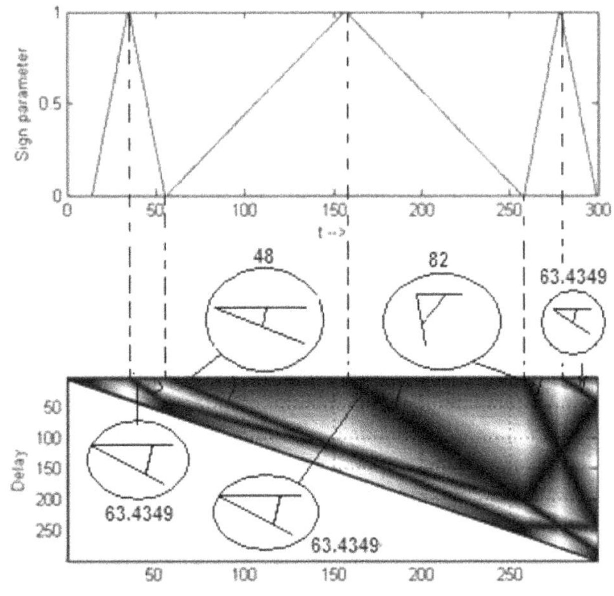

Fig. 21 *DAD signature directional variation segmentation*

the gesture related transitions are extracted for the model matching. As the reported segmentation scheme was proposed in conjunction with a recognition algorithm, no specific detail is found about its segmentation accuracy.

Research [26] presents a direct trajectory segmentation method on 27 SL sentences with minimal velocity and maximum directional angle change. The reported accuracy was 88% with 11.2% false alarm when initial segmentation was subjected to a 'Naive Bayesian' classifier.

3.5 *Segmentation signature*

Motion signatures relate the temporal variation of the signer's body contours with the segmentation of continuous gestures. As shown in Fig. 18, normalised 'Euclidean distances' are accumulated from every point belonging to the signer's body contour to a centroid. Recording of all the distances over a specific period of time generates a patterned surface called 'motion signature', shown in Fig. 19 (left). Like any other signature, motion signatures are distinct patterns, which occur at the boundary of two connected gestures (also called compound gestures) [27, 28, 34, 35].

3.5.1 Delayed absolute difference (DAD) signature: DAD signature [36] is similar to the motion signature but instead of using the morphological variations of a signer body, the spatiotemporal variations of the articulator are encoded as distinct segmentation features. DAD signature of a signal quantifies and localises the intra-signal variations which are candidates for the deterministic

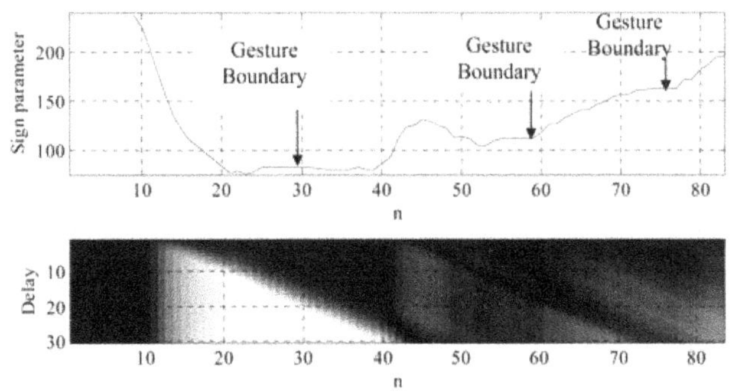

Fig. 22 *Annotated continuous stream*

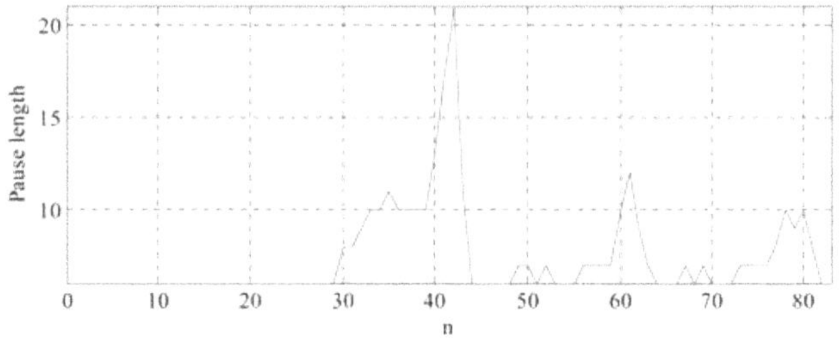

Fig. 23 *Every end point of a pause has a peak that shows its length*

boundary of a sign-like constancy or pauses, sudden changes and sign reduplication. Fig. 20 presents the segmentation features acquired over a signal with the simulated pauses, directional variations and repetitions. It shows inter-sign pauses, which are transformed into DAD pattern (a black inverted triangle) where the length of its base or height corresponds to the length of the pause segment. Similarly the sign repetitions are short termed so they are simulated as a slightly higher-frequency sinusoid. The resultant repetition patterns in the DAD signatures comprise of black horizontal lines that show the similarity (small differences) of a segment with its previous occurrences.

Another prominent segmentation feature is the abrupt change in the articulator trajectory shown in Fig. 21. DAD patterns for each significant directional variation are encoded as black slanted lines where angle of each line relates to the degree of change in direction.

DAD is a deterministic algorithm for sign boundary detection. Unlike the subjective transcription, which is inherently inconsistent, DAD results are far more robust and reliable. For example, Fig. 22 shows a real signal of three connected gestures of New Zealand SL in a natural sentence. The arrows there show the boundary annotation by an experienced signer. DAD signature extracts some prominent segmentation features at the candidate points (shown in Fig. 23). The length of an inter-sign pause in the every candidate frame can be helpful to retrieve the start of that pause, which can help in retrieving its temporal localisation. Once the 'start' and 'end' of a pause segment is decided, we can extract the actual gestures from end of previous pause to the start of the next one. By this means the sign components relating only to the linguistically significant units are processed for gesture modelling and recognition. A projection of the detected pauses over the actual signal is given in Fig. 24.

4 Real-time system implementation

In this research, the client/server approach between 'Windows7' and 'LINUX Fedora 16' was developed for real-time SL recognition. The main objective behind this approach was to utilise positive features of both operating systems. Under the 'Windows7', the 'Visual C++ 2010' combined with 'OpenCV 1.1pre' library can support video processing while providing a powerful graphical user interface. Meanwhile, the Gt^2k for gesture recognition can be

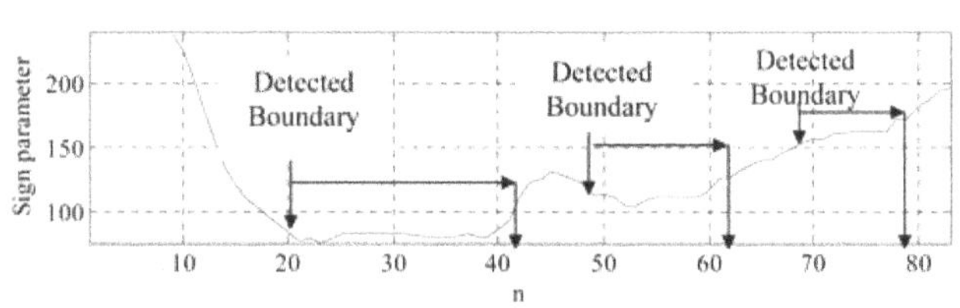

Fig. 24 *Overlaying the detected pause features over the sign parameter (vertical arrows indicate the start and end frames, whereas the horizontal arrows show the length of the quasi-stationary segments)*

Results from Gt^2k recogniser
#!MLF!#
"/mnt/javacode/test1.txt.rec"
0 172000 car -21368.128906
///
0 172000 noodles -11883843.000000
///
0 172000 dog -14732120.000000
///
0 172000 chicken -18813332.000000
///
0 172000 school -19096242.000000

Fig. 25 *MLF recognition results*

fully supported under 'LINUX'. Therefore a client/server technology saves significant time during the SL recognition system development.

It also helps to enhance the algorithms in such a way that it is not required to transfer one of the two programs: either the Gt^2k from 'LINUX Fedora 16' to 'Windows7' or the 'VC++ 2010' and 'OpenCV 1.1pre' library from 'Windows7' to 'LINUX Fedora 16'. Program 'VC++ 2010' cannot be transferred from 'Windows7' to 'LINUX Fedora 16' operating system because it is a Microsoft product. It offers powerful functionality for developers specifically in developing codes for video and image processing. However, it works only under 'Windows' operating system. Another open source product – *Qt* can execute image and video visualising under 'LINUX' and could be another alternative for 'VC++'. Unfortunately as a new programming product *Qt* is offering just limited resources for developers [37].

An alternative method to link the two different operating systems ('Windows7' and 'LINUX Fedora 16') involves the use of a physical link (cable) between the client and server computers. However, such an approach limits the speed of data transfer and may be affected by potential connection instability. In addition, the cost of this solution is higher since two different computers are required to build such a system.

The reported system was launched on just one powerful computer using two operating systems ('Windows7' and 'LINUX Fedora 16'). This was possible by using the 'VMware' workstation software.

4.1 Sharing files between the client and server

Client and server are two separate programming modules which communicate through the network while having explicitly distinctive tasks. Usually clients can be categorised into two types: 'thin client' and 'thick client'. A thin client is capable of achieving an acceptable computation performance over wide area networks. However, for the purpose of the reported study, a thick client ('VC++ 2010' and 'OpenCV 1.1pre' library, which resides in 'Windows7') was used. The use of the thick client is significant is this research during the pre-processing of sign videos, which requires high visualisation performance. Similarly, on the server side, Gt^2k is a tool for gesture recognition, which helps human

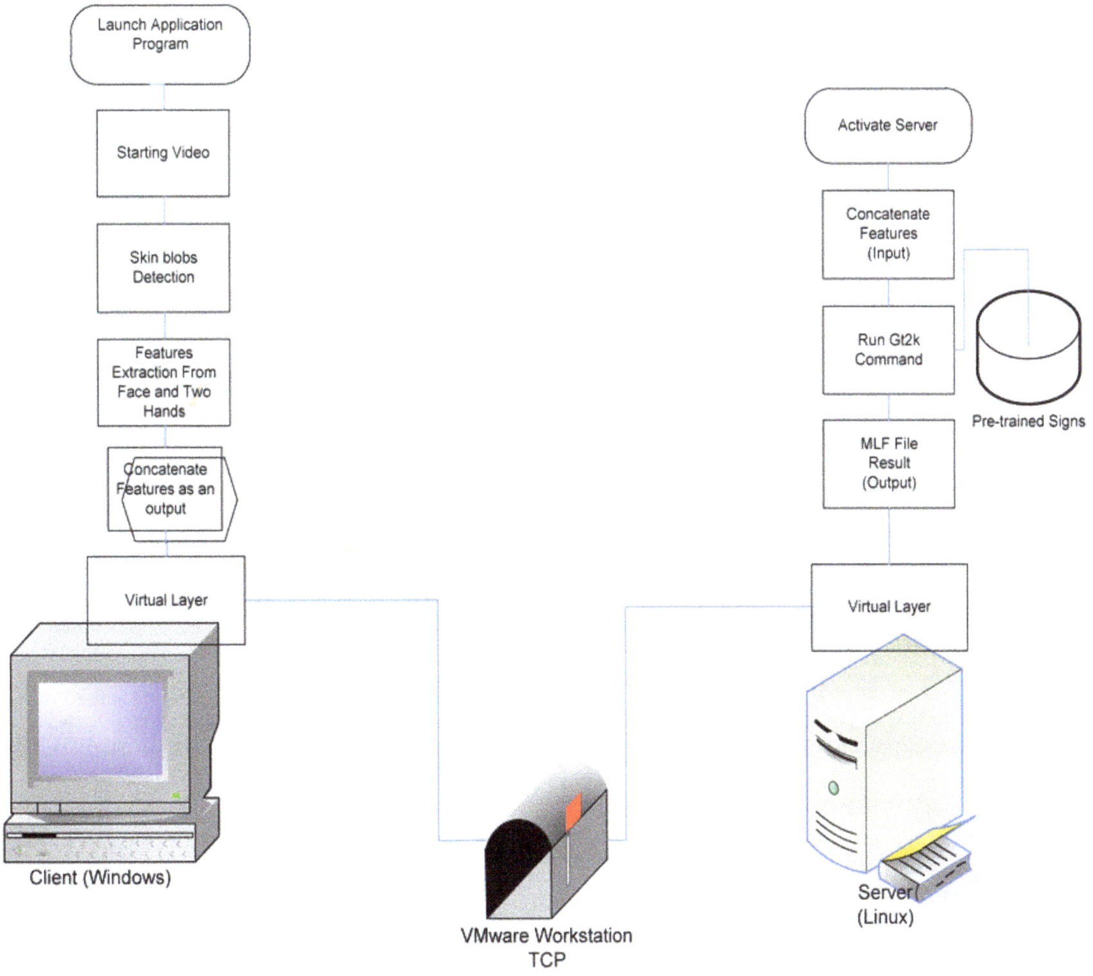

Fig. 26 *Real-time MSL recognition system*

computer interface developers to focus mainly on pre-recognition stages rather than spending time and efforts on building the HMM-based recogniser.

(a) *Thick client:* In this paper the algorithms for face and hands detection, skin blob tracking up to the feature extraction stage were implemented under 'Windows7' environment using 'VC++' 2010 with the help of 'OpenCV 1.1pre' library. To train HMM in an offline mode, feature files were transferred to the Gt^2k for training on a specific number of signs.
(b) *Server for SL recognition:* To achieve the SL recognition using a new non-trained sign, a 'recognize.sh' script can be used from Gt^2k software. Four arguments are to be passed: data files 'signs.txt'; a file to store the recognition results 'results.txt', which will be created automatically; 'options.sh' and the trained HMM model. HMM model is called 'newMacros' and the output is in the form of an MLF. The MLF file contains gestures ranked by their likelihood scores as shown in Fig. 25, which shows MLF for five isolated words. The command for recognition using Gt^2k is

Recognize.sh signs.txt result.txt options.sh NewMacros.

As stated earlier, establishing network connection is necessary to begin an interaction between the client and server. It is required to establish communication, which enables the feature collector client to send the features file to the analysing server.

4.2 Thick client and server interface

'VMware' workstation was lunched as a media on 'Windows7' to install 'LINUX Fedora 16'. Following that, a transmission control protocol (TCP) connection type is created between the client and server to transfer the processed video file from the client to the server. Similarly, TCP transfer, the resulted output from the server is inputted to the client as illustrated in Fig. 26.

5 Conclusions

ASLT is an example of a systems leading to a higher life quality. In this research, an automatic Malaysian SL recognition system was developed. The system utilises four stages including: face and hand detection and tracking; HUB detection; feature extraction; and real-time SL recognition. Hands and face are detected by the developed new hybrid method, which combines the appearance-based approach with the skin colour detection algorithm, and which is not restricted by a background. The HUB detection is achieved by applying a new face measurement approach. Various feature vectors from hand shape, hand motion trajectory and hand position are extracted and investigated. During the recognition stage, HMM is used to train and test the developed system using a newly developed Malaysian SL database and implementing new feature matching methods. Meanwhile a real-time SL recognition is implemented using client/server technology between 'Windows7' and 'LINUX Fedora 16'. These interfaces offer additional degrees of mobility and control, which can lead in the future to a betterment and development of portable SL translation devices (which is the ultimate aim of the presented research in the long term).

The overall system processing time for isolated signs into text and/or voice (in English) is <2 s on an upper mid-range commodity personal computer and is shorter on a more powerful machine, thus qualifying for real-time translation. Meanwhile, at this initial prototype implementation stage the average recognition accuracy for 20 isolated signs reached 80% and was at 55% for a total lexicon of 37 words in 20 sentences. It is expected that the accuracy will be significantly increased with the system further development and fine-tuning. Among the additional benefits of the systems is its offering a natural environment for a user where signing can be done by one or both hands, with or without pauses between signs, with flexible velocity of signing, without the burden of wearing any devices and while operating in variable illumination conditions with varying backgrounds.

6 References

[1] MFD, *Malaysian Sign Language*. 2012; Available at http://www.mfd.org.my/public/edu_eSign.asp
[2] MacKenzie I.S.: 'Input devices and interaction techniques for advanced computing, in virtual environments and advanced interface design' (Oxford University Press, Oxford, UK, 1995), pp. 437–470
[3] Pavlovic V.I., Sharma R., Huang T.S.: 'Visual interpretation of hand gestures for human–computer interaction: a review', *IEEE Trans. Pattern Anal. Mach. Intell.*, 1997, **19**, (7), pp. 677–695
[4] Yang M.H., Ahuja N.: 'Recognizing hand gesture using motion trajectories'. IEEE Conf. Computer Vision and Pattern Recognition, IEEE Computer Society, Fort Collins, CO, USA, 1999, vol. 1, pp. 466–483
[5] Grzeszczuk R., Bradski G., Chu M.H., Bouguet J.: 'Stereo based gesture recognition invariant to 3D pose and lighting'. IEEE Conf. Computer Vision and Pattern Recognition, IEEE Computer Society, Head Island, SC, USA, 2000, vol. 1, pp. 826–833
[6] Viola P., Jones M.J.: 'Robust real-time face detection', *Int. J. Comput. Vis.*, 2004, **57**, (2), pp. 137–154
[7] Bilal S., Akmeliawati R., Momoh J., Shafie A.A.: 'Dynamic approach for real-time skin detection', *J. Real-Time Image Process.*, 2012, p. 6
[8] Chang F., Chen C.-J., Lu C.-J.: 'A linear-time component-labeling algorithm using contour tracing technique', *Comput. Vis. Image Underst.*, 2004, **93**, (2), pp. 206–220
[9] Imagawa K., Lu S., Igi S.: 'Color-based hands tracking system for sign language recognition'. Third IEEE Int. Conf. Automatic Face and Gesture Recognition, Nara, Japan, 1998, pp. 462–467
[10] Bilal S., Akmeliawati R., Salami M.J.E., Shafie A.A., Bouhabba E.M., *ET AL.*: 'A hybrid method using Haar-like and skin-color algorithm for hand posture detection, recognition and tracking'. Int. Conf. Mechatronics and Automation (ICMA), Xi'an, China, 2010, pp. 934–939
[11] Kilian J.: Simple Image Analysis by Moments, 2001, 8 pp. Available at http://www.scribd.com/doc/39759766/Simple-Image-Analysis-by-Moments
[12] Jusko D.: Full Real Color Wheel Course, 2011. Available at http://www.realcolorwheel.com/human.htm
[13] Bradski G.R.: 'Computer vision face tracking for use in a perceptual user interface', *Intel Technol. J.*, 1998, **2**, (3), pp. 1–15
[14] Bilal S., Akmeliawati R., Shafie A.A., Salami M.J.E.: 'Modelling of human upper body for sign language recognition'. Fifth Int. Conf. Automation, Robotics and Applications (ICARA), Wellington, New Zealand, 2011, pp. 104–108
[15] Starner T.E., Pentland A.: 'Real-time American sign language recognition from video using hidden Markov models'. IEEE Int. Symp. Computer Vision, Coral Gables, FL, USA, 1995, pp. 265–270
[16] Segouat J., Braffort A.: 'Toward modeling sign language coarticulation', in Kopp S., Wachsmuth I. (Eds.): 'Gesture in embodied communication and human–computer interaction' (Springer Berlin Heidelberg, 2010), pp. 325–336
[17] Segouat J.: 'A study of sign language coarticulation', *Spec. Interest Group Accessible Comput. (SIGACCESS)*, 2009, **2009**, (93), pp. 31–38
[18] San-Segundo R., Pardo J.M., Ferreiros J., *ET AL.*: 'Spoken Spanish generation from sign language', *Interact. Comput.*, 2010, **22**, (2), pp. 123–139
[19] San-Segundo R., Barra R., Cordoba R., *ET AL.*: 'Speech to sign language translation system for Spanish', *Speech Commun.*, 2008, **50**, (11–12), pp. 1009–1020
[20] Alon J., Athitsos V., Quan Y., Sclaroff S.: 'A unified framework for gesture recognition and spatiotemporal gesture segmentation', *IEEE Trans. Pattern Anal. Mach. Intell.*, 2009, **21**, pp. 1685–1699
[21] Yang R., Sarkar S., Loeding B.: 'Handling movement epenthesis and hand segmentation ambiguities in continuous sign language recognition using nested dynamic programming', *IEEE Trans. Pattern Anal. Mach. Intell.*, 2009, **32**, pp. 462–477
[22] Viblis M.K., Kyriakopoulos K.J.: 'Gesture recognition: the gesture segmentation problem', *J. Intell. Robot. Syst.*, 2000, **28**, pp. 151–158
[23] Kahol K., Tripathi P., Panchanathan S., Rikakis T.: 'Gesture segmentation in complex motion sequences'. IEEE Int. Conf. Automatic Face and Gesture Recognition, Seoul, Korea, 2004, vol. 3, pp. II-105–8
[24] Ong S.C.W., Ranganath S.: 'A new probabilistic model for recognizing signs with systematic modulations', Third International Workshop on

Analysis and Modelling of Faces and Gestures, Rio de Janeiro, Brazil, 2007 (*LNCS*, **4778/2007**), pp. 16–30

[25] Ruiduo Y., Sarkar S.: 'Detecting coarticulation in sign language using conditional random fields'. 18th Int. Conf. Pattern Recognition, 2006, ICPR 2006, 2006, vol. 2, pp. 108–112

[26] Kong W.W., Ranganath S.: 'Sign language phoneme transcription with rule-based hand trajectory segmentation', *J. Signal Process. Syst.*, 2010, **59**, (2), pp. 211–222

[27] Li H., Greenspan M.: 'Segmentation and recognition of continuous gestures'. IEEE Int. Conf. Image Processing, 2007, ICIP 2007, 2007, vol. 1, pp. 365–368

[28] Li H., Greenspan M.: 'Continuous time-varying gesture segmentation by dynamic time warping of compound gesture models'. Int. Workshop on Human Activity Recognition and Modelling (HARAM2005), 2005, p. 8

[29] Starner T., Pentland A.: 'Real time American sign language recognition from video using hidden Markov model'. Int. Symp. Computer Vision, Florida, USA, 1995, pp. 265–270

[30] Vogler C.P.: 'American sign language recognition: reducing the complexity of the task with phoneme-based modeling and parallel hidden Markov models'. PhD dissertation, University of Pennsylvania, USA, p. 172

[31] Guerrero-Curieses A., Rojo-Álvarez J.L., Conde-Pardo P., Landesa-Vazquez I., Ramos-Lopez J., Alba-Castro J.L.: 'On the performance of kernel methods for skin color segmentation', *EURASIP J. Adv. Signal Process.*, 2009, **2009**, pp. 1–13

[32] Han J., Awad G., Sutherland A.: 'Modelling and segmenting subunits for sign language recognition based on hand motion analysis', *Pattern Recognit. Lett.*, 2009, **30**, (6), pp. 623–633

[33] Liang R.-H., Ming O.: 'A real-time continuous gesture recognition system for sign language'. IEEE Int. Conf. Automatic Face and Gesture Recognition, Japan, 1998, pp. 558–567

[34] Li H., Greenspan M.: 'Multi-scale gesture recognition from time-varying contours'. 10th IEEE Int. Conf. Computer Vision, ICCV 2005, 2005, vol. 1, pp. 236–243

[35] Li H., Greenspan M.: 'Model-based segmentation and recognition of dynamic gestures in continuous video streams', *Pattern Recognit.*, 2011, **44**, (8), pp. 1614–1628

[36] Khan S., Bailey D.G., Sen Gupta G.: 'Delayed absolute difference (DAD) signatures of dynamic features for sign language segmentation'. Fifth Int. Conf. Automation, Robotics and Applications (ICARA2011), Wellington, New Zealand, 2011, pp. 109–114

[37] Qt Project. Available at http://www.qt-project.org/

8

Stage-dependent minimum bit resolution maps of full-parallel pipelined FFT/IFFT architectures incorporated in real-time optical orthogonal frequency division multiplexing transceivers

Junjie Zhang[1,2], Wenyan Yuan[1,2], Kai Wang[2], Bingyao Cao[1,2], Roger P. Giddings[1], Min Wang[2], Jianming Tang[1]

[1]*School of Electronic Engineering, Bangor University, Bangor LL571UT, UK*
[2]*Key Laboratory of Specialty Fiber Optics and Optical Access Networks, Shanghai University, Shanghai 200072, People's Republic of China*
E-mail: j.tang@bangor.ac.uk

Abstract: Fast Fourier transform (FFT) and inverse FFT (IFFT) are the fundamental algorithms at the heart of optical orthogonal frequency division multiplexing (OOFDM) transceivers. The high digital signal processing (DSP) complexity has become one of the most significant obstacles to experimentally demonstrating real-time high-capacity OOFDM transceivers. In this study, extensive numerical explorations are undertaken, for the first time, of the impacts of each individual transceiver DSP element on the inverse error vector magnitude (IEVM) performance of the OOFDM transceivers incorporating full-parallel pipelined FFT/IFFT architectures. More importantly, FFT/IFFT stage-dependent minimum bit resolution maps are identified, based on which minimum bit resolutions of individual DSP elements of various FFT/IFFT stages can be easily selected according to chosen analogue-to-digital converter/digital-to-analogue converter resolutions. The validity and high accuracy of the identified maps are experimentally verified in field programmable gate array (FPGA)-based platforms. In addition to great ease of practical OOFDM transceiver designs, the maps also significantly reduce the FPGA logic resource usage without degrading the overall transceiver IEVM performance.

1 Introduction

Optical orthogonal frequency division multiplexing (OOFDM) is a promising technique for practically realising intelligent transceivers for future software-defined elastic optical networks [1]. The inherent digital signal processing (DSP)-rich OOFDM transceivers offer salient features including automatic awareness of channel spectral characteristics, excellent adaptability to component/system/network imperfections, dynamically variable capacity against reach performance and user-controlled transceiver capability of performing channel add/drop multiplexing [2]. Over the past several years, a number of real-time OOFDM implementations of transmitters [3–6], receivers [7–9] and transceivers [2, 10] have been reported using field programmable gate arrays (FPGAs). Technically speaking, the huge FPGA logic resource usage associated with the high OOFDM DSP complexity has become one of the most significant obstacles to experimentally demonstrating real-time OOFDM transceivers offering more advanced functionalities and flexibilities. For example, just the inverse fast Fourier transform (IFFT) DSP algorithm alone can take >82% of the total FPGA logic resources [5]. To address the challenge, architectures employing multiple currently available FPGAs have been utilised in [7, 8] to perform all necessary OOFDM DSP algorithms. Furthermore, when considering product development, reducing the logic resource usage is also critically important for achieving low-cost and low-power consumption DSPs implemented in application specific integrated circuits.

Therefore, from the practical application point of view, it is vital to explore effective approaches capable of minimising, via optimising the DSP designs, the FPGA logic resource usage without degrading the transceiver performance. Given the fact that the FFT/IFFT is the fundamental algorithm at the heart of the real-time OOFDM transceivers, in this paper, special attention is thus focused on the FFT/IFFT.

For traditional application scenarios involving digital-to-analogue converters (DACs) and analogue-to-digital converters (ADCs) with relatively low sampling rates in a regime of less than a few hundred megahertz, the well-known partial-parallel pipelined FFT/IFFT architecture has been widely adopted, where the total number of parallel samples input to the FFT/IFFT is smaller than the size of the FFT/IFFT. To reduce the DSP complexity associated with such an FFT/IFFT architecture, maximising the reuse of those complex multiplication functions is a preferred option, to realise which various FFT/IFFT DSP architectures have been proposed, these architectures include single-path delay feedback [11, 12], multi-path delay feedback or parallel feedback [13, 14] and multi-path delay commutator also known as feed-forward architectures [15–17].

However, for advanced high-speed OOFDM transceivers, ADCs/DACs with their sampling rates of tens of gigahertz are often employed [4, 9], which operate significantly faster than the processing speeds of typical FPGAs. This requires that one transfer operation per FPGA clock cycle has to be performed. Thus the full-parallel pipelined FFT/IFFT architecture is very promising, in which the total number of parallel samples input to the FFT/IFFT equals to the size of the FFT/IFFT. To reduce the DSP complexity of such FFT/IFFT architecture, it is, however, infeasible to employ any of the previously proposed approaches [11–17] because of the 100% utilisation of all complex multiplication functions in the full-parallel pipelined FFT/IFFT architecture. This suggests that the reduction in bit resolutions of the FFT/IFFT DSP elements is the dominant strategy for effectively reducing the FFT/IFFT DSP complexity and logic resource usage. Recently, investigations of the impacts of FFT/IFFT bit resolution on the overall OOFDM transceiver performance have been reported [18, 19], where the FFT/IFFT operation is, however, treated as a 'black-box' without taking into account bit resolution variations between different intermediate FFT/IFFT stages.

In this paper, over a broad variety of FFT/IFFT DSP elements, extensive numerical explorations are undertaken, for the first time, of the impacts of each individual transceiver DSP element on the inverse error vector magnitude (IEVM) performance of the OOFDM transceivers incorporating the full-parallel pipelined FFT/IFFT architectures. More importantly, FFT/IFFT stage-dependent minimum bit resolution maps are identified, based on which minimum bit resolutions of individual DSP elements of various FFT/IFFT stages can be easily selected according to chosen ADC/DAC resolutions. The validity and high accuracy of the identified maps are experimentally verified in FPGA-based platforms. In addition to great ease of practical OOFDM transceiver designs, the maps also significantly reduce the FPGA logic resource usage without degrading the overall transceiver IEVM performance.

2 Numerical identifications of FFT/IFFT-stage dependent minimum bit resolution maps

2.1 Transceiver DSP architectures and operation parameters

In Fig. 1*a*, the major DSP processes involved in a representative OOFDM transceiver are illustrated, which are very similar to those reported in [10]. Fig. 1*b* shows the considered FFT having the Cooley–Tukey radix-2 decimation-in-time architecture. In Fig. 1*b*, the integer numbers at the FFT input represent the index of the input data sequence, whereas the integer numbers at the FFT output correspond to the subcarrier frequencies. The red italic number, *n*, occurring between two consecutive FFT stages represents the twiddle factor defined as

$$W(n) = \exp\left(\frac{-j2\pi}{32}n\right) \quad (1)$$

As the IFFT can be easily achieved utilising the simply modified FFT, therefore, the similar architecture is also taken to realise the IFFT. For each FFT/IFFT stage, both the output bit resolution and the twiddle factor bit resolution are independently adjustable.

In the transmitter, an incoming pseudo-random bit sequence of $2^{15}-1$ is first truncated into parallel words, each of which is encoded using a variety of *M*-quadrature amplitude modulation (QAM) signal modulation formats including 16-QAM, 32-QAM, 64-QAM, 128-QAM and 256-QAM. Identical to that implemented in the real-time OOFDM transceivers [10], the size of the FFT/IFFT is taken to be 32; this results in the FFT/IFFT architecture consisting of five different stages, as shown in Fig. 1*b*. The bit resolutions of the encoded signals at the input of the IFFT are as follows: 3 bits for 16-QAM; 4 bits for 32-QAM and 64-QAM, as well as 5 bits for 128-QAM and 256-QAM. About 15 QAM-encoded complex information-bearing subcarriers are arranged to satisfy the Hermitian symmetry with respect to their conjugate counterparts to generate real-valued OFDM symbols after the IFFT.

After the IFFT, a 25% cyclic prefix is added to each OFDM symbol before the digital OFDM signal is converted to an analogue signal by a *k*-bit DAC ($k=5$–12). The receiver OOFDM DSP processes are just an inverse replica of the above-mentioned transmitter DSP processes. Unless stated explicitly in the corresponding text, the signal clipping effect is not included to ensure that the minimum bit resolution maps identified in this paper are capable of representing the worst-case scenarios, where the dynamic quantisation range of the transmitter DAC always spans from the maximum signal peak value to the minimum signal peak value, thus giving rise to the maximum signal quantisation noise effect for a given DAC bit resolution. Although for the ADC in the receiver, similar to the approach reported in [2, 10], the received signal amplitude is always adjusted appropriately to ensure that the signal covers the ADC's full-scale range and gives rise to the best bit error rate (BER) performance. For simplicity without losing any generality, the effects of electrical-to-optical (E–O) and optical-to-electrical (O–E) conversions, as well as optical fibre transmission are excluded to highlight the key aspects of interest of this paper. A total number of 5000 OFDM symbols are captured each time to evaluate the overall transceiver performance using the IEVM defined as

$$\text{IEVM} = \left[\frac{(1/N)\sum_{k=1}^{N}\sum_{r=1}^{15}\left(S_{\text{ideal}}(k,r)\right)^2}{(1/N)\sum_{k=1}^{N}\sum_{r=1}^{15}\left(S_{\text{ideal}}(k,r)-S_{\text{meas}}(k,r)\right)^2}\right]^{1/2} \quad (2)$$

where $S_{\text{meas}}(k,r)$ represents the actually received constellation point modulated onto the *r*th subcarrier and carried by the *k*th OFDM symbol; $S_{\text{ideal}}(k,r)$ is the ideal constellation point of the corresponding symbol, and *N* is the total number of OFDM symbols captured each time. Under the Gaussian noise distribution assumption, the relationship between the IEVM and the BER can be expressed by

$$\text{BER} = \frac{2}{\log_2(m)}\left(1-\frac{1}{\sqrt{m}}\right)\text{erfc}\left[\sqrt{\frac{3\,\text{IEVM}^2}{2(m-1)}}\right] \quad (3)$$

where *m* is the modulation level of the QAM signal. It is clear from

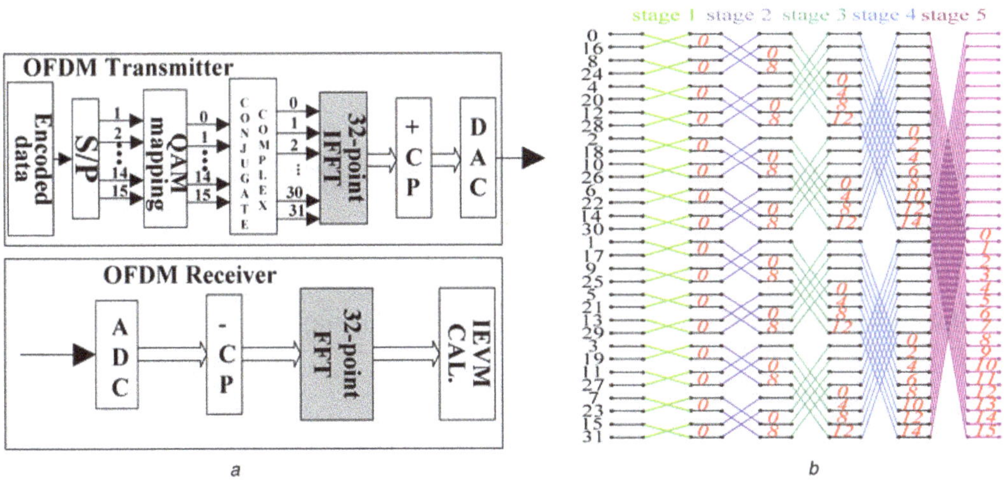

Fig. 1 *Transceiver DSP architectures and FFT architecture*
a Major OOFDM transceiver DSP architectures
CP: cyclic prefix and S/P: serial-to-parallel
b 32-point full-parallel radix-2 decimation-in-time FFT architecture

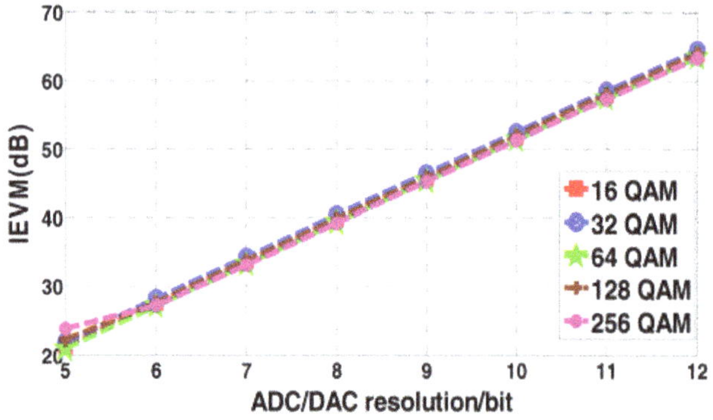

Fig. 2 *ADC/DAC resolution-dependent transceiver IEVM performance*

Fig. 1*b* that the overall transceiver IEVM performance is mainly determined by the bit resolution of the ADC/DAC and the bit resolution of each individual DSP element involved in the full-parallel pipelined FFT/IFFT architecture.

2.2 Impacts of ADC/DAC bit resolutions

For different signal modulation formats, the ADC/DAC resolution-dependent transceiver IEVM performance is plotted in Fig. 2, in obtaining which both the FFT and IFFT are performed using floating-point operations. It can be seen in Fig. 2 that an increase of the converter resolution by 1 bit improves the IEVM by ~6 dB, and that the transceiver IEVM performance is signal modulation format independent. For a high-speed OOFDM transceiver, the available signal converter resolution is often quite limited, the available ADC/DAC resolution, therefore, determines the maximum achievable IEVM performance of the OOFDM transceiver. As a direct result, Fig. 2 can be treated as the bench-marker results for evaluating the effects of the bit resolutions of all FFT/IFFT DSP elements discussed below.

2.3 Impacts of bit resolutions of FFT/IFFT DSP elements

In addition to the ADC/DAC resolutions discussed above, bit resolutions adopted in various FFT/IFFT stages also play a key role in determining the transceiver IEVM performance. Given the fact that the considered 32-point radix-2 FFT/IFFT consists of five stages in total, and that the first and second stages only have addition and subtraction operations because of their simple twiddle factors expressed in (1); thus in the following discussion, special attention is given to the impacts of the bit resolutions of the twiddle factors and output bit resolution of the FFT/IFFT for stage 3, stage 4 and stage 5 only.

For stage 3, stage 4 and stage 5 of the IFFT (or FFT) in the transmitter (or receiver) utilising different signal modulation formats, the simulated transceiver IEVM performances as a function of the output bit resolution are presented in Figs. 3*a* (or 3*b*), in computing which the floating-point FFT (or IFFT) in the receiver (or transmitter) is employed. In addition, in obtaining the above-mentioned two figures, except for the targeted transmitter IFFT (or receiver FFT) stage, floating-point computations are also considered for all other remaining transmitter IFFT (or receiver FFT) stages, all the twiddle factors and the ADC/DAC. As expected, Figs. 3*a* and *b* show that the signal modulation format-free IEVM developing trends are very similar to those observed in Fig. 2, and that there

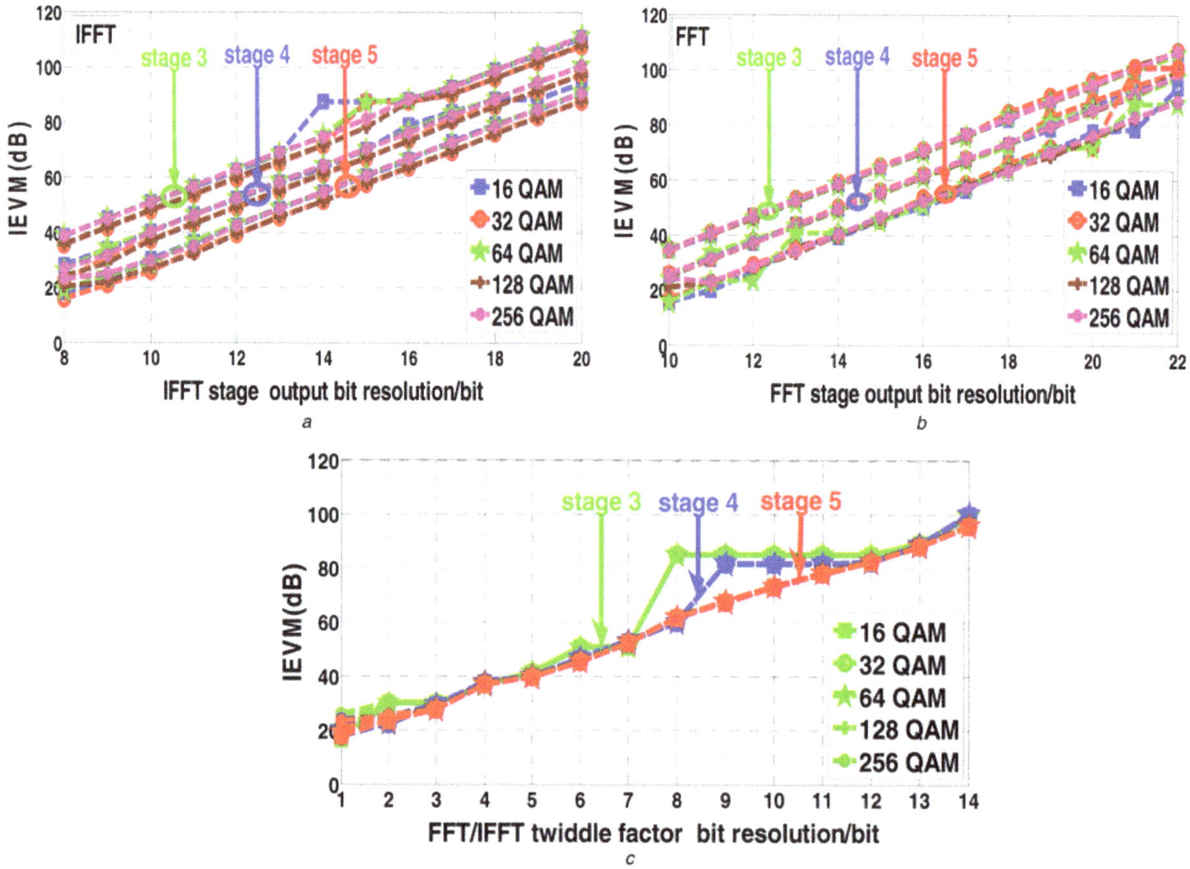

Fig. 3 *Simulated IEVM performances for stage 3, stage 4 and stage 5 and different signal modulation formats*
a Output bit resolution-dependent IEVMs for the IFFT in the transmitter
b Output bit resolution-dependent IEVMs for the FFT in the receiver
c Twiddle factor bit resolution-dependent IEVMs for the FFT/IFFT

Table 1 Twiddle factor value

Stage numbers	Twiddle factors	Twiddle factor values
3/4	*W*(0)	1
	W(4)	0.7071–0.7071j
	W(8)	–j
	W(12)	−0.7071–0.7071j
4	*W*(2)	0.9239–0.3827j
	W(6)	0.3827–0.9239j
	W(10)	−0.3827–0.9239j

also exists a 6 dB IEVM improvement for a 1 bit increase in output bit resolution for different FFT/IFFT stages. In addition, it is also very interesting to note in Figs. 3*a* and *b* that, to achieve the same IEVM performances, an ~1.5 bit increase in output bit resolution is required for a 1-step increase in FFT/IFFT stage. For a fixed output bit resolution, a specific stage of the IFFT has a better IEVM performance than the same stage of the FFT. For example, to achieve the same IEVM performance for the same stage, the minimum required output bit resolution increases by about 2 for the FFT compared with the IFFT. Such differences can be explained by considering the fact that the IFFT input signals are QAM-encoded and have a set of fixed amplitudes, whereas the OFDM signals input to the FFT have much high peak-to-average power ratios, hence high bit resolutions of the FFT are essential to enable the appropriate accommodation of the resulting large signal dynamic ranges.

Apart from the output bit resolution of each individual FFT/IFFT stage, numerical simulations are also undertaken to investigate the influence of the bit resolution of the twiddle factor of each IFFT (or FFT) stage on the transceiver IEVM performance. The numerically simulated results are plotted in Fig. 3*c*, where the floating-point output resolutions for both the IFFT and FFT are adopted. Moreover, the floating-point calculations are also employed in the following operations including analogue to digital/digital to analogue (AD/DA) conversion and all other non-targeted twiddle factors in both the IFFT and FFT. As expected, Fig. 3*c* shows that the bit resolutions of the twiddle factors of both the IFFT and the FFT have the same effects on the transceiver IEVM performance, and that the twiddle factor's impacts are independent of signal modulation format.

It is also seen in Fig. 3*c* that a 1 bit increase in the twiddle factor bit resolution gives rise to a ~6 dB IEVM improvement. The above statement holds well for various FFT/IFFT stages when the FFT/IFFT twiddle factor bit resolutions are <7. However, when the twiddle factor bit resolutions for stage 3 (stage 4) are >8 (>9), a sudden 30 dB (20 dB) IEVM improvement occurs, and the resulting IEVM performance remains constant until the bit resolutions reach ~12, as seen in Fig. 3*c*. This can be explained by considering the twiddle factor values of these two stages, which are explicitly listed in Tables 1 and 2. It is clear from these two tables that

Table 2 Binary number representation

Numbers	Binary representations (bit resolution = 16)
0.7071	1011 0101 0000 0101
0.9239	1110 1100 1000 0011
0.3827	0110 0001 1111 1000

0.7071 occurring in stage 3 remains constant for twiddle factor bit resolutions varying from 8 to 12. Similarly, 0.3827, 0.9239 and 0.7071 appearing in stage 4 also remain unchanged for twiddle factor bit resolutions altering from 9 to 12, respectively.

2.4 FFT/IFFT stage-dependent minimum bit resolution maps

Having investigated the effect of each individual DSP element separately in Sections 2.2 and 2.3, in this section effort is then given to explore a more important case where all the aforementioned DSP elements are present simultaneously within a single transceiver architecture. It has been shown [20] that it is sufficiently accurate to assume that the clipping and quantisation noises have Gaussian distributions when an OFDM signal is clipped by a clipping ratio not considerably smaller than the optimum clipping ratio. For all the cases considered here, the above-mentioned condition holds well. By taking into account (2), the overall transceiver IEVM performance can be expressed as (see (4)), where $IEVM_{ADC}$ is the IEVM associated with the finite bit resolution of the adopted ADC/DAC; $IEVM_{FFT\_output}(k)$ and $IEVM_{IFFT\_output}(k)$ are the IEVMs induced only by the output bit resolutions of the *k*th stage FFT and IFFT, respectively; and $IEVM_{FFT\_twiddle}(k)$ and $IEVM_{IFFT\_twiddle}(k)$ are the IEVMs corresponding to the twiddle factor bit resolutions of the *k*th stage FFT and IFFT, respectively. Considering the transceiver parameters outlined previously, the integer number, *k*, can vary from 3 to 5. For a given ADC/DAC bit resolution, the FFT/IFFT bit resolution map is considered to be valid only when the following equation is satisfied

$$IEVM_{total\_dB} \leq IEVM_{ADC\_dB} - 0.5 \tag{5}$$

where $IEVM_{total\_dB}$ is the bit resolution map-based overall transceiver IEVM performance and $IEVM_{ADC\_dB}$ is the bench-marker transceiver IEVM performance presented in Fig. 2. In (5), both $IEVM_{total\_dB}$ and $IEVM_{ADC\_dB}$ are in a unit of decibel.

To considerably shorten the computing execution time, it is necessary to identify an upper boundary of the FFT/IFFT stage-dependent bit resolutions. For a specific ADC/DAC bit resolution, if the noise effects associated with all the individual FFT/IFFT DSP elements are assumed to be equal, from (4) and (5) it can be worked out that the IEVM differences between these individual FFT/IFFT DSP elements and the $IEVM_{ADC\_dB}$ should be larger than 20 dB. This gives rise to an upper bit resolution boundary, here referred to as $\{b_U\}$. The optimum bit resolution map, $\{b_{OPT}\}$, should be applicable for all signal modulation formats up to 256-QAM to enable the use of adaptive bit loading on different subcarriers in the real-time OOFDM transceivers [2, 10].

Giving the fact that there exists a huge number of DSP elements, whose bit resolution variation ranges are relatively large, to further simplify the numerical exploration process of identifying the optimum maps, some classifications to the FFT/IFFT DSP elements are also made: for a given ADC/DAC bit resolution, use is made of Fig. 3 to prioritise these DSP elements according to their individual impacts on the overall transceiver IEVM performance, that is, an FFT/IFFT DSP element producing the lowest IEVM is assigned the highest priority and is subsequently ordered as 1, and an FFT/IFFT DSP element which has the relatively next least significant impact on the transceiver IEVM performance is ordered as 2. Such treatments repeat until all DSP elements are ordered appropriately. Generally speaking, the FFT of a specific stage has a higher priority than the IFFT of the same stage, and the FFT/IFFT output bit resolutions have higher priority than those

$$\frac{1}{IEVM_{total}^2} = \left\{ \frac{1}{IEVM_{ADC}^2} + \sum_{k=3}^{5} \left(\frac{1}{IEVM_{FFT\_output}^2(k)} \right) + \sum_{k=3}^{5} \left(\frac{1}{IEVM_{FFT\_twiddle}^2(k)} \right) + \sum_{k=3}^{5} \left(\frac{1}{IEVM_{IFFT\_output}^2(k)} \right) + \sum_{k=3}^{5} \left(\frac{1}{IEVM_{IFFT\_twiddle}^2(k)} \right) \right\} \tag{4}$$

corresponding to the twiddle factor bit resolution. In addition, it is also clear from Fig. 3 that, within both the FFT and IFFT, a late stage operation has a higher priority over its earlier stage operations.

Fig. 4 explicitly illustrates the computing procedure adopted in numerically identifying the FFT/IFFT stage-dependent minimum bit resolution maps. The procedure is outlined as follows:

1. In the initial simulation phase, for a given ADC/DAC bit resolution, a corresponding upper boundary bit resolution set, $\{b_U\}$, is produced along with an FFT/IFFT DSP element list. Each item contained in the parameter set is ordered according to its' priority.
2. The FFT/IFFT DSP element of the highest order in the un-processed DSP element list is selected and treated first. For different signal modulation formats taken on all subcarriers, numerical simulations are performed utilising the upper boundary bit resolution set for the first round simulations or the conditions specified in step 5 for all other round simulations.
3. If the simulated overall transceiver IEVM performance meets the constraint in (5), numerical simulations are performed once again, taking into account the following two conditions: (a) the bit resolution of the selected FFT/IFFT DSP element is reduced by 1 and (b) the previously adopted bit resolutions of all non-selected DSP elements still remain unchanged. Such procedures repeat until the minimum bit resolution of the selected DSP element is reached to satisfy (5).
4. The selected DSP element is removed from the un-processed DSP element list and its' bit resolution value is added to the optimum map.
5. The next FFT/IFFT DSP element of the highest order in the present un-processed DSP element list is selected; step 2, step 3 and step 4 are repeated. Here use is made of the obtained optimum map for the already processed DSP elements and the remaining part of the upper boundary bit resolution set for the un-processed DSP elements.
6. Step 5 repeats until a full optimum map containing all the FFT/IFFT DSP elements, $\{b_{OPT}\}$, is generated.
7. $\{b_U\}$ employed in the initial phase is replaced by the newly generated $\{b_{OPT}\}$, all the above-mentioned six steps are repeated to produce an updated version of $\{b_{OPT}\}$. Such iterative procedure continues until a stable $\{b_{OPT}\}$ is obtained.
8. The above-mentioned seven steps are repeated to generate another optimum $\{b_{OPT}\}$ corresponding to a different ADC/DAC bit resolution.

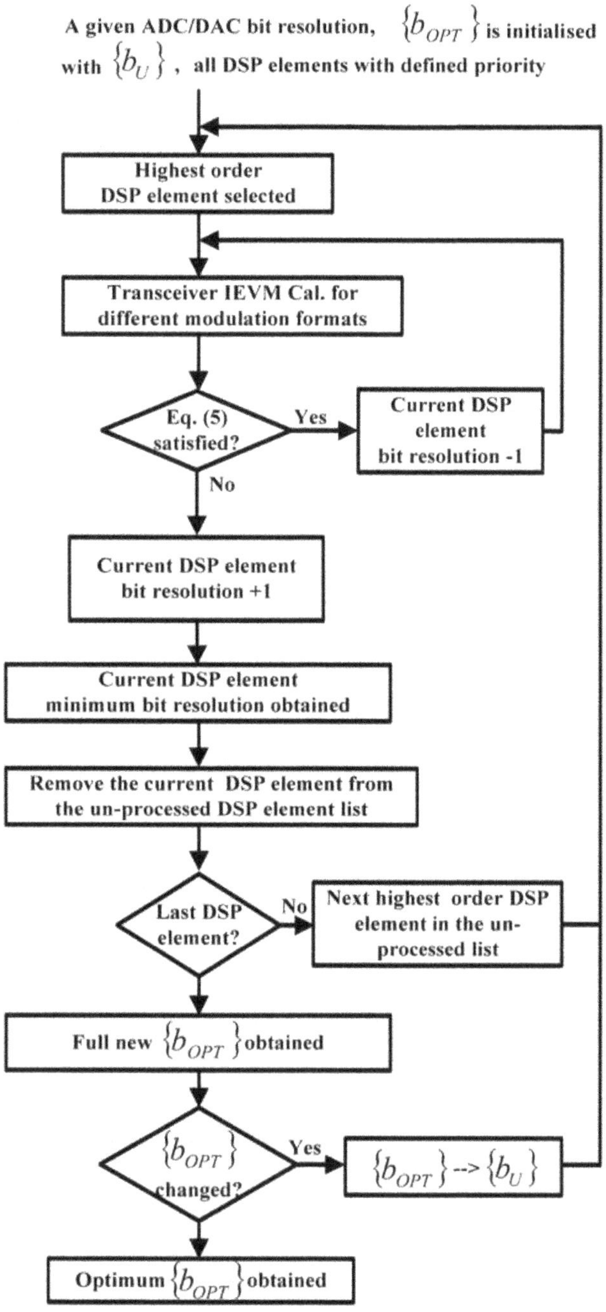

Fig. 4 *Computing procedure adopted in numerically identifying the FFT/IFFT stage-dependent minimum bit resolution maps*

The resulting FFT/IFFT stage-dependent minimum bit resolution maps against ADC/DAC bit resolution are presented in Figs. 5*a* and *b* for the IFFT output bit resolution and twiddle factor bit resolution in the transmitter, and in Figs. 5*c* and *d* for the FFT output bit resolution and twiddle factor bit resolution in the corresponding receiver. It can be seen in Figs. 5*a* and *c* that the minimum FFT/IFFT output bit resolution increases by about 1.5 bits for a one-step increase in FFT/IFFT stage, and the minimum FFT/IFFT output bit resolution grows almost linearly with ADC/DAC bit resolution. In addition, in comparison with the IFFT, the FFT requires approximately two more output bits to achieve the same transceiver IEVM performance. Furthermore, as expected from Fig. 3*c*, the FFT and IFFT have similar twiddle factor bit resolution maps, as shown in Figs. 5*b* and *d*.

To examine the identified FFT/IFFT stage-dependent minimum bit resolution maps, for different signal modulation formats Fig. 6*a* is plotted, which shows the resulting ADC/DAC bit resolution-dependent IEVM differences between the optimum map-based overall transceiver IEVM performance and the ADC/DAC-only bench-marker results presented in Fig. 2. It is very interesting to note in Fig. 6*a* that < 0.5 dB IEVM differences occur not only over a wide ADC/DAC bit resolution range from 5 to 12, but also for all practically adopted signal modulation formats from 16-QAM to 256-QAM. As such, it can be concluded that, for designing practical real-time OOFDM transceivers, Figs. 5*a*–*d* serve as valuable maps enabling the optimum selection of minimum bit resolutions for all the involved FFT/IFFT DSP elements according to the available ADC/DAC bit resolutions.

It is well known [10, 21] that the signal clipping effect may considerably affect the OOFDM transceiver performance. Such an effect is not considered in all the numerical simulations discussed above to ensure that the identified FFT/IFFT stage-dependent minimum bit resolution maps are applicable for the worst-case application scenarios, as mentioned in Section 2.1. For the cases where the signal clipping effect is included, the validity of the minimum bit resolution maps is examined in Fig. 6*b*, where the ADC/DAC bit resolution-dependent IEVM differences for different signal modulation formats are plotted between the following two cases: Case I: the ADC/DAC-induced signal clipping and

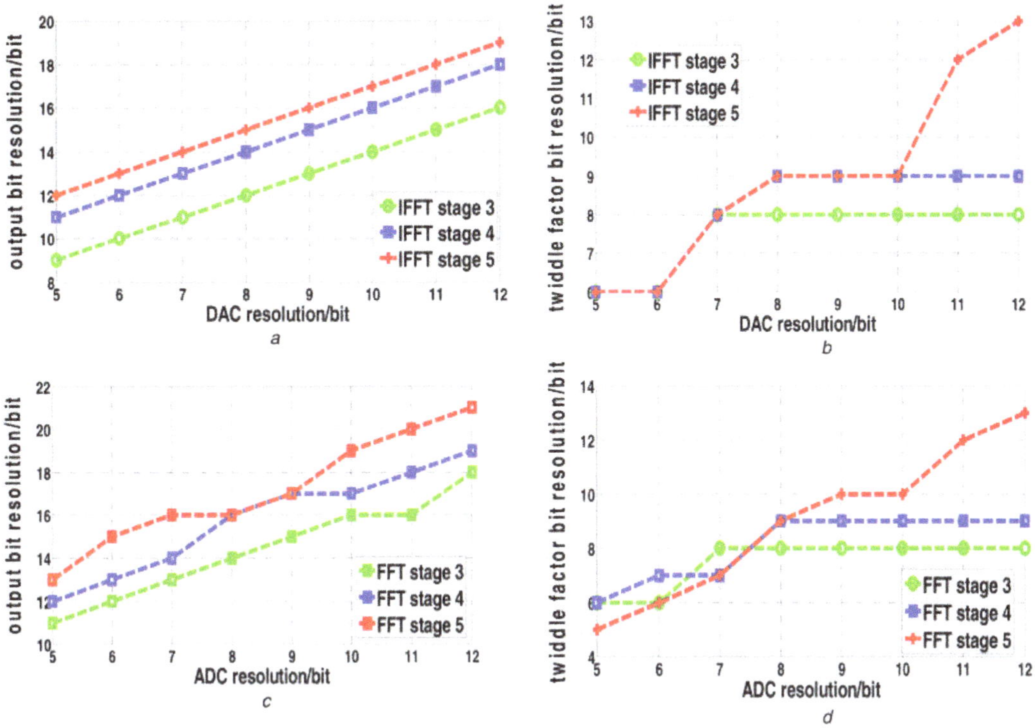

Fig. 5 *Identified FFT/IFFT stage-dependent minimum bit resolution maps*
a IFFT bit resolution map against DAC bit resolution
b IFFT twiddle factor bit resolution map against DAC bit resolution
c FFT bit resolution map against ADC bit resolution
d FFT twiddle factor bit resolution map against ADC bit resolution

quantisation effects and the floating-point FFT/IFFT operations are considered and Case II: the ADC/DAC-induced signal clipping and quantisation effects and the stage-dependent minimum bit resolution map-based FFT/IFFT operations are considered. In simulating Fig. 6*b*, the optimum signal clipping ratios listed in Table 3 are adopted for different ADC/DAC bit resolutions and signal modulation formats. These optimum values are numerically obtained using an approach identical to that reported in [21]. It can be seen in Fig. 6*b* that the IEVM difference developing trends are very similar to those represented in Fig. 6*a*, indicating that the identified minimum bit resolution maps are valid when the OOFDM signals are clipped at their optimum ratios. In particular, for a specific signal modulation format, in comparison with Fig. 6*a*, a much flatter IEVM difference curve is observed in Fig. 6*b*, confirming that the identified minimum bit resolution maps are capable of representing the worst-case scenarios.

In addition to the real-valued OOFDM signal cases discussed above, for complex-valued coherent OOFDM signals generated using asymmetric IFFT inputs, numerical simulations are also undertaken of the ADC/DAC bit resolution-dependent IEVM differences between the ideal case of including the floating-point IFFT/FFT operations and the case of including the identified minimum bit resolution map-based IFFT/FFT operations. In both cases, the ADC/DAC-induced quantisation noise effect is considered. It is shown that, for an ADC/DAC bit resolution range varying from 5 to 12, the IEVM differences of < 0.5 dB are still obtainable for various signal modulation formats up to 256-QAM. This indicates that the FFT/IFFT stage-dependent minimum bit

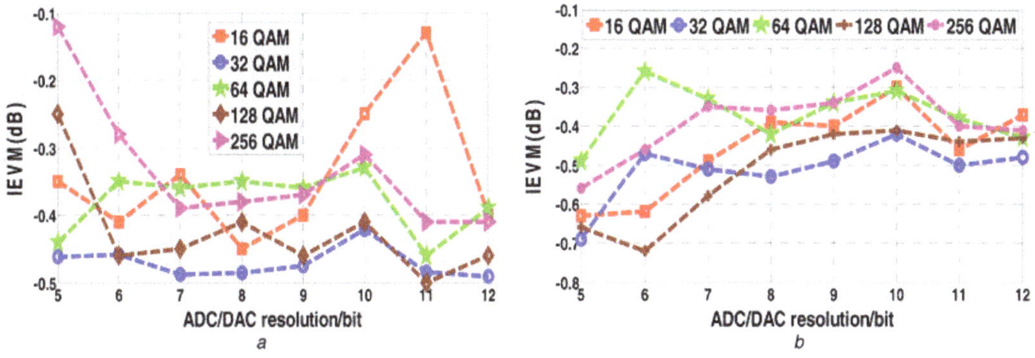

Fig. 6 *Simulated IEVM differences between the identified map-based transceiver IEVM performance and the ADC/DAC-only transceiver IEVM performance incorporating the floating-point FFT and IFFT*
a Without considering the signal clipping effect
b Optimum signal clipping ratios shown in Table 3 are employed

Table 3 Optimum signal clipping ratio (dB) for different ADC/DAC bit resolutions and signal modulation formats

Modulation formats	ADC/DAC bit resolutions							
	5	6	7	8	9	10	11	12
16-QAM	9	11	11	12	13	13	14	14
32-QAM	9	10	11	12	12	13	13	13
64-QAM	9	10	12	13	13	14	14	14
128-QAM	9	10	11	12	12	13	13	14
256-QAM	8	10	10	11	12	13	14	14

resolution maps identified above are also applicable for coherent OOFDM transceivers.

To illustrate the practical use of the identified minimum bit resolution maps, for the adopted signal modulation formats and targeted transceiver IEVM performance, use is first made of Fig. 2 to identify the minimum required ADC/DAC bit resolutions. According to this identified resolution, the bit resolution maps presented in Fig. 5 are utilised to determine the minimum stage-dependent output and twiddle-factor resolutions for minimising the FFT/IFFT logic resource usages. Compared with the floating-point FFT/IFFT case, the corresponding transceiver IEVM penalty is then found from Fig. 6.

3 Experimental map verifications and logic resource usages

To rigorously verify the validity and accuracy of the FFT/IFFT stage-dependent minimum bit resolution maps numerically identified in Section 2, in this section, experimental measurements are undertaken in a 150 MHz-clocked Altera FPGA-based experimental system, as illustrated in Fig. 7*a*. The detailed OOFDM transceiver DSP structures are identical to those illustrated in Fig. 1*a*. In each of the transceiver FPGAs, a self-developed full-parallel pipelined 32-point FFT/IFFT architecture is implemented and synthesised using Altera Quartus II. At the input of the IFFT, use is made of the signal bit resolutions identical to those adopted in numerical simulations, as explicitly specified in Section 2.1.

In the experimental measurements, a MATLAB-generated and QAM-encoded digital OFDM signal of length of 1000 symbols is first fed to the transmitter FPGA's input random access memory (RAM) having a depth of 1024. In the transmitter FPGA, the IFFT operation is performed taking into account the numerically identified stage-dependent minimum bit resolution maps. From the output RAM of the transmitter FPGA, the resulting time-domain digital signal is then passed to MATLAB to emulate the effect of quantisation noise associated with the DAC. After appropriately adjusting the signal amplitude to cover the full dynamic range of a corresponding ADC, the signal is transferred to the ADC, and the ADC-processed digital signal is fed to the receiver FPGA's input RAM with a depth of 1024 to perform the FFT operation employing the corresponding bit resolution maps numerically identified. Finally, from the output RAM of the receiver FPGA, the output digital signal is transferred back to MATLAB. For each case, the above-mentioned procedure is repeated five times, thus a total 5000 uncorrelated OFDM symbols are utilised to calculate the overall transceiver IEVM performance. Different ADCs/DACs of various bit resolutions varying from 5 to 12 are utilised.

Making use of the identified FFT/IFFT stage-dependent minimum bit resolution maps, for various signal modulation

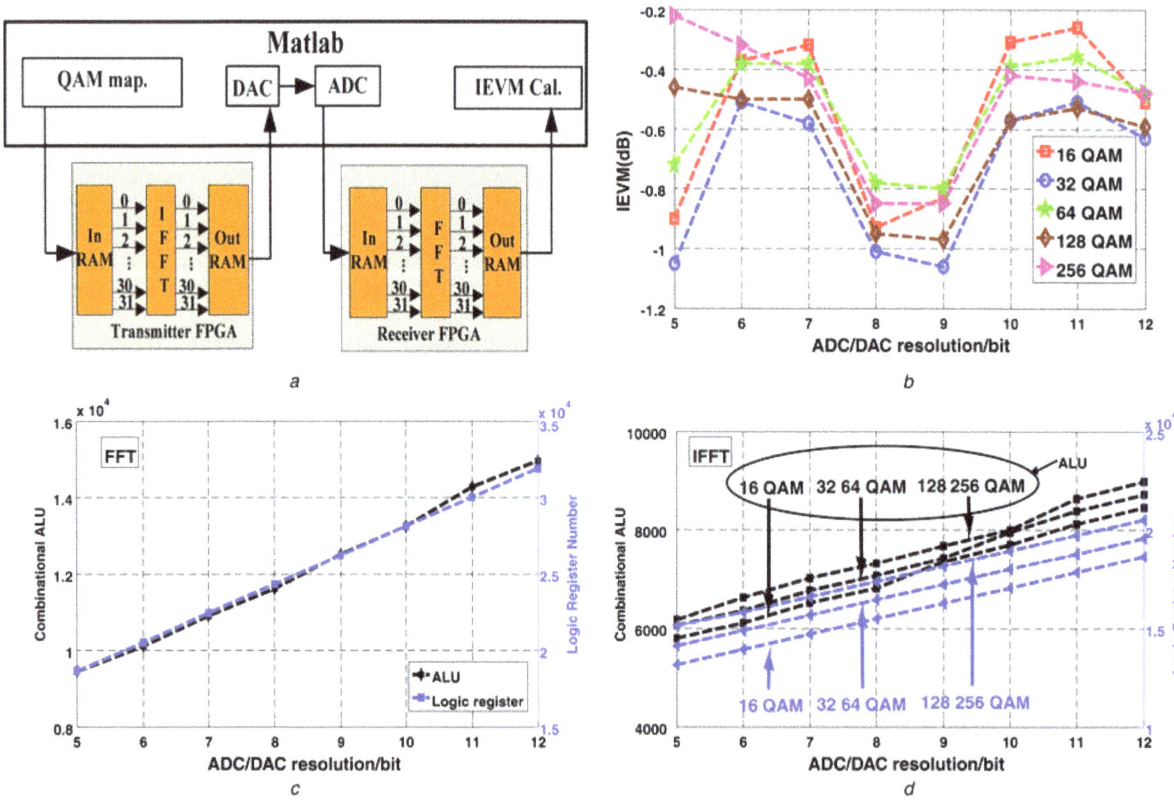

Fig. 7 *Experimental map verifications and logic resource usage*
a Experimental system setup for verifying the maps
All the parameters and internal RAMs are controlled via the FPGA's joint test action group interface
b IEVM differences between FPGA-generated/recovered signals obtained utilising the optimum bit resolution maps and MATLAB-based 'ideal' signals utilising the floating-point FFT/IFFT
c ALU and logic register usage for the receiver FFT
d ALU and register usage for the transmitter IFFT

formats, the ADC/DAC bit resolution-dependent IEVM differences between the digital signals generated and recovered by the transmitter and receiver FPGAs and the MATLAB-based 'ideal' signals using the floating-point FFT/IFFT are given in Fig. 7*b*, which agrees very well with the IEVM difference behaviours shown in Fig. 6. This strongly indicates that the FFT/IFFT minimum bit resolution maps are valid and have relatively high accuracy over the entire ADC/DAC bit resolution range practically adopted. In comparison with Fig. 6, the slightly large IEVM differences observed in Fig. 7*b* are mainly due to the fact that the finite bit resolution is adopted in the FPGA-based butterfly FFT/IFFT operations, whereas in obtaining Fig. 6 the corresponding operations are performed using the floating-point operations.

As a direct result of implementing the FFT/IFFT stage-dependent minimum bit resolution maps in the FPGAs, the consumed logic resources in terms of arithmetic logic unit (ALU) and logic register are plotted as a function of ADC/DAC bit resolution in Fig. 7*c* for the receiver FFT operation and in Fig. 7*d* for the transmitter IFFT operation. The ALU usage results from arithmetic operations such as addition, subtraction and multiplication, whereas the logic register usage results from the pipelined FFT/IFFT configurations. As expected, Figs. 7*c* and *d* show that the ALU and register usages almost linearly increase with ADC/DAC bit resolution. Moreover, the logic resource usage in the FFT is independent of signal modulation format (various ALU and logic register usage curves obtained for different signal modulation formats overlap completely in Fig. 7*c*). On the other hand, the bit resolutions of stage 1 and stage 2 of the IFFT are modulation format-dependent; this leads to the small logic resource usage variations shown in Fig. 7*d*. It can also be seen from Figs. 7*c* and *d* that, for the same ADC/DAC resolution, the ALU and logic register usages for the FFT are about 1.4 times higher than those for the IFFT.

4 Conclusions

Extensive numerical explorations have been undertaken of the impacts of each individual transceiver DSP element on the IEVM performance of the OOFDM transceivers incorporating the full-parallel pipelined FFT/IFFT architectures. The FFT/IFFT stage-dependent minimum bit resolution maps have been identified, for the first time, based on which the FPGA logic resource usage can be reduced significantly without degrading the transceiver IEVM performance. The validity and high accuracy of the identified bit resolution maps have also been experimentally verified in FPGA-based platforms.

Although the minimum bit resolution maps have been numerically identified and experimentally examined for the fixed 32-point FFT/IFFT, making use of the parameter developing trends seen in Fig. 5, the maps can be further extended to describe full-parallel pipelined FFT/IFFT architectures of arbitrary FFT/IFFT sizes. For available DACs/ADCs, the identified bit resolution maps offer an effective and simple approach of selecting minimum bit resolutions for individual DSP elements of various FFT/IFFT stages. This greatly eases practical designs of both high-capacity OOFDM transceivers and other FFT/IFFT-based DSP functions.

5 References

[1] Fischer J.K., Alreesh S., Elschner R., Frey F., Nölle M., Schubert C.: 'Bandwidth-variable transceivers based on 4D modulation formats for future flexible networks'. Proc. European Conf. Optical Communication (ECOC), London, 2013, Paper Tu.3.C.1

[2] Giddings R.P., Hugues-Salas E., Tang J.M.: '30 Gb/s real-time triple sub-band OFDM transceivers for future PONs beyond 10 Gb/s/λ'. Proc. European Conf. Optical Communication (ECOC), London, 2013, Paper P.6.7.5

[3] Buchali F., Dischler R., Klekamp A., Bernhard M., Efinger D.: 'Realisation of real-time 12.1 Gb/s optical OFDM transmitter and its application in a 109 Gb/s transmission system with coherent reception'. Proc. European Conf. Optical Communication (ECOC), Vienna, 2009, Paper PD2.1

[4] Schmogrow R., Bouziane R., Hillerkuss D., ET AL.: '85.4 Gbit/s real-time OFDM signal generation with transmission over 400 km and preamble-less reception'. CLEO 2012, San Jose, CA, USA, 6–11 May, 2012, Paper CTh1H.4

[5] Inan B., Adhikari S., Karakaya O., ET AL.: 'Real-time 93.8 Gb/s polarization-multiplexed OFDM transmitter with 1024-point IFFT', *Opt. Express*, 2011, **19**, (26), pp. B64–B68

[6] Inan B., Karakaya O., Kainzmaier P., ET AL.: 'Realization of a 23.9 Gb/s real time optical-OFDM transmitter with a 1024 point IFFT'. OFC/NFOEC2011, Los Angeles, CA, USA, 6–10 March 2011, Paper OMS2

[7] Qian D., Kwok T.T.-O., Cvijetic N., Hu J., Wang T.: '41.25 Gb/s real-time OFDM receiver for variable rate WDM-OFDMA-PON transmission'. OFC/NFOEC2010, San Diego, CA, USA, 21–25 March 2010, Paper PDPD9

[8] Kaneda N., Pfau T., Corteselli S., Yang Q., Leven A., Chen Y.K.: 'Real-time polarization division multiplexed coherent optical OFDM receiver at 9.83 GS/s for 28.6 Gb/s single-band intradyne detection'. ECOC 2011, Geneva, Switzerland, 18–22 September 2011, Paper We.9.A.6

[9] Chen S., Yang Q., William S.: 'Demonstration of 12.1 Gb/s single-band real-time coherent optical OFDM reception'. OECC 2010, Sapporo, Japan, 5–9 July 2010, Paper 8B4-2

[10] Giddings R.P., Jin X.Q., Hugues-Salas E., Giacoumidis E., Wei J.L., Tang J.M.: 'Experimental demonstration of a record high 11.25 Gb/s real-time optical OFDM transceiver supporting 25 km SMF end-to-end transmission in simple IMDD systems', *Opt. Express*, 2010, **18**, (6), pp. 5541–5555

[11] Wold E.H., Despain A.M.: 'Pipelined and parallel-pipeline FFT processors for VLSI implementation', *IEEE Trans. Comput.*, 1984, **c-33**, (5), pp. 414–426

[12] Yang L., Zhang K., Liu H., Huang J., Huang S.: 'An efficient locally pipelined FFT processor', *IEEE Trans. Circuits Syst. II, Express Briefs*, 2006, **53**, (7), pp. 585–589

[13] Liu H., Lee H.: 'A high performance four-parallel 128/64-point radix-24 FFT/IFFT processor for MIMO-OFDM systems'. Asia Pacific Conf. Circuits Systems/APCCAS2008, Macao, China, 30 November–3 December 2008, pp. 834–837

[14] Tang S.N., Tsai J.W., Chang T.Y.: 'A 2.4 Gs/s FFT processor for OFDM-based WPAN applications', *IEEE Trans. Circuits Syst. II, Express Briefs*, 2010, **57**, (6), pp. 451–455

[15] Ahmed T., Garrido M., Gustafsson O.: 'A 512-point 8-parallel pipelined feedforward FFT for WPAN'. Proc. of Forty Fifth Asilomar Conf. on Signals, Systems and Computers /ASILOMAR 2011, Pacific Grove, CA, USA, 6–9 November 2011, pp. 981–984

[16] Garrido M., Grajal J., Sánchez M.A., Gustafsson O.: 'Pipelined radix-2 k feedforward FFT architectures', *IEEE Trans. VLSI Syst.*, 2013, **21**, (1), pp. 23–32

[17] Yang K.J., Tsai S.H., Chuang G.C.H.: 'MDC FFT/IFFT processor with variable length for MIMO-OFDM systems', *IEEE Trans. VLSI*, 2013, **21**, (4), pp. 720–731

[18] Milder P., Bouziane R., Koutsoyannis R., Berger C.R., Benlachtar Y., Killey R.I., Glick M., Hoe J.C.: 'Design and simulation of 25 Gb/s optical OFDM transceiver ASICs'. The ECOC 2011, Geneva, Switzerland, 18–22 September 2011, Paper We.9.A.5

[19] Bouziane R., Koutsoyannis R., Milder P., ET AL.: 'Optimizing FFT precision in optical OFDM transceivers', *IEEE Photonics Technol. Lett.*, 2011, **23**, (20), pp. 1550–1552

[20] Dardari D.: 'Joint clip and quantization effects characterization in OFDM receivers', *IEEE Trans. Circuits Syst.*, 2006, **53**, (8), pp. 1741–1748

[21] Tang J.M., Shore K.A.: 'Maximizing the transmission performance of adaptively modulated optical OFDM signals in multimode fibre links by optimizing analogue-to-digital converters', *IEEE J. Lightwave Technol.*, 2007, **25**, (3), pp. 787–798

9

Ultra-wideband wireless receiver front-end for high-speed indoor applications

Zhe-Yang Huang[1], Chun-Chieh Chen[2], Chung-Chih Hung[1]

[1]*Department of Electrical and Computer Engineering, National Chiao Tung University, Hsinchu, Taiwan*
[2]*Department of Electronics Engineering, Chung Yuan Christian University, Chungli, Taiwan*
E-mail: cchung@mail.nctu.edu.tw

Abstract: Low-noise, ultra-wideband (UWB) wireless receiver front-end circuits were presented in this study. A two-stage common-source low-noise amplifier with wideband input impedance matching network, an active-balun and a double-balanced down-conversion mixer were adopted in the UWB wireless receiver front-end. The proposed wireless receiver front-end circuits were implemented in 0.18 μm radio-frequency-CMOS process. The maximum down-conversion power gain of the front-end is 25.8 dB; minimum single-sideband noise figure of the front-end is 4.9 dB over complete UWB band ranging from 3.1 to 10.6 GHz. Power consumption including buffers is 39.2 mW.

1 Introduction

The ultra-wideband (UWB) system frequency spectrum is allocated in the frequency ranging from 3.168 to 10.650 GHz. There are 6 band groups, and 14 band channels are defined in the frequency spectrum. Band Group 1–Band Group 4 and Band Group 6 are with three band channels. Band Group 5 is with two band channels. The UWB system adopts the orthogonal frequency-division multiplexing digital modulation. In the UWB receiver system, the minimum receiver sensitivity is −80.8, −70.4 and −63.5 dBm for 53.3, 480 and 1024 Mb/s data rate, respectively. The transmitted spectral mask for indoor applications is −41.3 dBm/Hz power spectrum density (PSD).

The receivers and front-ends in papers [1–5] adopted one single-balanced low-noise amplifier (LNA), and at least one single-balanced down-conversion mixer. The pros are lower power consumption, easy implementation and smaller silicon area. The cons are higher noise figure (NF), and high even-order harmonics. The receivers and front-ends in papers [6–8] adopted one differential LNA, and at least one double-balanced down-conversion mixer. The pros are lower NF and low even-order harmonics. The cons are higher power consumption, and a complicated and larger silicon area. In this paper, one single-balanced LNA is adopted to save the power consumption. The active-balun is adopted to convert the single-balanced signals to double-balanced signals with small area and with low power consumption. The double-balanced down-conversion mixer is adopted to compress the even-order harmonics in the receiver system (Table 1). In paper [9], the UWB down-conversion mixer is designed as passive mixer. The receiver sensitivity is not good while adopting the passive mixers. In paper [10], the UWB down-conversion mixer is designed as distributed mixer. The distributed mixer consumes much more power than other types of mixers. In papers [1–5, 11], the UWB down-conversion mixers are designed as single-balanced Gilbert mixer.

2 Circuit design

2.1 UWB receiver front-end design

Low-intermediate (IF) wireless receiver architecture is adopted in our UWB receiver front-end. The IF band frequency is from 0 Hz to 528 MHz. Low-noise and high conversion gain are important to zero-IF wireless receiver and low-IF wireless receiver. In Fig. 1, the LNA and down-conversion mixer are implemented on chip, and the local oscillator (LO) is external. The block diagram of the proposed UWB receiver front-end is shown in Fig. 2. The first stage of the UWB receiver front-end circuits is a wideband matching network for 3.1–10.6 GHz, and the input return loss is lower than −10 dB. The second stage is the LNA. The noise performance of the LNA determines the receiver sensitivity, so the low-noise characteristic is an important design consideration for the LNA. High power gain is also an important characteristic of the low-IF receiver. In this UWB receiver front-end circuit, the LNA power gain is designed as 15 dB. The third stage is an active-balun which converts the single-balanced signals to double-balanced signals. The fourth stage is a *Gm* stage which provides additional gain for the received signals before the signals enter the switching stage. The fifth stage is the switching stage, which converts the radio-frequency (RF) signals into the IF signals. In this UWB receiver front-end circuit, the down-conversion mixer provides 10 dB power gain. A LO with dual phase is adopted as a part of mixing signals. The loading of switching stage provides IF gain for the receiver chain. The sixth stage is the buffer stage, which is designed for 50 Ω output impedance. The gain contribution of the UWB receiver front-end circuits is shown in Fig. 3. From 3.1 to 10.6 GHz, the LNA provides 15 dB power gain and the mixer provides 10 dB power gain. A total power gain of the front-end circuits is 25 dB.

2.2 Wideband LNA circuit design

The most important characteristics of the UWB LNA are gain flatness and high power gain, low-noise contribution and wide bandwidth. The basic topologies of the single-stage metal–oxide semiconductor field effect transistor (MOSFET) amplifier are common-drain stage, common-gate stage and common-source stage. The maximum gain of the common-drain stage is close to 1, so it is not a good candidate to design the gain stage of the amplifier. The common-gate stage has high gain for the amplifier, but the noise is high. The common-source stage has high gain and low-noise characteristic. That is the reason why common-source amplifier is widely used in LNA designs [2, 4, 5, 8, 12–16]. The topology of conventional LNA is a single-stage amplifier with parallel inductor and capacitor (LC) tank. However, this topology can only provide a limited bandwidth. A two-stage low-noise amplifier with wideband input impedance matching network is shown in Fig. 4. The two-stage common-source amplifier could achieve wide bandwidth, high gain and flat gain [17–20]. The first-stage amplifier is composed of a MOSFET transistor M_{L_1}, a source-degeneration inductor L_s and a load inductor L_{L_1}. The second-stage amplifier comprises a MOSFET transistor M_{L_2} and a load inductor L_{L_2}.

Table 1 Performance comparison

References	3 dB BW, GHz	Conversion gain, dB	NF, dB	*S11*, dB	IIP3, dBm	Current, mA	Power, mW	Tech.
[1] JSSC'05	3.1–8.2	51.9–52	3.3–3.4	<− 9	−7.5	22	59.4	0.18 μm SiGe
[2] JSSC'05	3.1–4.8	34–37	6.5–8.4	<− 11	N/A	10	15	0.13 μm CMOS
[3] JSSC'05	3.1–4.8	<59	< 4.5	<− 9	−6	20	50	0.25 μm SiGe
[4] TCASII'10[a]	3.1–10.6	19.5–21.7	3–3.8	<− 11	−9.6	9	10.8	65 nm CMOS
[5] EL'11[a]	3–9	31–32.6	3.2–6.5	<− 9.4	N/A	10.8	19.4	0.18 μm CMOS
[6] JSSC'07[a]	3.1–8	15–18	5.2–6.6	N/A	−5.6	19.5	44.9	0.18 μm CMOS
[7] JSSC'09	3.1–8	<42	6.5–8.3	<− 13	−12	24	43.2	0.18 μm CMOS
[8] JSSC'08	3.4–10.3	<38	5–10	<− 8.5	N/A	11	27.5	0.25 μm SiGe
[12] JSSC'06[a]	3–10	29–39	3.3–5	<− 5	N/A	29.8	53.7	0.18 μm SiGe
this work[a]	3–11.5	22.8–25.8	4.9–6.9	<− 8.9	−26	21.8	39.2	0.18 μm CMOS

[a]Receiver front-end.

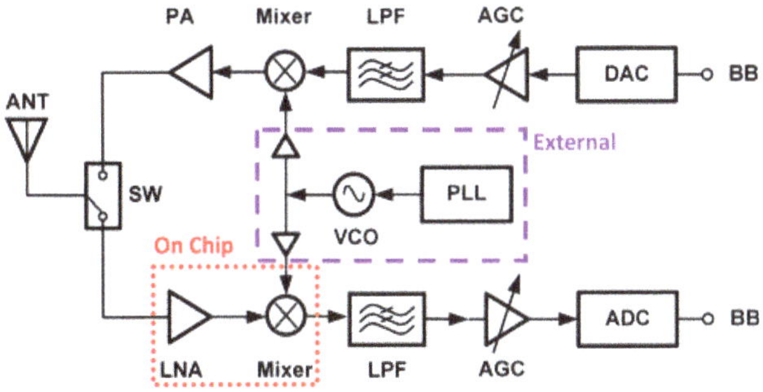

Fig. 1 *UWB transceiver architecture*

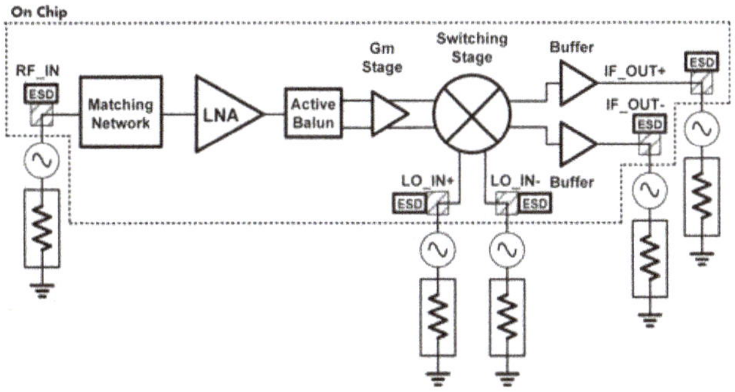

Fig. 2 *Double-balanced UWB receiver front-end*

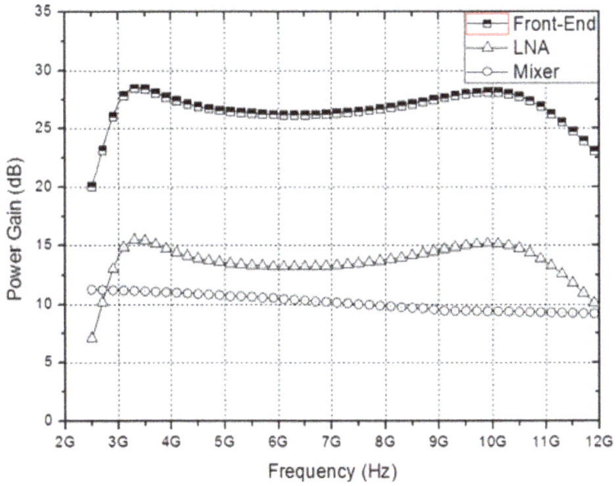

Fig. 3 *Gain combination*

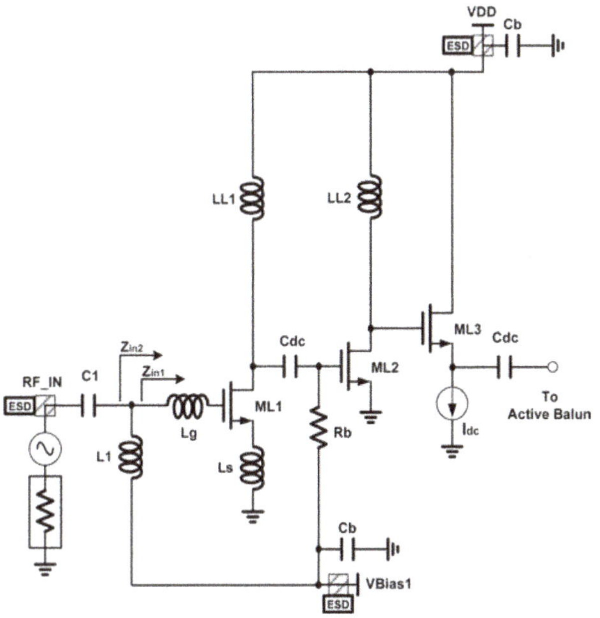

Fig. 4 *Two-stage cascade low-noise amplifier with wideband input impedance matching network*

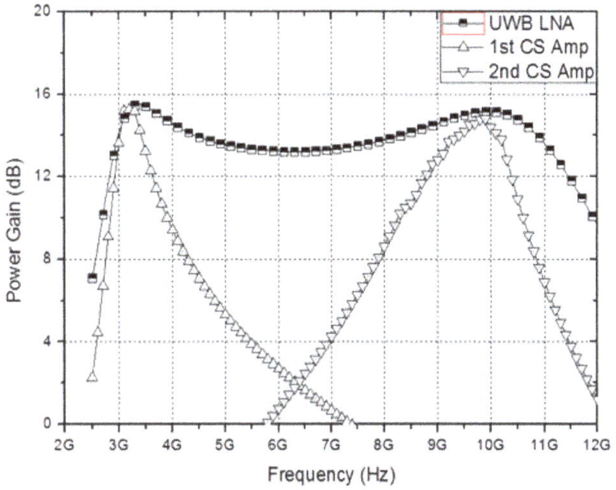

Fig. 5 *Superposition of the LNA power gain*

R_b is a biasing resistor and C_1 and C_{dc} are DC blockers. The source follower M_{L_3} is designed for inter-stage impedance matching and inter-stage isolation. In Fig. 5, the first-stage amplifier provides the

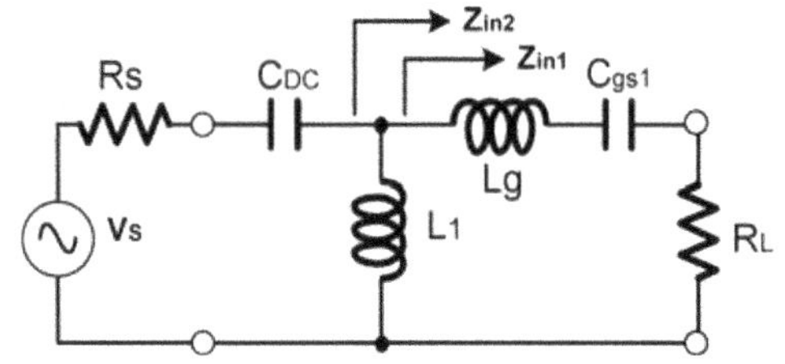

Fig. 6 *Small-signal model of the input stage*

low-frequency gain of the UWB receiver system, and the second-stage amplifier provides the high-frequency gain. The resonated centre frequency of the first-stage amplifier is designed at 3.5 GHz, and the resonated centre frequency of the second-stage amplifier is designed at 10 GHz. The overall gain of the UWB LNA follows the superposition theorem, so the bandwidth is extended as shown in Fig. 5. In conventional LNA circuit design, transistor M_{L_1}, inductor L_s and inductor L_g have 50 Ω impedance at one resonated frequency. However, the bandwidth is not wide enough for UWB system [21–23]. A shunt inductor L_1 provides another resonated frequency. Therefore, the bandwidth of the input impedance matching can be extended to several gigahertz [24–30]. The input matching network includes inductors L_1, L_g, L_s and transistor M_{L_1}. The inductances of L_g, L_s and L_{L_1} at 3.5 GHz are 1.45, 0.21 and 5.87 nH. The inductances of L_1 and L_{L_2} at 10 GHz are 2.19 and 1.29 nH. The Q-factors of L_g, L_s and L_{L_1} at 3.5 GHz are 9, 7.4 and 7. The Q-factors of L_1 and L_{L_2} at 10 GHz are 9.5 and 11.7.

The input impedance of the UWB LNA Z_{in_1} and Z_{in_2} are shown in Fig. 4 and the small-signal model is shown in Fig. 6. According to the small-signal model, the input impedance Z_{in_1} is expressed in (1) and the input impedance Z_{in_2} is expressed in (2), where g_{m_1} is the trans-conductance of the transistor M_{L_1} and C_{gs_1} is the gate–source parasitic capacitor of the transistor M_{L_1}. From (1), the resonated frequency can be derived as $\omega_r = 1/\sqrt{L_g C_{gs_1}}$, which determines one input impedance matching frequency of the UWB LNA

$$Z_{in_1}(s) = sL_g + \frac{1}{sC_{gs_1}} + \frac{g_{m1}}{C_{gs_1}} L_S \simeq sL_g + \frac{1}{sC_{gs_1}} + \omega_T L_S \quad (1)$$

In (2), sC_{L_1} is the parasitic capacitor of the inductor L_1. From (2), assuming s_{L_1} is neglected, a resonated frequency $\omega_{r2} = 1/\sqrt{(L_1 + L_g)C_{gs_1}}$ can be obtained which determines another input impedance matching frequency of the UWB LNA. If the above two resonated frequencies are carefully designed, the bandwidth of input impedance matching will be extended to several gigahertz

$$\begin{aligned} Z_{in_2}(s) &= \left(s_{L_1} || \frac{1}{sC_{L_1}}\right) || \left(sL_g + \frac{1}{sC_{gs_1}} + \frac{g_{m1}}{C_{gs_1}} L_S\right) \\ &\simeq s_{L_1} \cdot \frac{s^2(L_g C_{gs_1}) + sL_s C_{gs1}\omega_T + 1}{s^2(L_g C_{gs_1} + L_1 C_{gs_1}) + sL_s C_{gs_1}\omega_T + 1} \end{aligned} \quad (2)$$

The voltage gain of the single-stage common-source amplifier is expressed in (3), where r_{o1} is the output impedance of transistors M_{L_1}, C_{L_1} is the parasitic capacitor of the inductor L_{L_1} and R_{L_1} is the parasitic series resistance of the inductor L_{L_1}

$$\begin{aligned} A_{V1}(s) &= -g_{m_1}\left(r_{o1} || sL_{L_1} + R_{L_1} || \frac{1}{sC_{L_1}}\right) \\ &\simeq -g_{m_1} \cdot \frac{sL_{L_1} + R_{L_1}}{s^2 L_{L_1} C_{L_1} + sR_{L_1} C_{L_1} + 1} \end{aligned} \quad (3)$$

In the proposed UWB LNA, the overall voltage gain that includes two common-source amplifiers and one source follower is expressed in (4)

$$\begin{aligned} A_{V-LNA}(s) &= \left[-g_{m_1}\left(r_{o1} || sL_{L_1} + R_{L_1} || \frac{1}{sC_{L_1}}\right)\right] \\ &\quad \cdot \left[-g_{m_2}\left(r_{o1} || sL_{L_2} + R_{L_2} || \frac{1}{sC_{L_2}}\right)\right] \cdot \frac{1}{g_{m_5} + g_{mb_5}} \\ &= \frac{g_{m_1} g_{m_2}}{g_{m_5}} \cdot \frac{sL_{L_1} + R_{L_1}}{s^2 L_{L_1} C_{L_1} + sR_{L_1} C_{L_1} + 1} \\ &\quad \cdot \frac{sL_{L_2} + R_{L_2}}{s^2 L_{L_2} C_{L_2} + sR_{L_2} C_{L_2} + 1} \\ &\simeq \frac{g_{m_1} g_{m_2}}{g_{m_5}} \cdot \frac{(sL_{L_1} + R_{L_1})(sL_{L_2} + R_{L_2})}{s^4 N_1 N_2 + s^2(N_1 + N_2) + 1} \end{aligned} \quad (4)$$

where r_{o2} is the output impedance of transistor M_{L_2}, C_{L_2} is the parasitic capacitor of the inductor L_{L_2}, R_{L_2} is the parasitic series resistance of the inductor L_{L_2}, $N_1 = C_{L_1} \cdot L_{L_1}$ and $N_2 = C_{L_2} \cdot L_{L_2}$. The

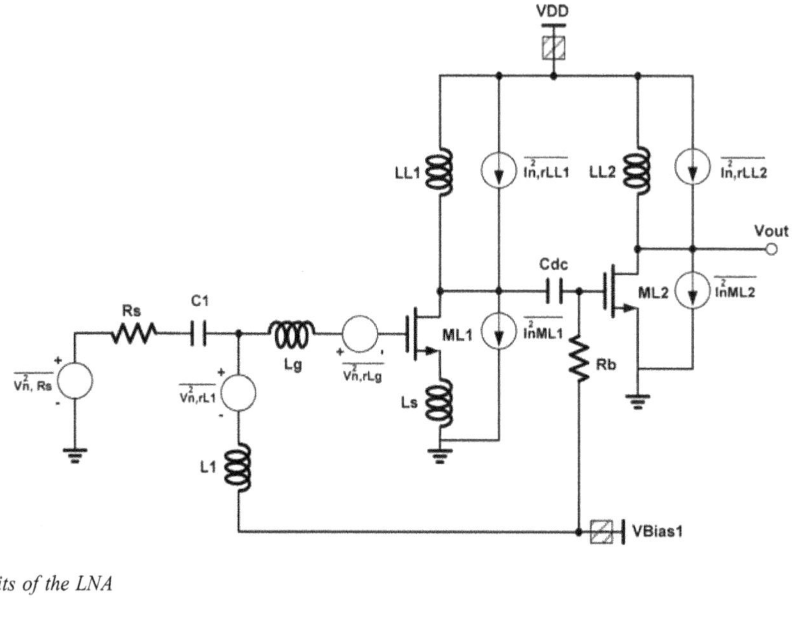

Fig. 7 *Noise equivalent circuits of the LNA*

noise equivalent circuits are shown in Fig. 7. The voltage noise sources $V_{n,Rs}$, $V_{n,\mathrm{r}L_1}$ and $V_{n,\mathrm{r}Lg}$ are from the source resistor R_s, the parasitic series resistance from inductor L_1 and the parasitic series resistance from inductor L_g. The current noise sources $I_{nM_{L_1}}$, $I_{nM_{L_2}}$, $I_{n,\mathrm{r}L_{L_1}}$ and $I_{n,\mathrm{r}L_{L_2}}$ are from the current noise of the MOSFET and the parasitic parallel resistance of the inductors L_{L_1} and L_{L_2}. The noise contribution of the parasitic resistance of the inductor L_s is very small. The total output noise $\overline{V_{n1,\mathrm{out}}^2}$ of the first-stage common-source amplifier is expressed in (5). The total output noise $\overline{V_{n2,\mathrm{out}}^2}$ of the second-stage common-source amplifier is expressed in (6)

$$\overline{V_{n1,\mathrm{out}}^2} = 4KTR_SA_1^2 + 4KTr_{L_1}A_1^2 + 4KTr_gA_1^2 + (\overline{I_{n,\mathrm{r}L_{L_1}}^2} + \overline{I_{nM_{L_1}}^2}) \cdot R_{\mathrm{out}}^2 \quad (5)$$

$$\overline{V_{n2,\mathrm{out}}^2} = 4KTR_SA_2^2 + (\overline{I_{n,\mathrm{r}L_{L_1}}^2} + \overline{I_{nM_{L_1}}^2}) \cdot R_{\mathrm{out}}^2 \quad (6)$$

where K is the Boltzmann constant and T is the absolute temperature. A_1 is the voltage gain of the first-stage common-source amplifier and A_2 is the voltage gain of the second-stage common-source amplifier. R_{out} is the equivalent output impedance of the amplifier. α_1 and α_2 are the short-channel factors for the current noise. $gm_{M_{L_1}}$ and $gm_{M_{L_2}}$ are the transconductances of the MOSFET transistor M_{L_1} and M_{L_2}, respectively. The drain current noise of M_{L_1} and M_{L_2} can be expressed as $\overline{I_{nM_{L_1}}^2} = 4\alpha_1 KTgm_{M_{L_1}}$ and $\overline{I_{nM_{L_2}}^2} = 4\alpha_2 KTgm_{M_{L_2}}$, respectively.

The noise factor of the first-stage common-source amplifier is expressed in (7)

$$F_1 = \frac{\overline{V_{n1,\mathrm{out}}^2}}{4KTR_SA_1^2} \quad (7)$$

The noise factor of the second-stage common-source amplifier is expressed in (8)

$$F_2 = \frac{\overline{V_{n2,\mathrm{out}}^2}}{4KTR_SA_2^2} \quad (8)$$

The calculated total noise factor of the low-noise amplifier is expressed in (9)

$$F_{\mathrm{Total}} = 1 + \frac{(r_g + r_{L_1})}{R_S} + \left(\frac{1}{r_{L_1}} + \alpha_1 gm_{M_{L_1}}\right) \cdot \frac{1}{gm_{M_{L_1}} R_S} + \frac{1 + \alpha_2 r_{L_2} gm_{M_{L_2}}}{A_1 r_{L_2} gm_{M_{L_2}} R_S} \quad (9)$$

2.3 Active-balun and double-balanced down-conversion mixer circuit design

In this paper, an active-balun and double-balanced mixer are adopted in the down-conversion mixer design (Fig. 8). In the third stage of the receiver front-end circuits, the active-balun converts the single-balanced signals to double-balanced signals. M_3 is a common-gate amplifier with the load resistor R_2. M_4 is a common-source amplifier with the load resistor R_3. R_1 is the biasing resistor and C_{dc} is the DC blocker. The single-balanced signals come from the LNA and go into the source of the M_3 transistor and the gate of the M_4 transistor. Assume the phase of the signals from the LNA is 0°. The signals separate into two paths, one is the common-gate amplifier and the other is the common-source amplifier. In one path, the input of the common-gate amplifier is from the source of the transistor M_3 and the output is to the drain of the transistor M_3. The phase of the signals is unchanged, from 0 to 0°. In the other path, the input of the common-source

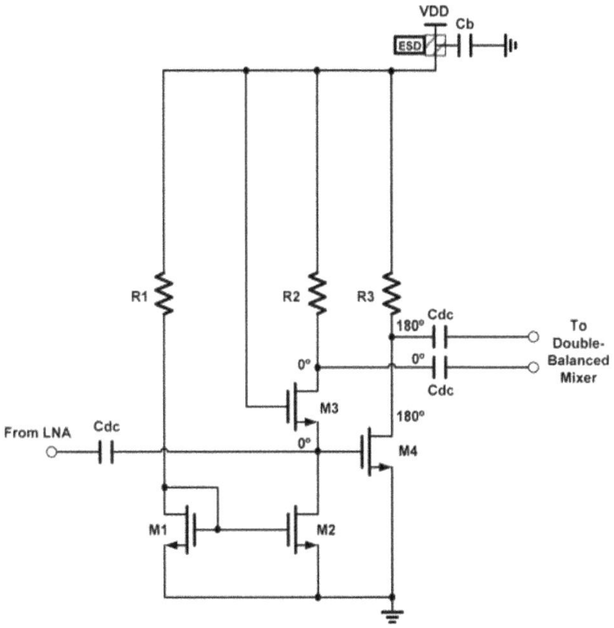

Fig. 8 *Active-balun circuit*

amplifier is from gate of the transistor M_4 and the output is to the drain of the transistor M_4. The phase of the signals is changed from 0 to 180°. The two paths signals are comprised of the differential signals. The gains of the two amplifiers can be optimised to be the same. The gain difference is < 0.3 dB. The phase differences of the common-gate amplifier and the common-source are <3°. To save the power consumption, the power gain of the active-balun is 3 dB.

The double-balanced down-conversion mixer is adopted in the UWB receiver front-end circuits. The advantages of the double-balanced down-conversion mixer are low power consumption, low input-referred noise and low even-harmonics [6–9, 31–34]. In this paper, the UWB receiver is a low-IF system and the LO feed-through problem can be solved by the following low-pass filter. The transistors M_5, M_6, M_7, M_8 and the resistor R_4 comprise the current mirror for the constant current source. The transistors M_9 and M_{10} comprise the differential *Gm* stage of the down-conversion mixer. The transistors M_{11}, M_{12}, M_{13} and M_{14} comprise the switching stage of the down-conversion mixer. The transistors M_{15}, M_{16} and the resistor R_L comprise the loading of the down-conversion mixer. R_b is a biasing resistor and C_{dc} is a DC blocker. The first stage of the down-conversion mixer is the *Gm* stage, which is common-source transistors M_9 and M_{10}. The *Gm* stage contributes additional gain for the received signals. The second stage of the down-conversion mixer is the switching stage, which is the transistors M_{11}, M_{12}, M_{13} and M_{14}. In the switching stage, differential LO signal comes in the gate of the transistor pairs M_{11}, M_{14} and M_{12}, M_{13}; and the RF signal comes in the source of the transistors. The mixing IF signals are at the drain of the switching transistors. When the RF signal is v_{RF} cos $\omega_{\mathrm{RF}}t$ and LO signal is v_{LO} cos $\omega_{\mathrm{LO}}t$, the cross-mixing signals are expressed in (10)

$$v_{\mathrm{corss}} = cv_{\mathrm{RF}}v_{\mathrm{LO}}[\cos(\omega_{\mathrm{RF}} - \omega_{\mathrm{LO}})t + \cos(\omega_{\mathrm{RF}} + \omega_{\mathrm{LO}})t] \quad (10)$$

The IF signals are the frequency subtraction of RF frequency and LO frequency. The loading of the switching stage is the active load with common-mode feedback. The diode connected transistors M_{15} and M_{16} extended the frequency bandwidth, and the resistor R_L stabilises the close loop. The fifth stage of the receiver front-end circuits is the buffer stage, which is the transistors M_{17} and M_{18}. To fit the measurement system impedance matching,

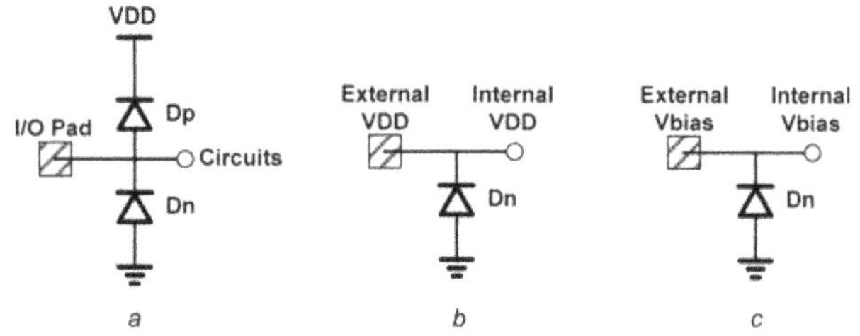

Fig. 9 *ESD diodes*
a ESD diodes for I/O pads
b ESD diodes for external V_{DD} pad
c ESD diodes for external V_{bias} pads

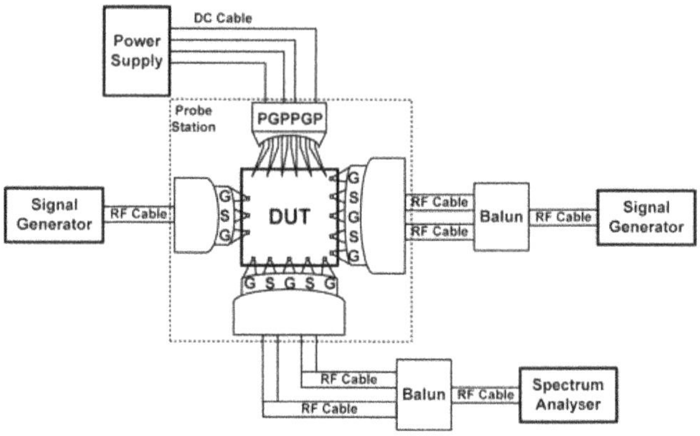

Fig. 11 *Diagram and photograph of measurement setup*

the output impedance of the buffers was designed to be 50 Ω, which consumes most power in the UWB receiver front-end. In wireless receiver, low-noise amplifier exposes to the electrostatic discharge (ESD), directly. The high-voltage static electricity from human body might destroy the transistors and paralyse the entire chip. In the proposed UWB receiver front-end circuits, the ESD protection is considered. The basic types of the ESD diodes are shown in Fig. 9. In Fig. 9*a*, the parallel p-type diode *Dp* and n-type diode Dn in the signal path are for the input–output (I/O) pads. In Fig. 9*b* and 9*c*, the parallel n-type diodes Dn are for the external voltage V_{DD}, V_{bias}.

The proposed UWB wireless receiver front-end circuits are shown in Fig. 10. The bias voltage of the two-stage common-source amplifiers is 0.7 V. The current consumption of first stage LNA is 3.9 mA, the current of second stage LNA is 2.7 mA and the current of the buffer stage is 2.3 mA. A total current consumption of the UWB LNA is 8.9 mA. The bias voltage of the *Gm* stage transistor is 0.7 V. The bias voltage of the switching stage is 0.9 V. The current of the active-balun is 2.3 mA, the current of mixer core is 4.6 mA and the current of buffer stage is 6 mA. The total current of the double-balanced down-conversion mixer is 10.6 mA. The V_{DD} of the UWB low-noise amplifier, active-balun and double-balanced down-conversion mixer is 1.8 V. The total DC current of the UWB wireless receiver front-end circuits, including LNA, active-balun, down-conversion mixer and output buffers is 21.8 mA. The total DC current consumption of the UWB wireless receiver front-end circuits without buffers is 13.5 mA.

2.4 Measurement consideration

The proposed UWB receiver front-end circuits were designed for on wafer test. The diagram of the measurement setup is shown in Fig. 11. On the probe station, there are four contact probes to probe the device under test. On left-hand side, the RF signals were generated from an external vector signal generator (VSG) as the received wireless signals, and the signals were transmitted through a microwave coaxial cable to a ground–signal–ground (GSG) microwave contact probe. The GSG microwave probe connected the single-balanced RF signals to the RF input of the UWB receiver front-end. On the right-hand side, the other external VSG was adopted as the LO. However, most of the instruments do not have a double-balanced system. Therefore, a passive-balun was adopted to convert the single-balanced signal from the VSG to double-balanced LO signals. The single-balanced LO signals were generated from the VSG and transmitted through the microwave coaxial cable to the passive-balun. Then the single-balanced LO signals were converted to double-balanced LO signals and transmitted through two microwave coaxial cables to a GSG–signal–ground (GSGSG) microwave contact probe. The GSGSG microwave probe connected the double-balanced LO signals to the LO input of the UWB receiver front-end. On the bottom side, the vector spectrum analyser (VSA) was adopted to analyse the down-conversion IF signals. In a similar way to the LO, a passive-balun was adopted to convert the double-balanced output IF signals to single-balanced IF signals. The GSGSG microwave probe connected the UWB receiver front-end double-balanced output IF signals to two microwave coaxial cables and transmitted to the balun. The double-balanced IF signals were converted to single-balanced IF signals and transmitted through the microwave coaxial cable to the VSA. On the upper side, a six pins (PGPPGP) DC probe was used to connect the power supplies. There are four individual power supplies that can be used in this DC probe. In the UWB receiver front-end, a universal V_{DD}, and two voltage biasing V_{bias1}, V_{bias2} were used.

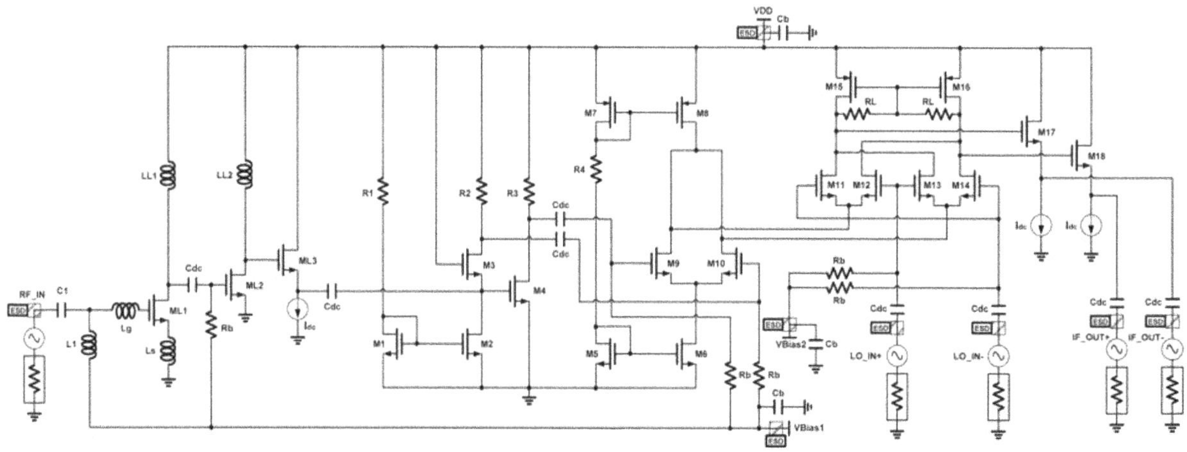

Fig. 10 *Proposed UWB wireless receiver front-end*

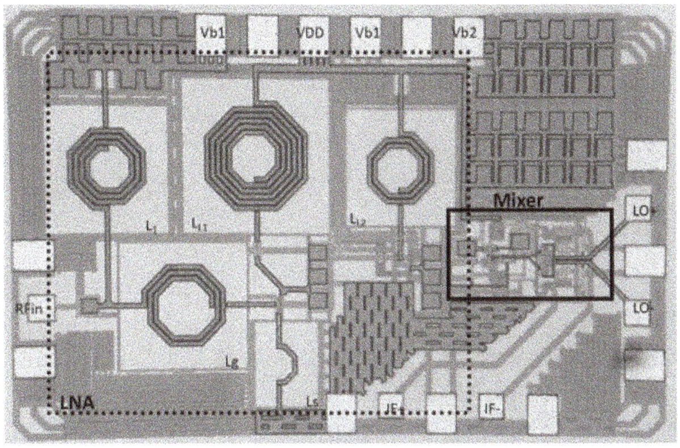

Fig. 12 *Micro-photography of the UWB receiver front-end*

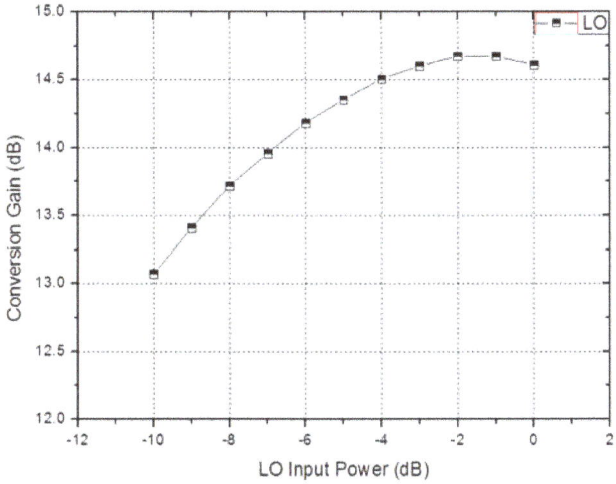

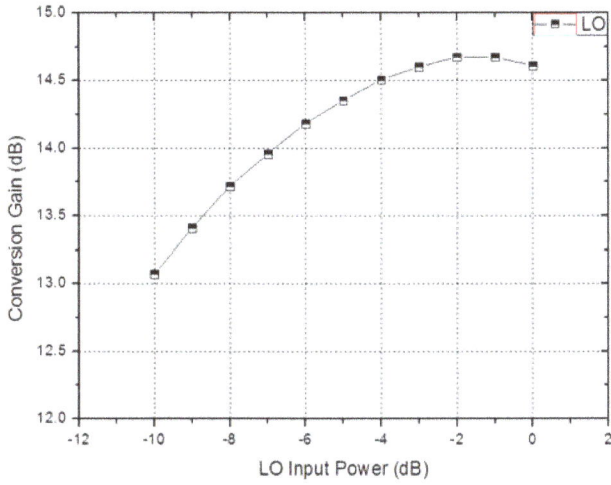

Fig. 13 *Input LO power against output IF power*

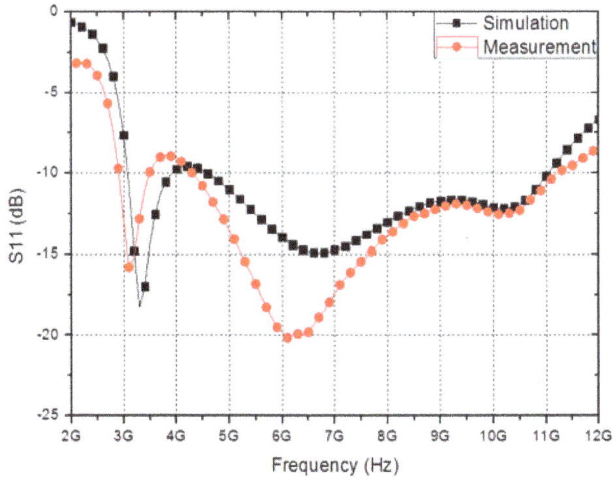

Fig. 14 *Measurement results of the input return loss (S11)*

3 Experimental results

Fig. 12 is the micro-photograph of the proposed UWB receiver front-end circuit, where the chip size including all pads and dummy is 1.28 mm × 0.81 mm [1.04 mm^2]. The red dash line indicates the area of the single-balanced LNA and the blue solid line indicates the area of the active-balun and the double-balanced down-conversion mixer. The experimental results of the receiver front-end circuit are shown in Figs. 14–19. The proposed UWB receiver front-end circuits were implemented in Taiwan semiconductor manufacturing company 0.18 μm RF-CMOS 1P6M process. The supply voltage is 1.8 V. The total DC current, including LNA, active-balun, mixer and output buffers is 21.8 mA and the power consumption is 39.2 mW. In Fig. 13, the input LO frequency is 5.808 GHz, and the maximum conversion gain occurs at the input LO power which is −2 dBm. Therefore, the external LO power is designed as −2 dBm. The frequency spectrum of all UWB bands is from 3.168 to 10.560 GHz. In Fig. 14, the input return loss (S_{11}) is lower than −8.9 dB over 3.1–10.6 GHz. The channel bandwidth of UWB is 528 MHz, and the transceiver IF is designed from DC to 528 MHz. In Fig. 15, the output return loss (S_{22}) is lower than −6.6 dB from DC to 528 MHz. In Fig. 16, the maximum down-conversion power gain (CG) of the front-end is 25.8 dB at 10 GHz and −3 dB frequency range is from 3 to 11.5 GHz. The minimum single-sideband NF is 4.9 dB at 9.6 GHz and the single-sideband NF is <6.9 dB over 3.1–10.6 GHz, as shown in Fig. 17. In Fig. 18, the input 1 dB compression point ($P_{1\,\mathrm{dB}}$) of the front-end is larger than −26 dBm. The maximum input $P_{1\,\mathrm{dB}}$ is −18 dBm. Fig. 19 shows the isolations of the front-end circuit. The isolations of LO-IF, LO-RF and RF-IF are −39, −56 and −22 dB, respectively. Fig. 19 shows the band-group #3, the two-tone frequencies are 7128 and 7138 MHz; the LO frequency is 7392 MHz; and the IIP3 is −19 dBm. The performance comparison with the state-of-the-art papers is shown in Table 1.

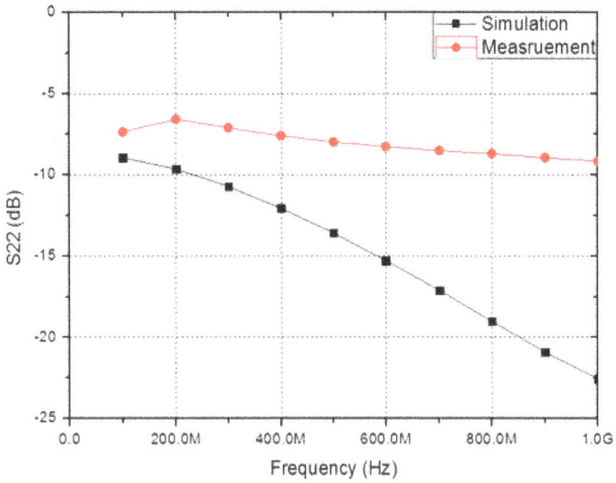

Fig. 15 *Measurement results of the output return loss (S22)*

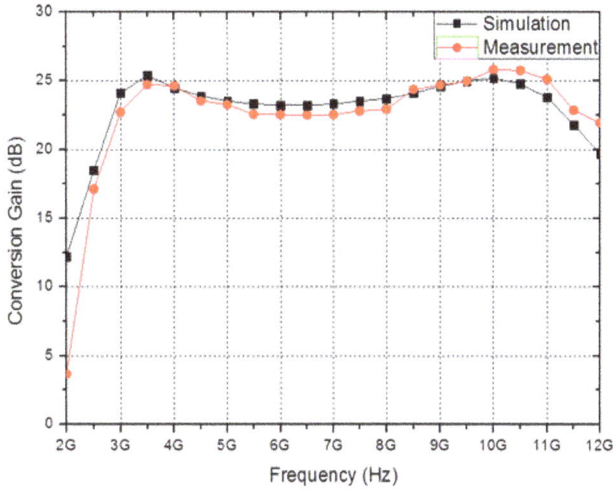

Fig. 16 *Measurement results of the down-conversion power gain (CG)*

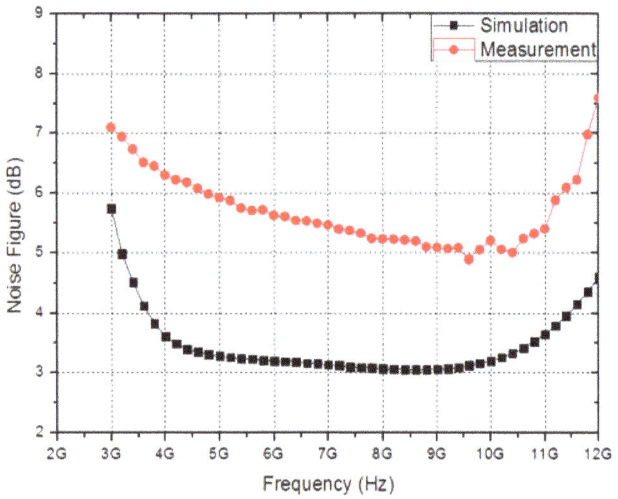

Fig. 17 *Measurement results of the down-conversion NF*

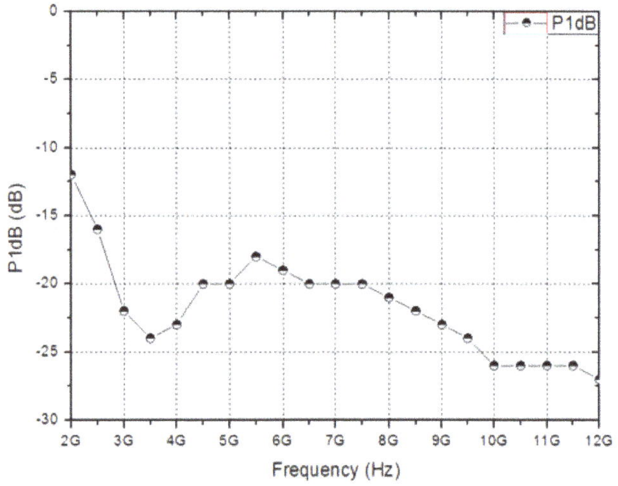

Fig. 18 *Input 1 dB compression point ($P_{1\ dB}$)*

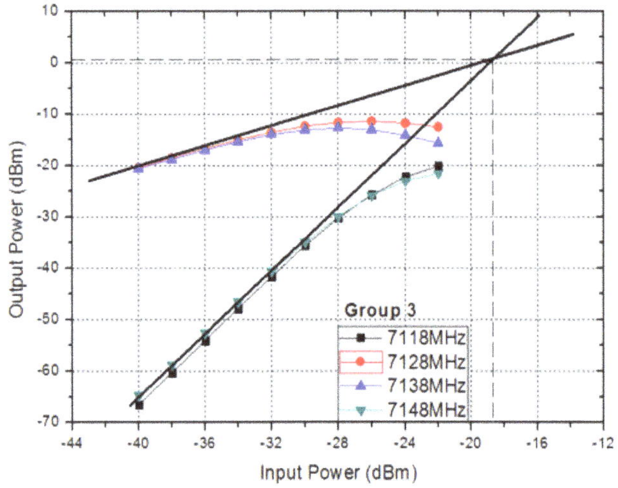

Fig. 19 *Input IIP3 of 7128 and 7138 MHz*

4 Conclusions

Low-noise, UWB wireless receiver front-end circuits were proposed in this paper. A two-stage common-source LNA with wideband input impedance matching network, active-balun and a double-balanced down-conversion mixer were presented. The wireless receiver front-end circuits were implemented in a 0.18 µm RF-CMOS process. The maximum down-conversion power gain of the front-end is 25.8 dB and the minimum single-sideband NF of the front-end is 4.9 dB over the 3.1–10.6 GHz. Power consumption including buffers is 39.2 mW.

5 Acknowledgments

This work was supported by the National Science Council (NSC), Taiwan, under the Contract NSC 101-2220-E-009-048. The authors thank the chip implementation center (CIC) for chip fabrication, technical and measurement support.

6 References

[1] Ismail A., Abidi A.A.: 'A 3.1- to 8.2 GHz zero-IF receiver and direct frequency synthesizer in 0.18 µm SiGeBi CMOS for mode-2 MB-OFDM UWB communication', *IEEE J. Solid-State Circuits*, 2005, **40**, (12), pp. 2573–2582

[2] Razavi B., Aytur T., Lam C., *ET AL.*: 'A UWB CMOS transceiver', *IEEE J. Solid-State Circuits*, 2005, **40**, (12), pp. 2555–2562

[3] Roovers R., Leenaerts D.M.W., Bergervoet J., *ET AL.*: 'An interference-robust receiver for ultra-wideband radio in SiGeBi CMOS technology', *IEEE J. Solid-State Circuits*, 2005, **40**, (12), pp. 2555–2562

[4] Simitsakis P., Papananos Y., Kytonaki E.-S.: 'Design of a low voltage-low power 3.1–10.6 GHz UWB RF front-end in a CMOS 65 nm technology', *IEEE Trans. Circuits Syst. II, Express Briefs*, 2010, **57**, (11), pp. 833–837

[5] Chang J.-F., Lin Y.-S.: '3.15 dB NF, 7.2 mW 3–9 GHz CMOS ultra-wideband receiver front-end', *IET Electron. Lett.*, 2011, **47**, (25), pp. 1401–1402

[6] Ranjan M., Larson L.E.: 'A low-cost and low-power CMOS receiver front-end for MB-OFDM ultra-wideband systems', *IEEE J. Solid-State Circuits*, 2007, **42**, (3), pp. 592–601

[7] Hui Z., Lou S., Lu D., Shen C., Chan T., Luong H.C.: 'A 3.1 GHz–8.0 GHz single-chip transceiver for MB-OFDM UWB in 0.18-µm CMOS process', *IEEE J. Solid-State Circuits*, 2009, **44**, (2), pp. 414–426

[8] Valdes-Garcia A., Mishra C., Bahmani F., Silva-Martinez J., Sanchez-Sinencio E.: 'An 11-band 3–10 GHz receiver in SiGeBi CMOS for multiband OFDM UWB communication', *IEEE J. Solid-State Circuits*, 2007, **42**, (4), pp. 935–948

[9] Han J., Cam N.: 'Coupled-slotline-hybrid sampling mixer integrated with step-recovery-diode pulse generator for UWB applications', *IEEE Trans. Microw. Theory Tech.*, 2005, **53**, (6), pp. 1875–1882

[10] Safarian A.Q., Yazdi A., Heydari P.: 'Design and analysis of an ultrawide-band distributed CMOS mixer', *IEEE Trans. Very Large Scale Integr. Syst.*, 2005, **13**, (5), pp. 618–629

[11] Syu J.-S., Meng C., Teng Y.-H.: 'UWB Gilbert down converter utilising a wideband LR-CR quadrature generator', *IET Electron. Lett.*, 2009, **45**, (10), pp. 514–515

[12] Lee F.S., Chandrakasan A.P.: 'A BiCMOS ultra-wideband 3.1–10.6-GHz front-end', *IEEE J. Solid-State Circuits*, 2006, **41**, (8), pp. 1784–1791

[13] Heydari P.: 'Design and analysis of a performance-optimized CMOS UWB distributed LNA', *IEEE J. Solid-State Circuits*, 2007, **42**, (9), pp. 1892–1905

[14] Bevilacqua A., Niknejad A.M.: 'An ultrawideband CMOS low-noise amplifier for 3.1-10.6-GHz wireless receivers', *IEEE J. Solid-State Circuits*, 2004, **39**, (12), pp. 2259–2268

[15] Ismail A., Abidi A.A.: 'A3-10-GHz low-noise amplifier with wide-band LC-ladder matching network', *IEEE J. Solid-State Circuits*, 2004, **39**, (12), pp. 2269–2277

[16] Liang C.-P., Rao P.-Z., Huang T.-J., Chung S.-J.: 'Analysis and design of two low-power ultra-wideband CMOS low-noise amplifiers with out-band rejection', *IEEE Trans. Microw. Theory Tech.*, 2010, **58**, (2), pp. 277–286

[17] Chen K.-H., Lu J.-H., Chen B.-J., Liu S.-I.: 'Design of a low voltage-low power 3.1–10.6 GHz UWB RF front-end in a CMOS 65 nm technology', *IEEE Trans. Circuits Syst. II, Express Briefs*, 2007, **54**, (3), pp. 217–221

[18] Wu C.-Y., Lo Y.-K., Chen M.-C.: 'A 3–10 GHz CMOS UWB low-noise amplifier with ESD protection circuits', *IEEE Microw. Wirel. Compon. Lett.*, 2009, **19**, (11), pp. 737–739

[19] Lin J.-Y., Chiou H.-K.: 'Power-constrained third-order active notch filter applied in IR-LNA for UWB standards', *IEEE Trans. Circuits Syst. II, Express Briefs*, 2011, **58**, (1), pp. 11–15

[20] Lo Y.-T., Kiang J.-F.: 'Design of wideband LNAs using parallel-to-series resonant matching network between common-gate and common-source stages', *IEEE Trans. Microw. Theory Tech.*, 2011, **59**, (9), pp. 2285–2294

[21] He K.-C., Li M.-T., Li C.-M., Tarng J.-H.: 'Parallel-RC feedback low-noise amplifier for UWB applications', *IEEE Trans. Circuits Syst. II, Express Briefs*, 2010, **57**, (8), pp. 582–586

[22] Chang J.-F., Lin Y.-S.: 'A 3-10 GHz low-power CMOS low-noise amplifier for ultra-wideband communication', *IEEE Trans. Microw. Theory Tech.*, 2011, **59**, (3), pp. 678–686

[23] Gharpurey R.: 'A broadband low-noise front-end amplifier for ultra wideband in 0.13 μm CMOS', *IEEE J. Solid-State Circuits*, 2005, **40**, (9), pp. 1983–1986

[24] Datta P.K., Xi F., Fischer G.: 'A transceiver front-end for ultra-wide-band application', *IEEE Trans. Circuits Syst. II, Express Briefs*, 2007, **54**, (4), pp. 362–366

[25] Lai Q.-T., Mao J.-F.: 'A 0.5–11 GHz CMOS low noise amplifier using dual-channel shunt technique', *IEEE Microw. Wirel. Compon. Lett.*, 2010, **20**, (5), pp. 280–282

[26] Reiha M.T., Long J.R.: 'A 1.2 V reactive-feedback 3.1–10.6 GHz low-noise amplifier in 0.13 μm CMOS', *IEEE J. Solid-State Circuits*, 2007, **42**, (5), pp. 1023–1033

[27] Chao F., Law C.L., Hwang J.: 'A 3.1–10.6 GHz ultra-wideband low noise amplifier with 13 dB gain, 3.4 dB noise figure, and consumes only 12.9 mW of DC power', *IEEE Microw. Wirel. Compon. Lett.*, 2007, **17**, (4), pp. 295–297

[28] Liao C.-F., Liu S.-I.: 'A broadband noise-canceling CMOS LNA for 3.1–10.6 GHz UWB receivers', *IEEE J. Solid-State Circuits*, 2007, **42**, (2), pp. 329–339

[29] Lin Y.-S., Chen C.-Z., Yang H.-Y., *ET AL.*: 'Analysis and design of a CMOS UWB LNA with dual-RLC-branch wideband input matching network', *IEEE Trans. Microw. Theory Tech.*, 2010, **58**, (2), pp. 287–296

[30] Rezaul Hasan S.M.: 'Analysis and design of a multistage CMOS band-pass low-noise preamplifier for ultrawideband RF receiver', *IEEE Trans. Very Large Scale Integr. Syst.*, 2010, **18**, (4), pp. 638–651

[31] Seo J.-B., Kim J.-H., Sun H., Yun T.-Y.: 'A low-power and high-gain mixer for UWB systems', *IEEE Microw. Wirel. Compon. Lett.*, 2008, **18**, (12), pp. 280–282

[32] Kim M.-G., An H.-W., Kang Y.-M., Lee J.-Y., Yun T.-Y.: 'A low-voltage, low-power, and low-noise UWB mixer using bulk-injection and switched biasing techniques', *IEEE Trans. Microw. Theory Tech.*, 2012, **60**, (8), pp. 2486–2493

[33] Chang F.-C., Huang P.-C., Chao S.-F., Wang H.: 'A low power folded mixer for UWB system applications in 0.18 μm CMOS technology', *IEEE Microw. Wirel. Compon. Lett.*, 2007, **17**, (5), pp. 367–369

[34] Rao P.-Z., Chang T.-Y., Liang C.-P., Chung S.-J.: 'An ultra-wideband high-linearity CMOS mixer with new wideband active baluns', *IEEE Trans. Microw. Theory Tech.*, 2009, **57**, (9), pp. 2184–2192

10

Approximate symbol error rate of cooperative communication over generalised κ–μ and η–μ fading channels

Brijesh Kumbhani, Lomada Nerusupalli Baya Reddy, Rakhesh Singh Kshetrimayum

Department of Electronics and Electrical Engineering, Indian Institute of Technology Guwahati, Guwahati, India
E-mail: krs@iitg.ernet.in

Abstract: Closed form expressions for approximate symbol error rate are obtained using moment generating function for a two branch cooperative communication system over generalised κ–μ and η–μ i.i.d. fading channels for BPSK and QAM modulation schemes. Selective decode and forward protocol is used at the relay transmitter. At the destination maximal-ratio combining is used. Monte Carlo simulations are performed to verify the analytical results.

1 Introduction

Cooperative communication [1] is a solution for reliable reception in single antenna systems. In cooperative communication, each user not only sends their information but also forward other user's information to the destination. Owing to single antenna system there is no inter-antenna interference and also there is diversity gain advantage in cooperative communication. Generally amplify and forward (AF) and decode and forward (DF) protocols are used at the relay node. In AF protocol, the received noisy signal at the relay will be amplified and forwarded to the destination, whereas in DF protocol noisy signal at the relay will be first decoded and forwarded to destination if the signal decodes correctly.

The performance analysis of cooperative communication over classical fading channels has been studied in the past. Performance analysis of cooperative communication over Rayleigh, Weibull and Nakagami-*m* fading channels is reported in [2–4]. In incremental relaying cooperative communication, the channel resources are used efficiently by not forwarding the information from relay to destination if the destination itself decodes it correctly [5].

The physical model of κ–μ and η–μ distributions are described in [6] and moment generating function (MGF) of κ–μ and η–μ distributions are given in [7]. The κ–μ and η–μ distributions are best suited to model practical small scale fading channels in line of sight (LOS) and non-LOS (NLOS) environments, respectively [6]. κ–μ and η–μ are generalised models which accommodate the best known fading models such as Rayleigh, Rician, Nakagami-*m*, Nakagami-*n* and Hoyt distributions etc. These models assume non-homogenous physical environment which is closer to the practical scenario.

In this paper, we derived, for the first time, closed-form expression for approximate SER for a two branch cooperative communication system over generalised κ–μ and η–μ fading channels for quadrature amplitude modulation (QAM) and binary phase shift keying (BPSK) modulation schemes. In our analysis, we considered DF protocol at the relay. We also show SER results obtained from Monte Carlo simulations to validate our analytical results.

Section 2 describes the system model and about κ–μ and η–μ fading distributions. In Section 3 approximate SER expression is derived for BPSK and QAM modulation schemes. In Section 4 simulation results were shown. The paper is concluded in Section 5.

2 System model

Let us consider a cooperative communication system with one source node (S) communicating with one destination node (D) and one relay node (R) as shown in Fig. 1 [8]. The transmission of information from the source to destination is performed in two orthogonal phases. In the first phase, the source broadcasts data to the relay and the destination. The received signals $y_{\text{s,d}}$ and $y_{\text{s,r}}$ at the destination and the relay, respectively, can be written as

$$y_{\text{s, d}} = \sqrt{p_1} f x + n_1 \tag{1}$$

$$y_{\text{s, r}} = \sqrt{p_1} g x + n_2 \tag{2}$$

in which p_1 is the transmitted power at the source, x is the transmitted information symbol, n_1 and n_2 are zero-mean complex Gaussian random variables with variance N_0 and channel gains f and g are κ–μ or η–μ distribution fading coefficients between source and destination, and between source and relay, respectively.

In second phase, for a DF cooperation protocol, if the relay is able to decode the transmitted symbol correctly, then it forwards the decoded symbol with power p_2 to the destination, otherwise the relay does not send or remains idle. The received signal $y_{\text{r,d}}$ at the destination can be written as

$$y_{\text{r, d}} = \sqrt{\tilde{p}_2} h x + n_3 \tag{3}$$

where $\tilde{p}_2 = p_2$ if the relay decodes the transmitted symbol correctly, otherwise $\tilde{p}_2 = 0$. n_3 is zero-mean complex Gaussian random variables with variance N_0 and channel gain h is κ–μ or η–μ distribution fading coefficients between relay and destination. The fading coefficients f, g and h are assumed to be known at the receiver, but not at the transmitter. The destination jointly combines the

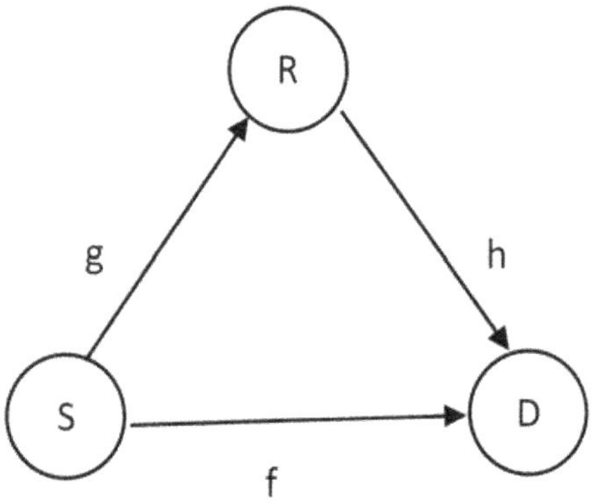

Fig. 1 *Cooperative communication system with single relay*

received signal from the source in phase 1 and that from the relay in phase 2, and detects the transmitted symbols by using maximum-ratio combining (MRC) [9]. The total transmitted power has to satisfy $p_1+p_2=p$.

2.1 κ−μ distribution

The probability density function of κ−μ distributed random variable (RV) is given as [6]

$$p_{x_{\kappa-\mu}}(x) = \frac{2\mu x^{\mu}}{\kappa^{(\mu-1/2)}e^{\mu\kappa}}\left(\frac{1+\kappa}{\Omega}\right)^{(\mu+1/2)} e^{-(\mu(1+\kappa)/\Omega)x^2} \times I_{\mu-1}\left(2\mu\sqrt{\frac{\kappa(1+\kappa)}{\Omega}}x\right) \quad (4)$$

where $\Omega=E[x^2]$, $E[\cdot]$ is the expectation operator, $\kappa>0$ and $\mu>0$ are the parameters of the distribution and $I_\nu(\cdot)$ is the modified Bessel function of the first kind and νth order. The parameter κ is the ratio of the total power because of dominant components to the total power because of scattered components and μ is the number of multipath clusters. This model includes Rice ($\mu=1$ and $\kappa=K$), Nakagami-m ($\kappa\to 0$ and $\mu=m$), Rayleigh ($\mu=1$ and $\kappa\to 0$) and one sided Gaussian distribution ($\mu=0.5$ and $\kappa\to 0$) fading models as special cases. This distribution is better suited for LOS propagation.

2.2 η−μ distribution

The probability density function (PDF) of η−μ distributed RV is given as [6]

$$p_{x_{\eta-\mu}}(x) = \frac{4\sqrt{\pi}\mu^{\mu+(1/2)}h^{\mu}x^{2\mu}}{\Gamma(\mu)H^{\mu-(1/2)}\Omega^{\mu+(1/2)}} e^{-(2\mu x/\Omega)x^2} \times I_{\mu-(1/2)}\left(\frac{2\mu H}{\Omega}x^2\right) \quad (5)$$

where $\Gamma(\cdot)$ is the gamma function, and h and H are functions of the parameter η defined for two formats in next subsections. μ denotes the number of multipath clusters. This fading model includes Hoyt ($\eta=q^2$, $\mu=0.5$), Nakagami-m ($\eta=1$, $\mu=m/2$), Rayleigh and one sided Gaussian distribution as special cases.

(1) *The η−μ distribution: format 1:* In this format $0<\eta<\infty$ is the power ratio of the in-phase and quadrature components of the scattered-waves of each multipath. It is assumed that the in-phase and quadrature phase components of fading signal within each cluster are independent of each other and have different average powers. In this case, $h=[(2+\eta^{-1}+\eta)/4]$ and $H=[(\eta^{-1}-\eta)/4]$. Within $0<\eta\le 1$, we have $H\ge 0$, on the other hand, within $0<\eta^{-1}\le 1$, we have $H\le 0$. Since $I_\nu(-z)=(-1)^\nu I_\nu(z)$, the distribution is symmetrical around $\eta=1$. Therefore as far as the envelope (or power) distribution is concerned, it is sufficient to consider η only within one of the ranges. In format 1, $H/h=(1-\eta)/(1+\eta)$.
(2) *The η-μ distribution: format 2:* In this format $-1<\eta<1$ is the correlation coefficient between the in-phase and quadrature components of the scattered-waves of each cluster. It is assumed that the in-phase and quadrature phase components of fading signal within each cluster are correlated and have identical powers. In this case, $h=(1/1-\eta^2)$ and $H=(\eta/1-\eta^2)$. Within $0\le\eta<1$, we have $H\ge 0$, on the other hand, within $-1<\eta\le 0$, we have $H\le 0$. Since $I_\nu(-z)=(-1)^\nu I_\nu(z)$, the distribution is symmetrical around $\eta=0$. Therefore as far as the envelope (or power) distribution is concerned, it is sufficient to consider η only within one of the ranges. In format 2, $H/h=\eta$.

3 Analysis of probability of error

The combined signal at the MRC detector can be written as in [8]

$$y = \frac{\sqrt{p_1}f^*}{N_0}y_{s,d} + \frac{\sqrt{\tilde{p}_2}h^*}{N_0}y_{r,d} \quad (6)$$

where f^* and h^* are the complex conjugates of f and h, respectively.
The signal-to-noise ratio (SNR) of the MRC output is

$$\gamma = \frac{p_1|f|^2+\tilde{p}_2|h|^2}{N_0} \quad (7)$$

3.1 BPSK modulation

The conditional SER can be calculated as in [8]

$$P^{\text{cond}}_{\text{bpsk}}(e) = \psi_{\text{bpsk}}(\gamma)\big|\tilde{p}_2 = 0 \times \psi_{\text{bpsk}}\left(\frac{p_1|g|^2}{N_0}\right) + \psi_{\text{bpsk}}(\gamma)\big|\tilde{p}_2 = p_2\times\left[1-\psi_{\text{bpsk}}\left(\frac{p_1|g|^2}{N_0}\right)\right] \quad (8)$$

where γ is SNR, $\psi_{\text{bpsk}}(\gamma)$ is conditional SER and is given by [10]

$$\psi_{\text{bpsk}}(\gamma) = Q(\sqrt{2\gamma}) \quad (9)$$

and $Q(.)$ is Q function and is defined as

$$Q(x) = \frac{1}{\sqrt{2\pi}}\int_x^\infty \exp\left(-\frac{y^2}{2}\right)\mathrm{d}x \quad (10)$$

by using approximation on $Q(.)$ function [11]

$$Q(x) \simeq \frac{1}{12}\exp\left(-\frac{x^2}{2}\right)+\frac{1}{4}\exp\left(-\frac{2x^2}{3}\right) \quad (11)$$

The final approximated average SER can be given by

$$P_{\text{bpsk}}(e) \simeq \varepsilon(p_1)\times\xi_1(p_1)+(1-\varepsilon(p_1))\times\xi_2(p_1+p_2) \quad (12)$$

where $\varepsilon(p_1)$ is the probability of error at the relay after the first phase and is given by

$$\varepsilon(p_1) \simeq \frac{1}{12}\text{MGF}_{p_1}(1)+\frac{1}{4}\text{MGF}_{p_1}\left(\frac{4}{3}\right) \quad (13)$$

$\xi_1(p_1)$ is the probability of error at the destination after the first phase and is given by

$$\xi_1(p_1) \simeq \frac{1}{12}\text{MGF}_{p_1}(1)+\frac{1}{4}\text{MGF}_{p_1}\left(\frac{4}{3}\right) \quad (14)$$

and $\xi_2(p_1+p_2)$ is the probability of error at the destination after the second phase and is given by

$$\xi_2(p_1+p_2) \simeq \frac{1}{12}\text{MGF}_{p_1}(1)\times\text{MGF}_{p_2}(1)+\frac{1}{4}\text{MGF}_{p_1}\left(\frac{4}{3}\right) \times\text{MGF}_{p_2}\left(\frac{4}{3}\right) \quad (15)$$

and MGF(.) is given by [7]

$$\text{MGF}_{\gamma_{\kappa-\mu}}(s) = \left[\frac{\mu(1+\kappa)}{\mu(1+\kappa)+s\tilde{\gamma}}\right]^{\mu}\exp\left[\frac{\mu^2\kappa(1+\kappa)}{\mu(1+\kappa)+s\tilde{\gamma}}-\mu\kappa\right] \quad (16)$$

$$\text{MGF}_{\gamma_{\eta-\mu}}(s) = \left(\frac{4\mu^2 h}{(2(h-H)\mu+s\tilde{\gamma})(2(h+H)\mu+s\tilde{\gamma})}\right)^{\mu} \quad (17)$$

3.2 4-QAM modulation

The conditional SER can be calculated as [8]

$$P_{\text{qam}}^{\text{cond}}(e) = \psi_{\text{qam}}(\gamma)\Big|_{\tilde{p}_2=0} \times \psi_{\text{qam}}\left(\frac{p_1|g|^2}{N_0}\right) + \psi_{\text{qam}}(\gamma)\Big|_{\tilde{p}_2=p_2} \times \left[1 - \psi_{\text{qam}}\left(\frac{p_1|g|^2}{N_0}\right)\right] \quad (18)$$

where $\psi_{\text{qam}}(\gamma)$ is conditional SER and is given by [10]

$$\psi_{\text{qam}}(\gamma) = 2Q\left(\sqrt{\frac{3}{4}\gamma}\right) - Q^2\left(\sqrt{\frac{3}{4}\gamma}\right) \quad (19)$$

by using $Q^2(.)$ and is given by

$$Q^2(x) \simeq \frac{1}{144}\exp(-x^2) + \frac{1}{16}\exp\left(-\frac{4x^2}{3}\right) + \frac{1}{24}\exp\left(-\frac{7x^2}{6}\right) \quad (20)$$

The final approximated average SER can be given by

$$P_{\text{qam}}(e) \simeq \varepsilon(p_1) \times \xi_1(p_1) + (1 - \varepsilon(p_1)) \times \xi_2(p_1 + p_2) \quad (21)$$

In the above equation, $\varepsilon(p_1)$, $\xi_1(p_1)$ and $\xi_2(p_1+p_2)$ are defined as follows

$$\varepsilon(p_1) \simeq 2\left\{\frac{1}{12}\text{MGF}_{p_1}\left(\frac{1}{2}\right) + \frac{1}{4}\text{MGF}_{p_1}\left(\frac{2}{3}\right)\right\} - \left\{\frac{1}{144}\text{MGF}_{p_1}(1) + \frac{1}{16}\text{MGF}_{p_1}\left(\frac{4}{3}\right) + \frac{1}{24}\text{MGF}_{p_1}\left(\frac{7}{6}\right)\right\} \quad (22)$$

$$\varepsilon(p_1) \simeq 2\left\{\frac{1}{12}\text{MGF}_{p_1}\left(\frac{1}{2}\right) + \frac{1}{4}\text{MGF}_{p_1}\left(\frac{2}{3}\right)\right\} - \left\{\frac{1}{144}\text{MGF}_{p_1}(1) + \frac{1}{16}\text{MGF}_{p_1}\left(\frac{4}{3}\right) + \frac{1}{24}\text{MGF}_{p_1}\left(\frac{7}{6}\right)\right\} \quad (23)$$

$$\xi_2(p_1+p_2) \simeq 2\left\{\frac{1}{12}\text{MGF}_{p_1}\left(\frac{1}{2}\right) \times \text{MGF}_{p_2}\left(\frac{1}{2}\right) + \frac{1}{4}\text{MGF}_{p_1}\left(\frac{2}{3}\right) \times \text{MGF}_{p_2}\left(\frac{2}{3}\right)\right\} - \left\{\frac{1}{144}\text{MGF}_{p_1}(1) \times \text{MGF}_{p_2}(1) + \frac{1}{16}\text{MGF}_{p_1}\left(\frac{4}{3}\right) \times \text{MGF}_{p_2}\left(\frac{4}{3}\right) + \frac{1}{24}\text{MGF}_{p_1}\left(\frac{7}{6}\right) \times \text{MGF}_{p_2}\left(\frac{7}{6}\right)\right\} \quad (24)$$

4 Simulation results

The derived approximate expressions of SER for different modulation schemes have been evaluated numerically and plotted with respect to average SNR, and are compared with the simulation results for different values of κ–μ and η–μ distributions.

Fig. 2 curves represent approximate average SER against average SNR for BPSK modulation scheme showing the results for Rayleigh ($\kappa = 0$, $\mu = 1$), Rice ($\kappa = 1$, $\mu = 1$) and Nakagami-*m* ($\kappa = 0$, $\mu = 2$, 4) as special cases of κ–μ distribution.

Fig. 3 curves represent approximate average SER against average SNR for 4-QAM modulation scheme showing the results for Rayleigh ($\kappa = 0$, $\mu = 1$), Rice ($\kappa = 2$, $\mu = 1$) and Nakagami-*m* ($\kappa = 0$, $\mu = 2$, 3) as special cases of κ–μ distribution. Fig. 4 curves represent approximate average SER against average SNR for BPSK

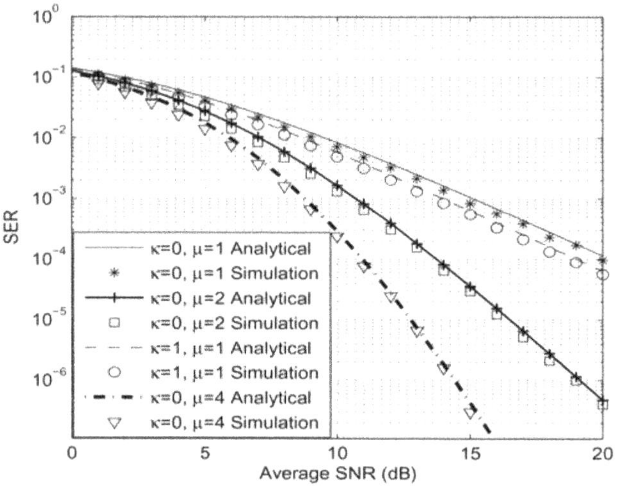

Fig. 2 *SNR against SER of BPSK for different values of* κ *and* μ *with* $p_1 = p_2 = p/2$

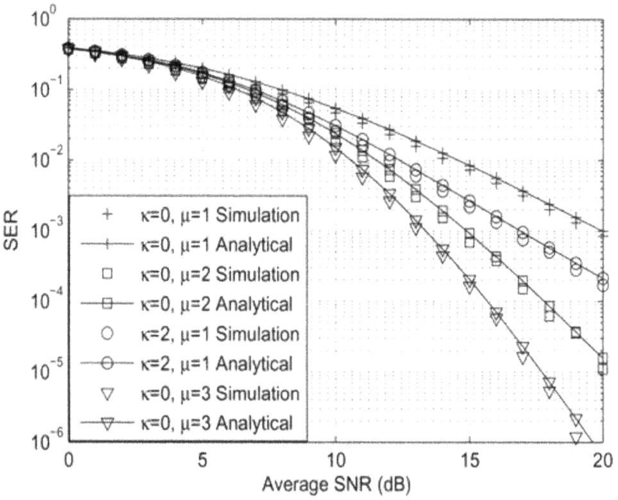

Fig. 3 *SNR against SER of QAM for different values of* κ *and* μ *with* $p_1 = p_2 = p/2$

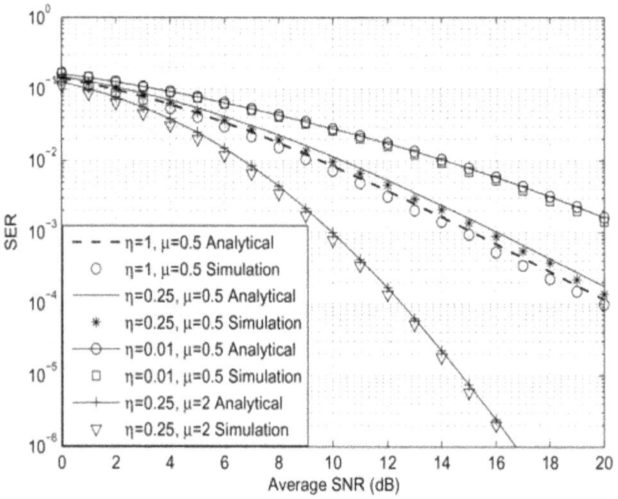

Fig. 4 *SNR against SER of BPSK for different values of* η *and* μ *with* $p_1 = p_2 = p/2$

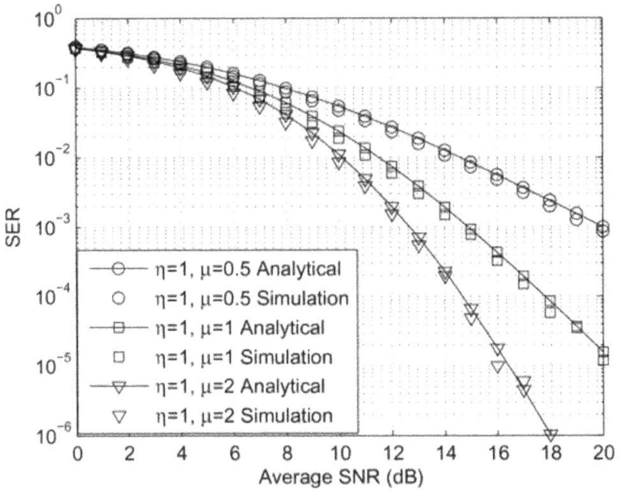

Fig. 5 *SNR against SER of QAM for different values of* η *and* μ *with* $p_1 = p_2 = p/2$

modulation scheme showing the results for Rayleigh ($\eta = 1, \mu = 0.5$) and Nakagami-*q*/Hoyt ($\eta = 0.01$, 0.25, $\mu = 0.5$) as special cases of $\eta-\mu$ distribution. Fig. 5 curves represent approximate average SER against average SNR for 4-QAM modulation scheme showing the results for Rayleigh ($\eta = 1$, $\mu = 0.5$), and Nakagami-*m* ($\eta = 1$, $\mu = 1$, 2) as special cases of $\eta-\mu$ distribution.

5 Conclusion

In this paper, approximate SER expression is derived for a cooperative communication system over the generalised $\kappa-\mu$ and $\eta-\mu$ distributed fading channels for BPSK and QAM modulation schemes using DF protocol. Representing Q function as sum of exponentials makes the analysis simpler at the cost of a small deviation between the analytical and simulation results. The advantage of $\kappa-\mu$ and $\eta-\mu$ distribution is that we can model Rayleigh, Rice, Nakagami-*m*, Hoyt and one sided Gaussian channel models as special cases.

6 References

[1] Nosratinia A., Hunter T., Hedayat A.: 'Cooperative communication in wireless networks', *IEEE Commun. Mag.*, 2004, **42**, (10), pp. 74–80

[2] Anghel P., Kaveh M.: 'Exact symbol error probability of a cooperative network in a rayleigh-fading environment', *IEEE Trans. Wirel. Commun.*, 2004, **3**, (5), pp. 1416–1421

[3] Ikki S., Ahmed M.: 'Performance analysis of cooperative diversity wireless networks over nakagami-m fading channel', *IEEE Commun. Lett.*, 2007, **11**, (4), pp. 334–336

[4] Lei Y., Cheng W., Zeng Z.: 'Performance analysis of selection combining for amplify-and-forward cooperative diversity networks over Weibull fading channels'. Proc. IEEE Int. Conf. Communications Technology and Applications, 2009, pp. 648–651

[5] Ikki S., Ahmed M.: 'Performance analysis of incremental-relaying cooperative-diversity networks over rayleigh fading channels', *IET Commun.*, 2011, **5**, (3), pp. 337–349

[6] Yacoub M.: 'The $\kappa-\mu$ distribution and the $\eta-\mu$ distribution', *IEEE Antennas Propag. Mag.*, 2007, **49**, (1), pp. 68–81

[7] Ermolova N.: 'Moment generating functions of the generalized $\eta-\mu$ and $\kappa-\mu$ distributions and their applications to performance evaluations of communication systems', *IEEE Commun. Lett.*, 2008, **12**, (7), pp. 502–504

[8] Liu K.J.R., Kwasinski A., Su W., Sadek A.K.: 'Cooperative communications and networking' (Cambridge University Press, New York, 2009, 1st edn.)

[9] Brennan D.G.: 'Linear diversity combining techniques', *IEEE Proc.*, 2003, **91**, (2), pp. 331–356

[10] Simon M.K., Alouini M.-S.: 'Digital communication over fading channels' (Wiley Interscience, 2005, 2nd edn.)

[11] Chiani M., Dardari D., Simon M.K.: 'New exponential bounds and approximations for the computation of error probability in fading channels', *IEEE Trans. Wirel. Commun.*, 2003, **2**, (4), pp. 840–845

11

Vedic division methodology for high-speed very large scale integration applications

Prabir Saha[1], *Deepak Kumar*[2], *Partha Bhattacharyya*[3], *Anup Dandapat*[1]

[1]*Department of Electronics and Communication Engineering, National Institute of Technology, Shillong, Meghalaya 793 003, India*
[2]*Department of Computer Science and Engineering, National Institute of Technology, Shillong, Meghalaya 793 003, India*
[3]*Department of Electronics and Telecommunication Engineering, Bengal Engineering and Science University, Shibpur, Howrah 711 103, India*
E-mail: anup.dandapat@gmail.com

Abstract: Transistor level implementation of division methodology using ancient Vedic mathematics is reported in this Letter. The potentiality of the 'Dhvajanka (on top of the flag)' formula was adopted from Vedic mathematics to implement such type of divider for practical very large scale integration applications. The division methodology was implemented through half of the divisor bit instead of the actual divisor, subtraction and little multiplication. Propagation delay and dynamic power consumption of divider circuitry were minimised significantly by stage reduction through Vedic division methodology. The functionality of the division algorithm was checked and performance parameters like propagation delay and dynamic power consumption were calculated through spice spectre with 90 nm complementary metal oxide semiconductor technology. The propagation delay of the resulted (32 ÷ 16) bit divider circuitry was only ~300 ns and consumed ~32.5 mW power for a layout area of 17.39 mm^2. Combination of Boolean arithmetic along with ancient Vedic mathematics, substantial amount of iterations were reduced resulted as ~47, ~38, 34% reduction in delay and ~34, ~21, ~18% reduction in power were investigated compared with the mostly used (e.g. digit-recurrence, Newton–Raphson, Goldschmidt) architectures.

1 Introduction

Division is a fundamental operation in many scientific and engineering applications, like arithmetic computation, signal processing, artificial intelligence, computer graphics etc. [1–3]. Generally, computations of such division operations are calculated in sequential manner, thereby costlier in terms of propagation delay (latency) compared with other mathematical operations like addition, subtraction and multiplication [4].

Substantial amount of works have so far been investigated by various researchers to implement the high-speed divider [1–15] like digit recurrence (DR) methodology (restoring [1, 3, 5], non-restoring [2, 6, 9]), division by convergence (Newton–Raphson (N–R) method [10–12]), division by series expansion (Goldschmidt (G–S) algorithm [13, 14]) etc. Generally, division architectures can be classified into two categories: namely (i) iteration based and (ii) multiplication based. Iterative divisions consist of shift-and-subtract operations, generates one quotient bits, in each of the iterations, like radix-2 restoring and non-restoring division. Thereby, in iterative division, after each subtraction cycle, it should require to check whether the resulting remainder is lesser than the divisor or negative. The cost in terms of computational complexity of DR algorithms [1–3, 5, 6, 9] is low because of the large number of iterations; therefore latency becomes high. Although, some of the researcher rely on higher radix implementation of DR algorithm [6, 7, 10] to reduce the iterations, therefore the latency becomes improved from earlier reports [1–3, 5, 9], but these schemes additionally increases the hardware complexity. Some other attractive ideas are based on functional iterations, like N–R [10–12] and G–S [13–15] algorithm, utilises multiplication techniques along-with the series expansion, where the amount of quotient bits obtained in each of the iterations is doubled. These methods converge quadratically towards the quotient when the number of iterations is increased, thereby latency becomes high. Each iterations of N–R and G–S methods involve two dependent multiplications; namely, the product of the first multiplication is one of the operands of the second multiplication thereby it cannot be optimised like a parallel multiplier [13]. The drawback of these methods is operands should be previously normalised, most used primitive are multiplications and the remainder is not directly obtained.

In algorithmic and structural levels, substantial amount of division techniques has so far been developed to reduce the propagation delay and power consumption of the divider circuitry; by reducing the iteration, aiming towards high-speed operations, but principle behind division techniques are same in all cases. Vedic mathematics [16] is the ancient system of mathematics which has unique computation techniques based on 16 sutras (formulae). Recently, we [17] reported on a Vedic divider based on 'Nikhilam Navatascaramam Dasatah' for some specific number system, like, the divisor was chosen very close to the base of operations. The implementation reduces the number of iterations, if the divisor is closer to the base of operation, otherwise increases the iterations, a serious bottleneck of the algorithm. In this Letter, we report on a division technique and its transistor level implementation of such circuitry based on such ancient mathematics. 'Dhvajanka' is a Sanskrit term indicating 'on top of the flag', is adopted from Vedas; formula is encountered to implement the division circuitry. In this approach, divider implementation was transformed into just small division instead of actual divisor, subtraction and few multiplication, thereby reduces the iterations, owing to the substantial reduction in propagation delay. Transistor level (application specific integrated circuit (ASIC)) implementation of such division circuitry was carried out by the combination of Boolean arithmetic with Vedic mathematics, performance parameters like propagation delay, dynamic switching power consumption calculation of the proposed method was calculated by using spice spectre in 90 nm complementary metal oxide semiconductor (CMOS) technology and compared with other designs like DR- [9], N–R- [11], and G–S [15]-based implementation. The calculated results revealed (32 ÷ 16) bit divider circuitry has propagation delay ~300 ns with ~32.53 mW dynamic switching power for a layout area of 17.39 mm^2.

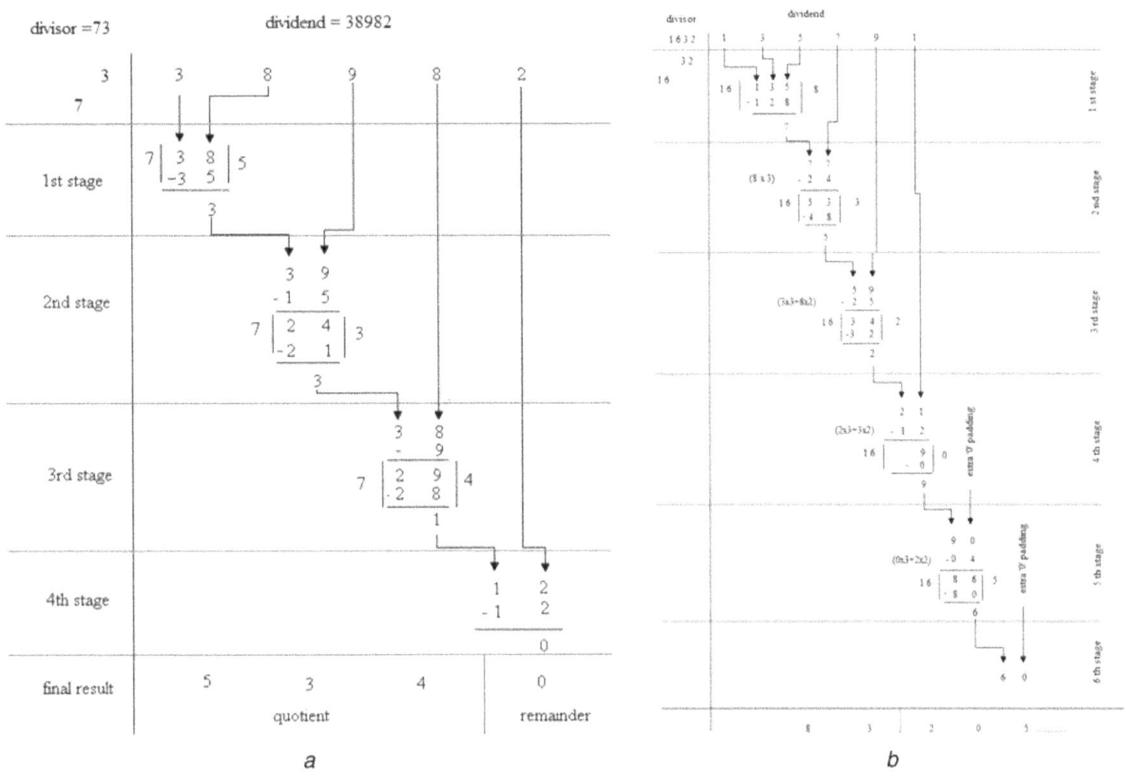

Fig. 1 *Illustration of 'Dhvajanka' sutra*
a Small divisor with exact division (remainder '0')
b Large divisor (remainder ≠ 0), have been considered for illustration purpose

2 Vedic division methodology

The gifts of the ancient Indian mathematics in the world history of mathematical science are not well recognised. The contributions of mathematician in the field of number theory, 'Sri Bharati Krsna Thirthaji Maharaja', in the form of Vedic sutras (formulae) [16] are significant for calculations. He had explored the mathematical potentials from Vedic primers and showed that the mathematical operations can be carried out mentally to produce fast answers using the sutras (formulae). In this Letter, we report only 'Dhvajanka' formula to implement the division algorithm and its architecture.

2.1 Numerical example of 'Dhvajanka' sutra

With the help of example, shown in Fig. 1*a*, dividend has been considered as 38 982 (five digit number) and divisor is equals to 73 (two digit number). Out of divisor 73, we put down only the first digit (i.e. 7) in the divisor column and put the other digit (i.e. 3) 'on top of the flag'. On the other hand, shown in Fig. 1*b*, dividend has been considered as 135 791 and divisor has been considered as 1632. The entire division for Fig. 1*a* is to be set by 7; and for Fig. 1*b* is to be set by 16. The diagram implementation procedure has been described in Table 1.

Table 1 Chart implementation procedure, the example has been considered from Fig. 1

Implementation steps of Fig. 1*a*	Implementation steps of Fig. 1*b*
1. One digit of divisor has been put on top; we allot one place (at the right end of the dividend) to the remainder portion of the answer and mark it off from the digit by a vertical line	1. Two digits have been put on top; we allot two places (at the right end of the dividend) to the floating point portion of the answer and mark it off from the digit by a vertical line
2. 38 is divided by the most significant digit (MSD) of the divisor (i.e. 7). Quotient is 5 and remainder is 3. This remainder will be used for next step division	2. 135 is divided by the MSD part of the divisor (i.e. 16). Quotient is 8 and remainder is 7. This remainder will be used for next step division
3. In this step, our actual gross dividend is 39 is subtracted by the value obtained by multiplying previous quotient (i.e. 5) with least significant digit (LSD) of divisor (i.e. 3). $\{39-(5\times 3)=24\}$. After subtraction, the result is again divided by 7. Quotient becomes 3 and remainder becomes 3	3. In this step, our actual gross dividend is 77 is subtracted by the value obtained by multiplying previous quotient (i.e. 8) with LSD part of divisor (i.e. 3). $\{77-(8\times 3)=53\}$. After subtraction, the result is again divided by 16. Quotient becomes 3 and remainder becomes 5
4. In the third stage, the gross dividend is 38. Again it is subtracted by 9 (3×3) similar to the previous step. Thus, the result $(38-9=29)$ is divided by 7. Quotient is 4 and remainder is 1	4. In the third stage, the gross dividend is 59. Again it is subtracted by cross-multiplication of LSD part of the divisor and the obtained quotient, that is, $(59-(3\times 3+8\times 2)=34)$. Thus, the result is divided by 16. Quotient is 2 and remainder is 2
5. This is the final stage for this example. Here, the final remainder is calculated. Actual gross dividend is subtracted by 12 and the result in the final remainders	5. In the fourth stage gross dividend is 21, again it is subtracted by the cross-multiplication of the digits of the quotients and LSD parts of the divisor and results become 9. This result is divided again by 16, quotient becomes '0' and remainder becomes '9'
6. Thus, we say quotient is equals to 534 and remainder is equals to 0	6. The process continues until the number of iterations 7. Thus, the results become 83.205

```
7x+3 ) 38x³ + 9x² + 8x + 2 ( 5x² + 3x + 4
       35x³ + 15x²
        3x³ − 6x²
        = 24 x² + 8x
         21 x² + 9x
          3 x² - x
           = 29 x + 2
             28 x + 12
                x - 10
             =10-10=0
```

Fig. 2 *Algebraical proof of the formula*

2.2 *Algebraic proof of 'Dhvajanka' sutra*

Algebraic proof of the formula is shown in Fig. 2, where x stands for 10. To understand the steps taken from Fig. 1*a*; by means of which 38 982 is sought to be divided by 73. Algebraically, the dividend is represented as $38x^3+9x^2+8x+2$; and the divisor is $7x+3$. Now, let us proceed with the division in the usual manner.

1. If we try to divide $38x^3$ by $7x$, our first quotient digit is $5x^2$. In the first step of the multiplication of the divisor by $5x^2$, we obtain the product $35x^3+15x^2$ and this gives us the remainder $3x^3+9x^2-15x^2$. Which is actually $30x^2+9x^2-15x^2=24x^2$.
2. The first step remainder term (i.e. $24x^2$) plus $8x$ being our second-step dividend, we multiply the divisor by second quotient and $21x^2+9x$ there from and then obtain $3x^2-x$ as the remainder.
3. However, this $3x^2$ is equals to $30x$ which (with $-x+2$) gives us $29x+12$ as the last step dividend. Again multiplying the divisor by 4, we obtain the product $28x+12$; and subtract this $28x+12$, thereby obtaining $x-10$ as the remainder. However, x is being 10, thus the remainder vanishes.

2.3 *Mathematical modelling of 'Dhvajanka' sutra*

Let us assume the numbers $A=\sum_{i=0}^{n-1}a_ix^i$ is dividend, and $B=\sum_{i=0}^{m-1}b_ix^i$is divisor, where x is the radix of the number. So 'A' can be expressed in terms of 'B' as

$$\begin{aligned} A = a_{n-1}x^{n-1} + a_{n-2}x^{n-2} + a_{n-3}x^{n-3} \\ + a_{n-4}x^{n-4} + \cdots + a_3x^3 + a_2x^2 + a_1x^1 + a_0 \end{aligned} \tag{1}$$

(see (2) and (3))

2.4 *Illustration of Dhvajanka sutra*

Consider dividend $f(x)=a_3x^3+a_2x^2+a_1x+a_0$ and divisor $g(x)=b_1x+b_0$, where 'x' is radix. We have to compute $f(x)/g(x)$ with the help of 'on top of the flag' sutra. Mathematically, $f(x)/g(x)=(a_3x^3+a_2x^2+a_1x+a_0)/(b_1x+b_0)$ can be represented as (see (4) and (5))

(see equation (6) at bottom of the next page)

Then $f(x)=Q(x)g(x)+R$, where, $Q(x)$ is quotient and R is

$$\begin{aligned} = &\left[a_{n-1}x^{n-1} + \left(\frac{a_{n-1}}{b_{m-1}}\right)b_{m-2}x^{n-2} + \cdots + \left(\frac{a_{n-1}}{b_{m-1}}\right)b_2x^{(n/2)+2} + \left(\frac{a_{n-1}}{b_{m-1}}\right)b_1x^{(n/2)+1} + \left(\frac{a_{n-1}}{b_{m-1}}\right)b_0x^{(n/2)}\right] \\ &+ \left[a_{n-2} - \left(\frac{a_{n-1}}{b_{m-1}}\right)b_{m-2}\right]x^{n-2} + \cdots + \left[a_{(n/2)+2} - \left(\frac{a_{n-1}}{b_{m-1}}\right)b_2\right]x^{(n/2)+2} \\ &+ \left[a_{(n/2)+1} - \left(\frac{a_{n-1}}{b_{m-1}}\right)b_1\right]x^{(n/2)+1} + \left[a_{n/2} - \left(\frac{a_{n-1}}{b_{m-1}}\right)b_0\right]x^{(n/2)} + \cdots + \left[a_0 - \frac{(a_1 - ((a_2 - (\ldots))/b_{m-1}))}{b_{m-1}}\right] \end{aligned} \tag{2}$$

$$\begin{aligned} = &\left(b_{m-1}x^{(n/2)-1} + b_{m-2}x^{(n/2)-2} + \cdots + b_0\right) \\ &\times \left(\frac{a_{n-1}}{b_{m-1}}x^{n/2} + \frac{(a_{n-2} - ((a_{n-1})/(b_{m-1})))b_{m-2}}{b_{m-1}}x^{(n/2)-1} + \cdots + \frac{a_0 - ((a_1 - (a_2 - (\ldots))/b_{m-1})/b_{m-1})b_0}{b_{m-1}}\right) \end{aligned} \tag{3}$$

$$= \frac{\frac{a_3}{b_1}x^2(b_1x+b_0) + \frac{\left(a_2 - \frac{a_3}{b_1}b_0\right)}{b_1}x(b_1x+b_0) + \frac{\left(a_1 - \frac{\left(a_2 - \frac{a_3}{b_1}b_0\right)}{b_1}b_0\right)}{b_1}(b_1x+b_0) + \left(a_0 - \frac{\left(a_1 - \frac{\left(a_2 - \frac{a_3}{b_1}b_0\right)}{b_1}b_0\right)}{b_1}b_0\right)}{b_1x+b_0} \tag{4}$$

$$= \frac{\left(\frac{a_3}{b_1}x^2 + \frac{(a_2 - (a_3/b_1)b_0)}{b_1}x + \frac{\left(a_1 - \frac{\left(a_2 - \frac{a_3}{b_1}b_0\right)}{b_1}b_0\right)}{b_1}\right)(b_1x+b_0)}{b_1x+b_0} + \frac{\left(a_0 - \frac{\left(a_1 - \frac{\left(a_2 - \frac{a_3}{b_1}b_0\right)}{b_1}b_0\right)}{b_1}b_0\right)}{b_1x+b_0} \tag{5}$$

remainder. Through the algebraic identity the equations can be re-written as

$$Q(x) = \left(\frac{a_3}{b_1}x^2 + \frac{(a_2 - (a_3/b_1)b_0)}{b_1}x + \frac{(a_1 - ((a_2 - (a_3/b_1)b_0)/b_1)b_0)}{b_1}\right)$$

$$\text{and } R = \left(a_0 - \frac{(a_1 - ((a_2 - (a_3/b_1)b_0)/b_1)b_0)}{b_1}b_0\right)$$

2.5 *Flowchart diagram of the algorithm*

In this section, divider implementation algorithm has been discussed leading towards high-speed operation. The flowchart of the algorithm is shown in Fig. 3. Where, dividend (A) and divisor (B) considered as n-bit and m-bit, respectively. The implementation procedure using the flowchart diagram has been described in Table 2, where two examples have been considered. Example 1 has been considered for perfect division (remainder = 0), Example 2 has been considered for imperfect division (remainder $\neq 0$). For simplicity purpose $(8 \div 4)$ bit divider example has been considered, example of higher order bit can be implemented in similar manner.

3 Divider implementation technique

Proposed divider implementation technique is shown in Fig. 4. The architecture has been implemented via (3). For simplicity purpose, let us assume dividend has greater length than divisor. Divisor has been broken into two parts, that is, most significant part (L) and least significant part (R). L is compared with equal number of bits of dividend taken from most significant bit (MSB) side. If the dividend is greater than L, directly divide the dividend bits by L, otherwise concatenation with next significant bit of dividend. Divide procedure has been implemented through subtractor. Difference is acting here as remainder, and borrow has been working as the selector input of the multiplexer. If the borrow is equal to '0' hence quotient '1' else '0'. The remainder is again concatenated of next MSD of the dividend and subtracted from the cross-multiplication result of the quotient bits and least significant bits of divisor. If result is negative, the quotient is reduced by '1' and set the new quotient bits, otherwise for positive result it is promoted to the next stage. Similarly, the division algorithm has been implemented.

Consider the number $A = \sum_{i=0}^{n-1} a_i 2^i$ to be divided by $B = \sum_{i=0}^{m-1} b_i 2^i$, where $(a_i, b_i \in 0, 1)$. To execute the division operation easily through 'Dhvajanka (on top of the flag)' methodology, it has been assumed that the length of dividend is greater than length of divisor.

3.1 *Implementation procedure*

Step 1: Consider the most significant part of dividend $\left(\sum_{i=n-(m/2)}^{n-1} a_i 2^i\right)$ and divisor $\left(\sum_{i=(m/2)}^{m-1} b_i 2^i\right)$.

Step 2: Determine $\left(\sum_{i=n-(m/2)}^{n-1} a_i 2^i - \sum_{i=(m/2)}^{m-1} b_i 2^i\right)$. Suppose the first borrow '0', then through multiplexer it will set the quotient (Q_n) '1' and the remainder is 'R'.

Step 3: Determine $Q_n \times b_{(m/2)-1}$. Concatenate R and $a_{n-(m/2)-1}$ and subtract $Q_n \times b_{(m/2)-1}$. Again divide in similar procedure (step 1). Set the quotient bit Q_{n-1} and remainder 'R'.

Step 4: Determine $Q_{n-1} \times b_{(m/2)-2} + Q_n \times b_{(m/2)-1}$. Concatenate R and $a_{n-(m/2)-2}$ and subtract $Q_{n-1} \times b_{(m/2)-2} + Q_n \times b_{(m/2)-1}$. Again divide in similar procedure (step 1). Set the quotient bit Q_{n-1} and remainder 'R'.

3.2 *Latency of the divider*

The hardware cost of the architecture can be computed based on the number of complex operations performed in its critical path, hence total propagation delay can be estimated. The reported architecture for division using Vedic mathematics can be computed in five steps shown in Fig. 5, with maximum 'n' (for imperfect division) iterations. So the total latency can be computed in terms of the propagation delay of summation the individual subsection, with 'n' iterations. The total propagation delay of the proposed architecture (t_{pd}) can be computed as

$$t_{pd} = t_{stage1} + t_{stage2} + t_{stage3} + t_{stage4} + t_{stage5} \tag{7}$$

where t_{stage1} is the propagation delay of stage1; t_{stage2} is the propagation delay of stage2; t_{stage3} is the propagation delay of stage3; t_{stage4} is the propagation delay of stage4; and t_{stage5} = propagation delay of stage5.

Stage 1 contains only comparator [18], and comparator has been implemented through '2' stage parallel adder and '2' stage XOR gates. For m bit divisor maximum, $m/2$ bit comparator is required. Thereby, maximum $m/2$ bit parallel adder is required in each case. Critical path to implement a full adder is equal to 2 XOR gate delay; thereby critical path for to implement m/over2 bit parallel adder is equal to $(m/2) \times 2$ XOR $= m$XOR gate delay. '2' stage parallel adders and '2' XOR stage are required to implement a comparator, thus total propagation delay equals to $(2m + 2)$ XOR gate delay. Second stage contains only $m/2$ bit parallel subtractor, and critical path of 1 bit subtractor equals to 3 XOR gate delay, thereby, total critical path delay for $m/2$ bit subtractor maybe estimated as $(m/2) \times 3$ XOR gate delay. Third stage contains only parallel adder of n bit, assuming one full adder may require 2 XOR gate delay, thereby total propagation delay of n bit parallel adder requires $n \times 2$ XOR gate delay. Fourth stage contains $m/2$ bit multiplier, and n bit subtractor in feedback path. Assume critical path delay of n bit subtractor equals to $3 \times n$ XOR gate delay. To implement multiplier, three stages are required, namely (i) partial product generation, (ii) partial product addition and (iii) final addition [18]. In partial product generation stage, maximum depth in a column of the partial product is equal to $m/2$. For generation of partial product, it requires $m/2$ XOR (let us assume XOR gate delay and 'AND' gate delays are equal) delays. For addition, it may require $(m/(2 \times 3)) \times 2$ XOR gate, that is, $m/3$ XOR gate for

$$= \left(\frac{a_3}{b_1}x^2 + \frac{\left(a_2 - \frac{a_3}{b_1}b_0\right)}{b_1}x + \frac{\left(a_1 - \frac{\left(a_2 - \frac{a_3}{b_1}b_0\right)}{b_1}b_0\right)}{b_1}\right) + \frac{(a_0 - \frac{\left(a_1 - \frac{\left(a_2 - \frac{a_3}{b_1}b_0\right)}{b_1}b_0\right)}{b_1}b_0)}{b_1 x + b_0} = Q(x) + \frac{R}{g(x)} \tag{6}$$

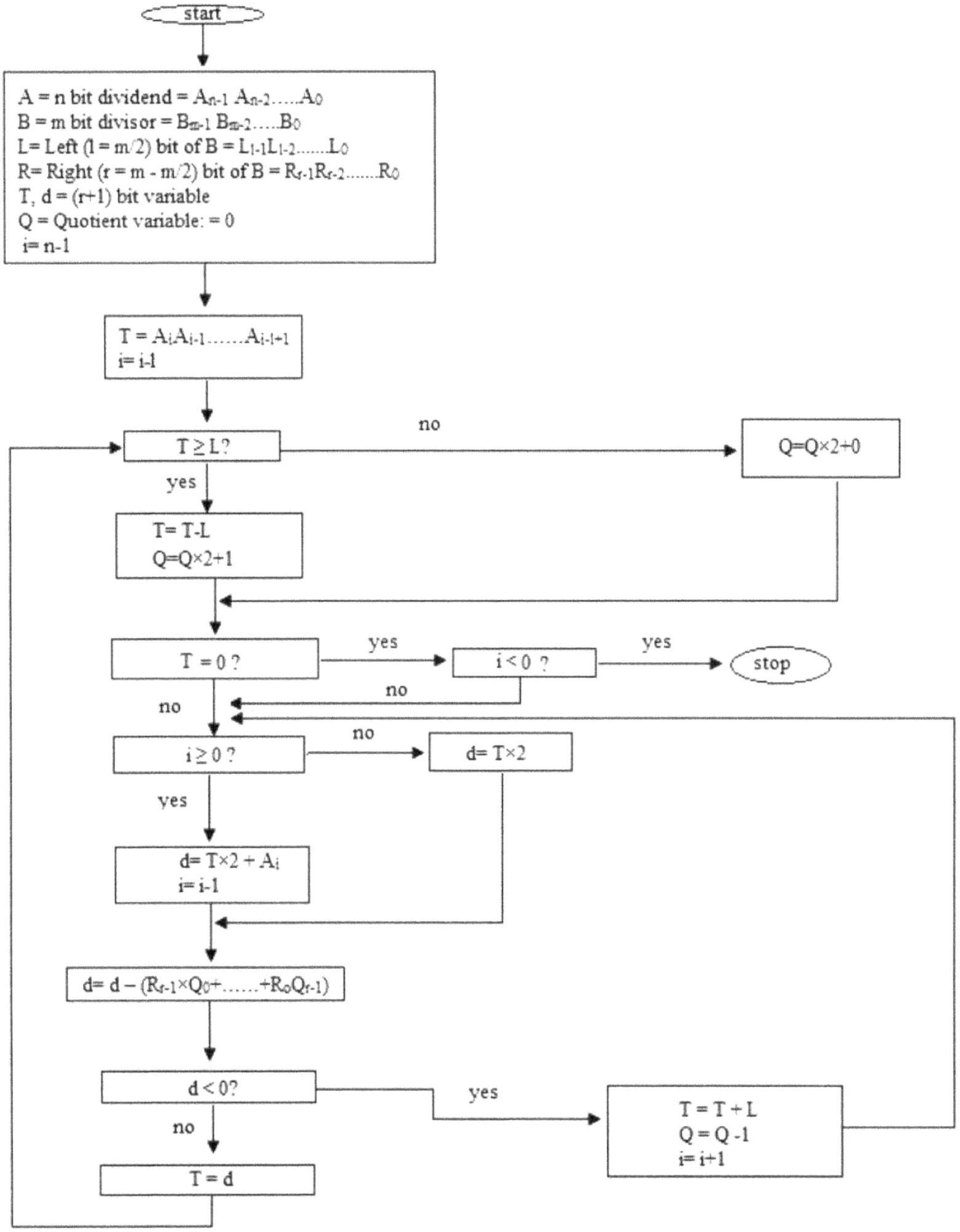

Fig. 3 *Flowchart representation of divider using dhvajanka formula*

partial product addition in first stage. For second stage requires $m/6$ XOR gate and so on, thus total addition purpose may be approximated as $m+(m/2)=(3m/2)$ XOR gate delay. Also for multiplication approximated, maximum XOR gate delay equals to $3m/2$. In the fifth stage, $m/2$ bit subtractor is required, thereby critical path delay of $m/2$ bit subtractor equals to $(m/2)\times 3$ XOR gate delay.

Thus, total propagation delay for each of the iterations may be approximated as

$$\begin{aligned} t_{\mathrm{pd}} &= t_{\mathrm{stage1}} + t_{\mathrm{stage2}} + t_{\mathrm{stage3}} + t_{\mathrm{stage4}} + t_{\mathrm{stage5}} \\ &= (2m+2) + (3m/2) + \ 2n + \big(3n + (3m/2)\big) \\ &\quad + \ (3m/2) = [5n + (13m/2) + 2] \end{aligned}$$

XOR gate delay. Thereby n iteration may consume $n(5n+(13m/2)+2)$ XOR gate delay.

4 Results and discussion

The advantages of CMOS transmission gate (TG) logic over conventional CMOS and complementary pass transistor logic (CPL) [19, 20] logic are well established. As the CMOS TG consists of one p-channel MOSFET (PMOS) and one n-channel MOSFET (NMOS), connected in parallel, the 'ON' resistance is smaller than even a single NMOS. Proper modifications at the device, circuit and architectural levels of design hierarchy have been implemented to reduce the energy delay product (EDP) and power delay product (PDP) for the proposed design. TGs are used for the design of different modules for faster operation and better logic transformation. Dual threshold voltage (V_{T}) operating mode was considered for simulation to determine the performance parameters. The proper choice of threshold voltages for a particular transistor in the circuit is based on a number of logics as described below:

(i) Placement of high-V_{T} transistors on the leakage path directly between supply and ground reduces the subthreshold leakage current and hence static power.

Table 2 Illustration of flowchart with the help of the examples. Example 1 has been considered for complete division (remainder = 0), Example 2 has been considered for incomplete division (remainder ≠ 0)

Steps	Example 1	Example 2
initialisation	$A = 10000100$	$A = 10101010$
	$B = 1011$	$B = 1111$
	$L = 10; l = 2$	$L = 11; l = 2$
	$R = 11; r = 2$	$R = 11; r = 2$
	$Q: = 0$	$Q: = 0$
	$i = 7$	$i = 7$
step 1	$T = 10$	$T = 10$
	$i = (i - 1) = (7 - 2) = 5$	$i = (i - 1) = (7 - 2) = 5$
	$T = T - L = 10 - 10 = 00$	$Q = Q \times 2 + 0 = 0$
	$Q = 1$	$d = 101$
	$d = 00$	$i = 4$
	$i = 4$	$d = 101 - 0 = 101$
	$d = 00 - (1 \times 1 + 1 \times 0) = 00 - 01 = -$ Ve	$T = d$
	$T = T + L = 00 + 10 = 10$	$T = 101$
	$Q = Q - 1 = 0$	$T = T - L = 101 - 11 = 10$
	$i = 5$	$Q = 1$
step 2	$d = 100$	$i = 4 \geq 0$
	$i = 4$	$d = 100; i = 3$
	$d = 100 - (1 \times 0 + 0 \times 1) = 100 - 00 = 100$	$d = 100 – (1 \times 1 - 1 \times 0) = 11$
	$T = d = 100$	$T = d = 11$
step 3	$T = T - L = 100 - 10 = 10$	$T = T – L = 11 - 11 = 0$
	$Q = 1$	$Q = 11$
	$d = 100$	$i = 3 \geq 0$
	$i = 3$	$d = 01; i = 2$
	$d = 100 - (1 \times 1 + 1 \times 0) = 100 - 01 = 11$	$d = d - (1 \times 1 + 1 \times 1) = 01 - 10 = -$ Ve
	$T = d = 11$	$T = T + L = 0 + 11 = 11$
		$Q = Q - 1 = 11 - 1 = 10$
		$i = i + 1 = 3$
step 4	$T = T - L = 11 - 10 = 01$	$d = 111$
	$Q = 11$	$i = i - 1 = 3 - 1 = 2$
	$d = 10$	$d = 111 - (1 \times 1 + 1 \times 0) = 111 - 01 = 110$
	$i = 2$	$T = d = 110$
	$d = 10 - (1 \times 1 + 1 \times 1) = 10 - 10 = 00$	
	$T = d = 00$	
step 5	$Q = 110$	$T = T – L = 110 - 11 = 11$
	$d = 01$	$Q = 101$
	$i = 1$	$d = 110, i = 1$
	$d = 01 - 01 = 00$	$d = 110 - (1 \times 1 + 1 \times 0) = 110 - 1 = 101$
	$T = d = 00$	$T = d = 101$
step 6	$Q = 1100$	$T = 101 - 11 = 10$
	$d = 00$	$Q = 1011$
	$i = 0$	$d = 101; i = 0$
	$d = 00 - 00 = 00$	$d = 101 - (1 \times 1 + 1 \times 1) = 101 - 10 = 11$
	$T = d - 00$	$T = d = 11$
step 7	$Q = 11000$	$T = T - L = 11 - 11 = 0$
	$d = 00$	$Q = 10111$
	$i = -1$	$d = 00$
	$d = 00 - 00 = 00$	$i = -1$
	$T = d = 00$	$d = 00 - (1 \times 1 + 1 \times 1) = 00 - 10 = -$Ve
		$T = T + L = 00 + 11 = 11$
		$Q = Q - 1 = 10110, i = 0$
step 8	$Q = 110000$	$d = 110, i = -1$
		$d = 110 - (1 \times 0 + 1 \times 1) = 110 - 1 = 101$
		$T = d = 101$
step 9		$T = T - L = 101 - 11 = 10$
		$Q = 101101$
		$d = 100$
		$d = 100 - (1 \times 1 + 1 \times 0) = 11$
		$T = d = 11$
result	when $i = r - 1$, then floating point (bit) start	when $i = r - 1$, then floating point (bit) start
	$Q = 1100.00$	$Q = 1011.01$

(ii) Placement of low-V_T transistors on the signal propagation path from the input node to the output improves the performance substantially.
(iii) A logical intersection of the conditions illustrated in (a) and (b) requires an optimised choice that leads to the minimum EDP.

The entire algorithm in this Letter was simulated and their functionality was examined by spice spectre simulator. Performance parameters like propagation delay and dynamic power consumptions analysis of this Letter was calculated using standard 90 nm CMOS technology with 1 V power supply, operated at 250 MHz. As shown, the application of the Vedic division methodology reduces the iteration resulted the reduction of propagation delay and dynamic switching power consumptions.

To implement the Vedic divider like (4 ÷ 4), (4 ÷ 8), (4 ÷ 16), (8 ÷ 4), (8 ÷ 8), (8 ÷ 16) etc. bits, all the individual modules such as subtractor, adder, cross-multiplier etc. were implemented through TG to make the circuit faster. The individual performance parameters such as propagation delay, dynamic switching power consumption, EDPs and PDPs for different circuit modules have been computed. With the help of all the modules, the final simulation has been carried out and performance parameters have been calculated. Comparative study between different architectures and proposed architecture like (4 ÷ 4), (4 ÷ 8), (4 ÷ 16), (8 ÷ 4), (8 ÷ 8), (8 ÷ 16) etc., bit divider is shown in Table 3. Proper modifications at the device, circuit and architectural levels of design hierarchy have been analysed in terms of propagation delay, average power dissipation and their products. The values of delay, power, EDP and PDP of different architectures are measured and tabulated in Table 3. The EDP (10^{-21}) J s and PDP (10^{-12}) J are quantitative measures of the efficiency and a compromise between speed and power dissipations. EDPs and PDPs are particularly important when high-speed operation is needed and its comparison at 1 V supplies voltage with 90 nm CMOS technology. Input data were taken in a regular fashion for experimental purpose. For each transition, the delay is measured from 50% of the input voltage swing to 50% of the output voltage swing.

It is worth mentioning here that we have taken the implementation methodology from different references [9, 11, 15] and implemented in the same technological environments (spice spectre with standard 90 nm CMOS technology) and then compared the performance parameters. The propagation delay and switching power are the worst-case delay and power of all possible bit combinations. It can be observed from Table 3 (32 ÷ 16) bit squarer requires ~300 ns to propagate a signal and consumes ~32.53 mw power for a layout area of ~17.39 mm^2. Proposed architecture offered ~47.3, ~38.4, ~34% faster operation (propagation delay) than DR [9], N–R [11] and G–S [15] architecture, respectively. On the other hand corresponding reduction of power consumption

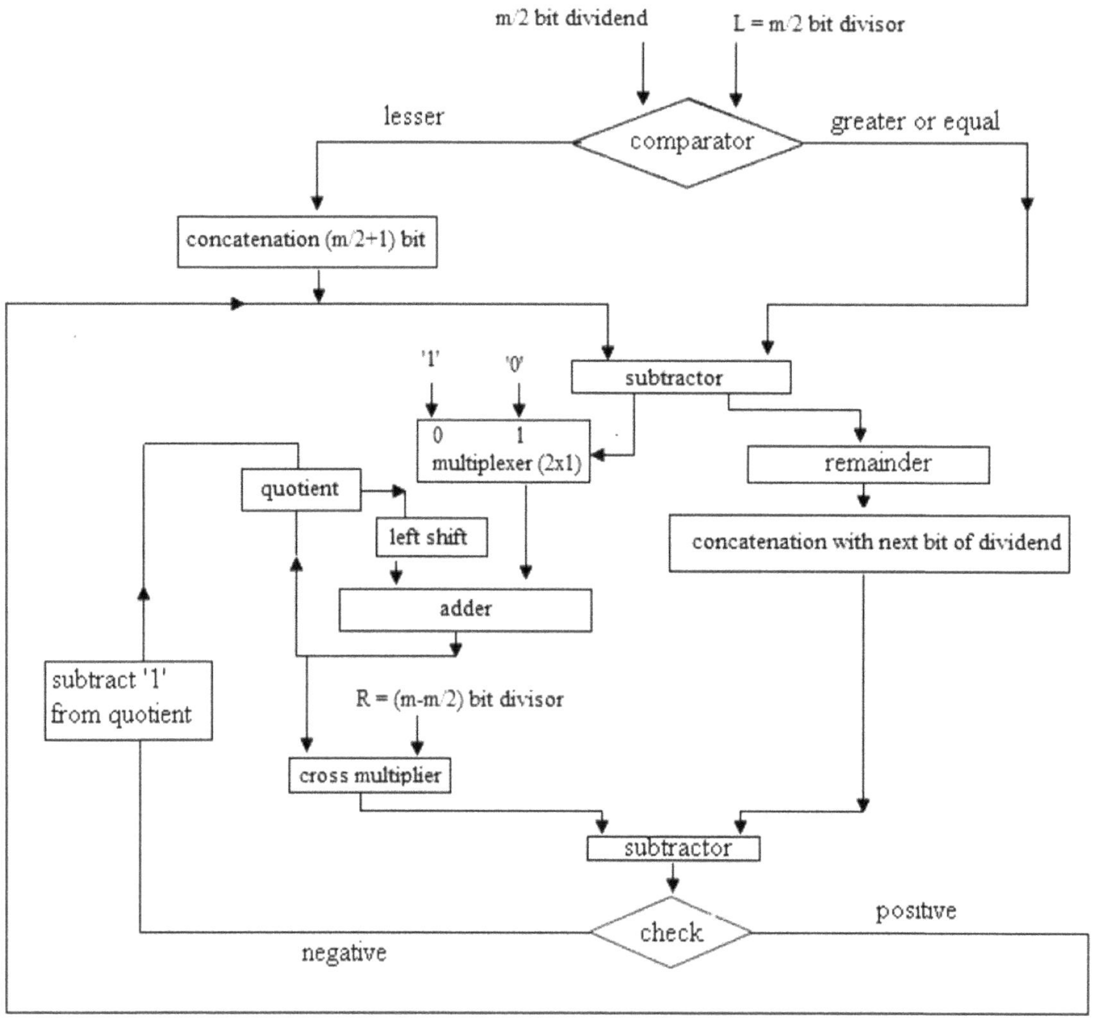

Fig. 4 *Hardware implementation of divider using dhvajanka formula*

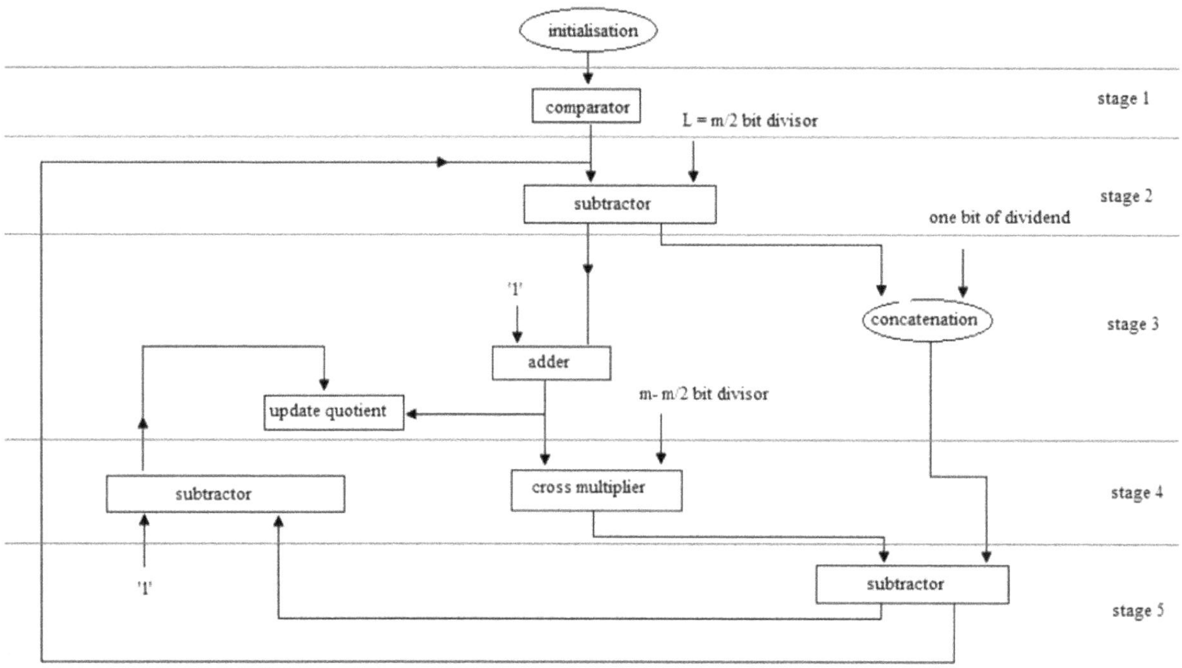

Fig. 5 *Latency analysis of divider using dhvajanka formula*

Table 3 Performance parameters like propagation delay (ns), dynamic switching power consumption (mW), EDP (10^{-24} J s), PDP (10^{-12} J), % savings in terms of propagation delay and dynamic switching power consumption compared with proposed methodology, as a function of input number of bits. The architecture has been implemented through spice spectre (T-Spice V13) simulator, with 90 nm CMOS technology. For each transition, the delay is measured from 50% of the input voltage swing to 50% of the output voltage swing

Input no. of bits	Architectures	Delay, nS	Power, mW	EDP (10^{-21}) J S	PDP (10^{-12}) J	Improvement in delay, %	Improvement in power, %
4 ÷ 4	DR [9]	12.8	0.99	162.2016	12.672	47.65	37.3
	N–R [11]	11.18	0.88	109.9933	9.8384	40	29.5
	G–S [15]	10.44	0.78	85.01501	8.1432	35.8	20.5
	proposed	6.7	0.62	27.83	4.15		
4 ÷ 8	DR [9]	19.79	1.59	622.7141	31.4661	48.86	37.7
	N–R [11]	17.3	1.34	401.0486	23.182	41.5	26.1
	G–S [15]	16.16	1.25	326.432	20.2	37.3	20.8
	proposed	10.12	0.99	101.39	10.01		
4 ÷ 16	DR [9]	33.7	2.97	3372.999	100.089	47	42
	N–R [11]	29.41	2.47	2136.422	72.6427	39.3	32.03
	G–S [15]	27.55	2.44	1851.966	67.222	35.2	29.5
	proposed	17.84	1.72	547.41	30.68		
8 ÷ 4	DR [9]	36.34	3.17	4186.288	115.1978	47	39.1
	N–R [11]	31.78	2.66	2686.516	84.5348	40	27.4
	G–S [15]	29.7	2.5	2205.225	74.25	36	22.8
	proposed	19.0	1.93	696.73	36.67		
8 ÷ 8	DR [9]	50.3	4.42	11 183	222.326	48.11	37.7
	N–R [11]	43.4	3.68	6931.501	159.712	39.86	25.2
	G–S [15]	41.0	3.45	5799.45	141.45	36.3	20.28
	proposed	26.1	2.25	1873.32	71.77		
8 ÷ 16	DR [9]	78.07	6.93	42 237.83	541.0251	34.8	35
	N–R [11]	68.25	5.775	26 900.31	394.1438	25.4	22
	G–S [15]	63.8	5.46	22 224.6	348.348	20.12	17
	proposed	50.9	4.5	11 658.65	229.05		
16 ÷ 4	DR [9]	115.5	10.01	133 535.9	1156.155	47.7	38
	N–R [11]	101.0	8.47	86 402.47	855.47	40.21	26.8
	G–S [15]	94.3	8.02	71 317.77	756.286	35.9	22.6
	proposed	60.38	6.2	22 603.6	374.35		
16 ÷ 8	DR [9]	143.3	12.77	262 230.5	1829.941	46.9	37.9
	N–R [11]	125.31	10.66	167 389.7	1335.805	39.3	25.7
	G–S [15]	117.0	9.59	131 277.5	1122.03	35.04	17.4
	proposed	76.0	7.92	45 745.92	601.92		
16 ÷ 16	DR [9]	198.91	17.67	699 116.9	3514.74	42.1	34.7
	N–R [11]	173.5	14.05	422 936.6	2437.675	33.7	17.95
	G–S [15]	162.3	13.95	367 461	2264.085	29.1	17.3
	proposed	115	11.53	152 484.3	1325.95		
32 ÷ 4	DR [9]	402.0	35.79	5 783 807	14 387.58	45.29	36
	N–R [11]	351.6	29.86	3 691 370	10 498.78	37.45	23.3
	G–S [15]	328.4	28.08	3 028 331	9221.472	33.0	18.4
	proposed	219.9	22.9	1 107 353	5035.71		
32 ÷ 8	DR [9]	457.8	40.78	8 546 707	18 669.08	45.19	37.4
	N–R [11]	400.28	34.82	5 579 002	13 937.75	37.3	26.7
	G–S [15]	375.9	31.98	4 518 800	12 021.28	33.2	20.2
	proposed	250.9	25.5	1 605 246	6397.95		
32 ÷ 16	DR [9]	570.01	49.5	16 083 114	28 215.5	47.3	34.2
	N–R [11]	487.52	41.25	9 804 125	20 110.2	38.4	21.18
	G–S [15]	454.75	39.7	8 209 863	18 053.58	34	18.06
	proposed	299.92	32.53	2 926 139	9756.398		

were ~34.2, ~21.18 and ~18.06%, respectively, compared with the same architectures. The layout of the proposed (32 ÷ 16) bit divider shown in Fig. 6, was implemented using L-Edit (T-Spice V-13) and the corresponding area was found to be ~17.39 mm^2.

5 Conclusions

A new division approach based on Vedic mathematics has been proposed for ultra-high-speed and low-power very large scale integration applications. Proposed approach is applied in (32 ÷ 16) division and it is found that it involves minimum memory space of the processor as compared with the conventional method. In addition to that the proposed algorithm is used efficiently so that it takes minimum stages for the division, which eventually reduces significant operational time.

In division circuitry, an (32 ÷ 16) bit divider implementation was transformed into just small division instead of actual divisor, subtraction and few multiplications, thereby reduces the iteration, owing to the substantial reduction in propagation delay. The propagation delay for (32 ÷ 16) bit division was only ~300 ns, whereas the power consumption of the same was 32.53 mW for a layout area of ~17.39 mm^2. Improvement in speed were found to be

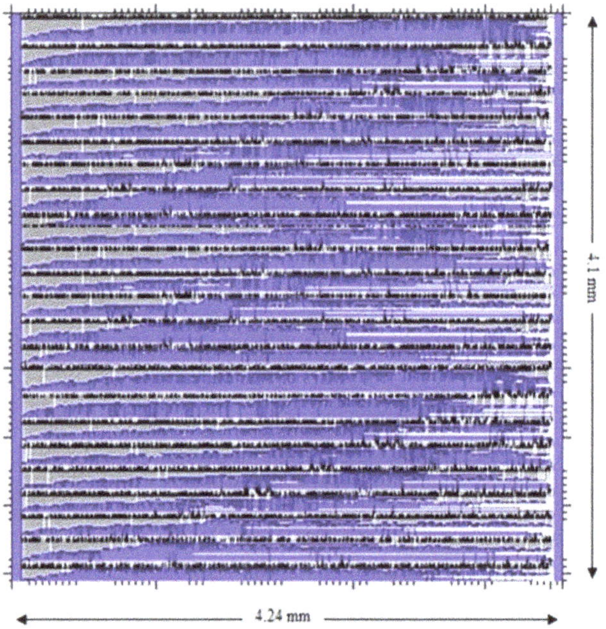

Fig. 6 *Layout of the proposed (32 ÷ 16) bit Vedic divider. Layout was implemented through L-Edit (T-Spice V-13) simulator the corresponding area was found to be ~17.39 mm²*

~47.3, ~38.4 and 34% for (32 ÷ 16) bit division circuitry, whereas corresponding reduction of power consumption were ~34.2, ~21.18 and ~18.06% compared with DR and N–R- and G–S-based implementation, respectively.

6 References

[1] Juang T.-B., Chen S.-H.H., Li S.M.: 'A novel VLSI iterative divider architecture for fast quotient generation'. Proc. IEEE Int. Symp. Circuits and Systems 2011, Seattle, WA, USA, May 2008, pp. 3358–3361

[2] Oberman S.F., Flynn M.J.: 'Division algorithms and implementations', *IEEE Trans. Comput.*, 1997, **46**, (8), pp. 833–854

[3] Deschamps J.-P., Bioul G.J.A., Sutter G.D.: 'Synthesis of arithmetic circuits, FPGA, ASIC and embedded system' (John Wiley & Sons, Inc., 2006)

[4] Hagglund R., Lowenborg P., Vesterbacka M.: 'A polynomial-based division algorithm', *Proc. IEEE Int. Symp. Circuits Syst.*, 2002, **3**, pp. 571–574

[5] Aggarwal N., Asooja K., Verma S.S., Negi S.: 'An improvement in the restoring division algorithm (needy restoring division algorithm)'. Proc. IEEE Int. Conf. Computer Science and Information Technology, Beijing, August 2009, pp. 246–249

[6] Sutter G., Deschamps J.P.: 'High speed fixed point divider for FPGAS'. Proc. IEEE Int. Conf. Field Programmable Logic and Applications, Prague, August 2009, pp. 448–452

[7] Sutter G., Deschamps J.P.: 'Fast radix 2^k divider for FPGAs'. Proc. IEEE Int. Conf. Programmable Logic, Sao Carlos, April 2009, pp. 115–122

[8] Jun K., Swartzlander E.E.Jr.: 'Modified non-restoring division algorithm with improved delay profile and error correction'. Proc. IEEE Int. Conf. Signals System and Computer, 2012, pp. 1460–1464

[9] Liu W., Nannarelli A.: 'Power efficient division and square root unit', *IEEE Trans. Comput.*, 2012, **61**, (8), pp. 1059–1070

[10] Louvet N., Muller J.M., Panhaleux A.: 'Newton–Raphson algorithms for floating-point division using an FMA'. Proc. IEEE Int. Conf. Application Specific Systems Architectures and Processors, Rennes, France, July 2010, pp. 200–207

[11] Piso D., Bruguera J.D.: 'Simplifying the rounding for Newton–Raphson algorithm with parallel remainder'. Proc. IEEE Int. Conf. Signals Systems and Computers, Pacific Grove, CA, USA, November 2009, pp. 921–925

[12] Nenadic N.M., Mladenovic S.B.: 'Fast division on fixed-point DSP processors using Newton–Raphson method'. Proc. IEEE Int. Conf. Computers as a Tool, Belgrade, November 2005, pp. 705–708

[13] Guy E., Seidel P.-M.M., Warren E.F.Jr.: 'A parametric error analysis of Goldschmidt's division algorithm'. Proc. IEEE Int. Conf. Computer Arithmetic, June 2003, pp. 165–171

[14] Ercegovac M.D., Imbert L., Matula D.W., Muller J.-M., Wei G.: 'Improving Goldschmidt division, square root, and square root reciprocal', *IEEE Trans. Comput.*, 2000, **49**, (7), pp. 759–763

[15] Kong I., Swartzlander E.E.Jr: 'A rounding method to reduce the required multiplier precision for Goldschmidt division', *IEEE Trans. Comput.*, 2010, **59**, (12), pp. 1703–1708

[16] Maharaja J.S.S.B.K.T.: 'Vedic mathematics' (Motilal Banarsidass Publishers Pvt Ltd, Delhi, 2001)

[17] Saha P., Banerjee A., Bhattacharyya P., Dandapat A.: 'Vedic divider: novel architecture (ASIC) for high speed VLSI applications'. Proc. IEEE Int. Symp. System Design, Kochi, India, December 2011, pp. 67–71

[18] Saha P., Banerjee A., Dandapat A., Bhattacharyya P.: 'ASIC design of a high speed low power circuit for calculation of factorial of 4-bit numbers based on ancient vedic mathematics', *Microelectron. J. (Elsevier)*, 2011, **42**, (12), pp. 1343–1352

[19] Uyemura J.P.: 'CMOS logic circuit design' (Kluwer Academic Publishers, 2001)

[20] Chang C.H., Gu J., Zhang M.: 'Ultra low-voltage low-power CMOS 4-2 and 5-2 compressors for fast arithmetic circuits', *IEEE Trans. Circuits Syst, I*, 2004, **51**, (10), pp. 1985–1997

12

A 65 nm CMOS broadband self-calibrated power detector for the square kilometre array radio telescope

Ge Wu, Leonid Belostotski, James W. Haslett

Department of Electrical and Computer Engineering, University of Calgary, Calgary, Alberta, Canada
E-mail: lbelosto@ucalgary.ca

Abstract: In this study, a 65 nm complementary metal oxide semiconductor (CMOS) broadband self-calibrated high-sensitivity power detector for use in the Square Kilometre Array (SKA), the next-generation high-sensitivity radio telescope, is presented. The power detector calibration is performed by adjusting voltages at the bulk terminals of the input transistors to compensate for mismatches in the output voltages because of process, voltage and temperature variations. Measurements show that the power detector, preceded by an input power-match circuit with 6 dB gain, has an input signal range from −48 to −11 dBm over which a 0.95 dB maximum error in the detected power is observed when the calibration rate is 20 kHz. The proposed broadband power detector has a 3 dB upper band edge of 1.8 GHz, which adequately covers the midband SKA frequency range from 0.7 to 1.4 GHz. The settling time and the calibration time are both <5 μs. The circuit consumes 1.2 mW from a 1.2 V power supply and the input-match circuit consumes another 5.8 mW. The presented power detector achieves the best combination of the detection range and sensitivity of previously published circuits.

1 Introduction

Measurement of received signal strength is common in many wireless communication receivers. Usually this measurement is performed by a power detector (PD) in an automatic gain control (AGC) loop that provides the optimum signal level for an analog-to-digital converter (ADC) [1–3]. In contrast to conventional AGCs, in this paper the PD, for use in a Square Kilometre Array (SKA) telescope receiver, does not form a part of an AGC, but rather is used to measure the total received power 'before' the AGC's variable-gain amplifier (VGA) in order to preserve the total received power information, required for some astronomical observations. The receiver will still have an AGC loop however, but the power measurement will be performed at the SKA control centre by analysing the ADC output during the telescope calibration cycle.

The SKA is the next-generation low-noise and high-sensitivity radio telescope, which has an effective collecting area of one square kilometre [4]. The scientific goals for the SKA require a wideband telescope with operating frequencies from 0.2 to 30 GHz and implemented with a number of different antenna technologies. Our work focuses on the midband SKA frequency range from 0.7 to 1.4 GHz [5–8].

The conceptual direct-sampling SKA receiver architecture proposed by our research group is shown in Fig. 1. Input signals, dominated by the receiver thermal noise, are collected by telescope antenna elements, amplified by low-noise amplifiers and gain stages, identified as 'LNA' in Fig. 1, filtered by a bandpass filter, and fed into a VGA [3, 7], before entering the ADC [9]. In this receiver, a large broadband gain from LNAs is required to provide sufficient signal amplitude for the ADC to sample. Even with large gain the power levels seen by the VGA and the PD are still low [6] and require a sensitive PD. Although the sensitivity of the PD can be traded-off against the gain of the LNA, the broadband LNA's power consumption has already been optimised to achieve the optimum gain-bandwidth performance. An increase in the LNA gain would therefore require an additional gain stage, which would consume more power and affect the LNA bandwidth. Therefore this study is seeking to improve the PD sensitivity so that an additional gain stage can be avoided. Depending on the final configuration of the SKA, the telescope may require a few millions of receivers and the PDs. If that is indeed the case, fabrication costs and power consumption of each receiver and the PD circuit are important considerations in the receiver design.

There are several methods to implement power detection, such as by using thermal detection or by rectifying incoming signals with a diode or a non-linear transconductor. The thermal detection method is widely used in radio frequency (RF) measurement equipment [10]. In this method, heating, which is generated by the measured RF signal absorbed by a resistive component, is compared with heating because of a calibrated DC signal. This method is broadband and accurate [10, 11]. However, a chip-level realisation is complicated by thermal coupling among adjacent circuits through the substrate and requires special attention to packaging. Therefore this method is not suitable for integration with low-cost complementary metal oxide semiconductor (CMOS) [12, 13]. Rectification by using non-linear dependence of currents in semiconductor components (diodes, bipolar transistors and metal oxide semiconductor field effect transistor (MOSFET) transistors) on their voltages can be used for power detection as well. For example, Schottky diodes and bipolar transistors are widely employed for power detection because of their non-linear small-signal characteristics [10, 14]. However, diode detectors have temperature-dependent performance and limited dynamic range because their square-law characteristics are only observed over a limited operating range [10, 14, 15]. This requires elaborate compensation techniques to operate these detectors as true root-mean-square detectors [12]. Moreover, Schottky diodes are not standard components in most CMOS technologies, and therefore custom layouts are needed for their implementation, which in turn require additional modelling steps as they are not modelled in CMOS design kits.

Non-linear behaviour of MOSFETs can also be used for power detection. For example, recent works demonstrate that MOSFETs operating in the deep triode region can be used to realise PDs [16, 17]. In this situation, the non-linearity of transistor channel resistance can produce an average current proportional to the amplitude of the RF signal. Owing to the low channel resistance of a MOSFET in triode, the instantaneous small-signal drain current is divided between its channel resistance and the load resistance. The accuracy of the current divider determines the average output voltage. Owing to their low intrinsic gains associated with low channel resistances, even very small mismatches in transistors can

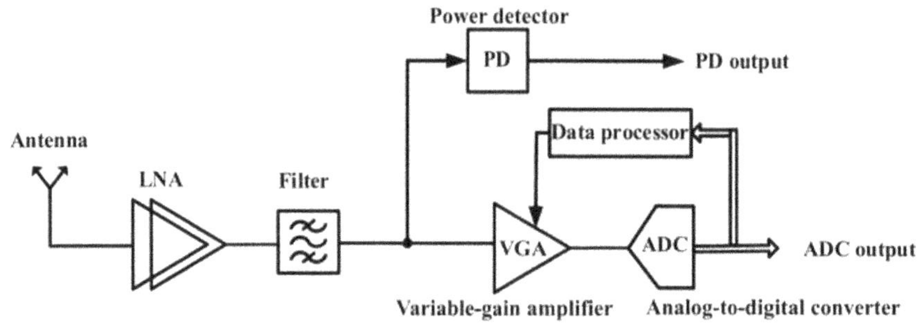

Fig. 1 *Conceptual SKA receiver block diagram*

significantly affect the PD performance [17]. In addition, since the output of a PD is a DC signal, process, voltage and temperature (PVT) related offsets affect the output and require calibration to remove these offsets to improve the sensitivity of the PD.

MOSFET square-law characteristics can be used for power detection as well. A differential-input single-ended-output PD circuit using the MOS transistor's square-law characteristics in the saturation region is presented in [18–20]. That circuit overcomes the input range limitation of diodes and bipolar transistors. In contrast to MOSFETs in triode, the circuit does not suffer from low channel resistance and overcomes the disadvantages of load-dependent performance present in [17] and described above. However, since mismatches in circuit parameters because of PVT were not addressed in [18–20], the minimum detected input powers of those circuits were high and their sensitivities were limited by the transistor mismatches. In [17], a calibration circuit was used to improve the PD sensitivity. In that circuit, an additional offset-subtractor amplifier was periodically connected before the main PD for offset cancellation. The switches used to connect the offset-subtractor introduce charge injection on the differential input of the PD. Although the differential charge injection is reduced by the differential circuit, it cannot be completely removed and the difference between the amounts of charge injection is amplified by the PD during its normal operation thus limiting the PD sensitivity.

In this paper, an improved differential CMOS PD based on the MOS transistor's square-law characteristic in the saturation region is presented in Section 2. To improve the PD sensitivity, an analogue self-calibration loop is proposed in Section 3 to compensate circuit component mismatches because of PVT variations. Section 4 analyses stability of the calibration loop and describes circuit configurations used in the calibrated PD. Experimental verification of the PD operation is demonstrated in Section 5. Section 6 concludes this paper.

2 PD architecture

2.1 Background: power detector cell

The basic differential-input single-ended-output PD using MOS transistor square-law characteristics in strong inversion is shown in Fig. 2 [18, 21]. This circuit uses a pair of same-sized N-channel MOSFET (NMOS) transistors M_1 and M_2 biased in saturation to convert the input signals v_{rf+} and v_{rf-} into currents $I_1(t)$ and $I_2(t)$, which have a square-law relationship with the input voltages. Assuming that RF input voltages are

$$v_{rf+}(t) = -v_{rf-}(t) = \frac{1}{2} V_{rf} \cos(\omega t) \tag{1}$$

the total current $I_D(t)$, shown in Fig. 2, can be expressed as

$$\begin{aligned} I_D(t) = I_1(t) + I_2(t) = 2k_n (V_{\text{bias}} - V_t)^2 \left(\frac{W}{L}\right)_{1,2} \\ + \underbrace{\frac{1}{4} k_n V_{rf}^2 \left(\frac{W}{L}\right)_{1,2}}_{\text{input signal power term}} + \frac{1}{4} k_n V_{rf}^2 \cos(2\omega t) \left(\frac{W}{L}\right)_{1,2} \end{aligned} \tag{2}$$

where the gate–source voltage $V_{GS} = V_{\text{bias}}$ and V_t are the gate bias voltage and threshold voltage for transistors $M_{1,2}$, respectively, $(W/L)_{1,2}$ is the width-to-length ratios of $M_{1,2}$, $k_n = (1/2)\mu_n C_{ox}$, μ_n is the carrier mobility and C_{ox} is the gate oxide capacitance per unit area. In (2), $I_D(t)$ includes three terms: the first term is the DC bias current, the second term is the DC current generated by the input signal, which has a squared relationship with the input RF signal amplitude, V_{rf}, and the third term is the second harmonic of the input signal. Therefore if the DC bias current and the second harmonic of the input signal in (2) are removed, then $I_D(t)$ represents the input signal power. Note that the fundamental components of the input signals are not present in (2) since they are out of phase and cancel at the output.

The basic PD circuit described above suffers from limited dynamic range because of the MOSFETs entering triode when input power is high and by circuit noise and an offset because of mismatch between the PD transistors when input power is low. These limitations will be addressed in the rest of this paper.

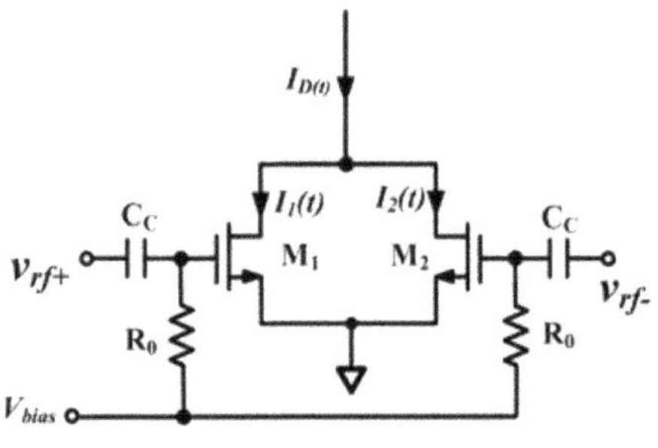

Fig. 2 *Differential-input single-ended-output CMOS PD cell*
R_0 and C_C provide biasing

2.2 Proposed differential power detector circuit

A differential CMOS PD is proposed as depicted in Fig. 3, which includes the main power detection circuit, based on the circuit in Fig. 2, and its duplicate circuit. The main power detection circuit consists of M_1, M_2, M_5, R_L and C_L, where C_L and R_L filter out the RF signals and convert current to voltage at node V_{O1}, and M_5 works as a current source, reducing voltage drop across R_L and thereby increasing the upper limit of the detection range. Note that while a similar circuit has been used by others [21], in this work the diode-connected transistor loads used in [21] are replaced by $M_{5,6}$ and R_L to increase the PD gain and reduce its noise. In addition, low-pass filters at the outputs V_{o1} and V_{o2} are added to filter out the second harmonic components of the input signal and capacitance C_s at the source of $M1$–$M4$, shown in Fig. 3, is added to remove the second harmonic components at the sources of $M1$–$M4$, which when coupled through gate–source

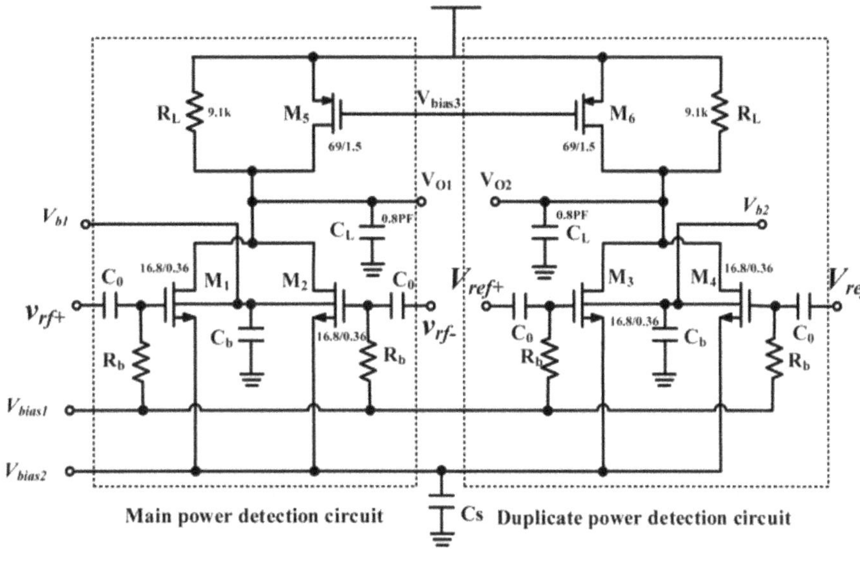

Fig. 3 *Differential PD schematic*

capacitances of $M3$ and $M4$ cause power-dependent fluctuations at the PD output.

In Fig. 3, when a differential-input signal $V_{rf}\cos(\omega t)$ is applied at the gate terminals of M_1 and M_2, the total current flowing through the load formed by R_L and M_5 is I_{D0}

$$I_{D0} = 2k_n\left(\frac{W}{L}\right)_{1,2}(V_{GS} - V_t)^2 + \underbrace{\frac{1}{4}k_n\left(\frac{W}{L}\right)_{1,2}V_{rf}^2}_{\text{input signal power term}} \tag{3}$$

and the second-order harmonic, previously seen in (2), is largely removed by C_L and R_L. To remove the DC bias component, the duplicate PD circuit, which consists of transistors M_3, M_4, M_6, R_L and C_L and is biased at the same condition as the main power detection circuit, ideally generates the same DC bias component in the current flowing through M_3, M_4 as M_1, M_2. By measuring the differential output voltage V_{out}

$$V_{out} = V_{O2} - V_{O1} = \underbrace{\frac{1}{4}k_n R_L\left(\frac{W}{L}\right)_{1,2}}_{\xi_p} V_{rf}^2 \tag{4}$$

the DC bias components cancel out and the output is only dependent on V_{rf}^2. In the ideal situation, when all transistors in the main PD circuit and its duplicate are perfectly matched and noise generated in the circuit is low, the output voltage is exactly related to the input power.

In practice, some amount of mismatch is inevitable and needs to be addressed in order to increase the dynamic range of the PD.

3 PD calibration and sensitivity

To compensate the offset at the PD differential output and to improve the sensitivity, this paper proposes a calibration process that uses MOSFET source-to-bulk voltage, V_{SB}, to tune MOSFET threshold voltages by making use of the well-known dependence of V_t on V_{SB}

$$V_t = V_{TO} + \gamma_{th}\left(\sqrt{|2\phi_F + V_{SB}|} - \sqrt{|2\phi_F|}\right) \tag{5}$$

where γ_{th} denotes the body effect coefficient, ϕ_F is the Fermi voltage and V_{TO} is the intrinsic threshold voltage. The proposed calibrated-PD circuit block diagram is illustrated in Fig. 4*a*. It is formed by a differential-input power-match amplifier Amp0, a PD circuit from Fig. 3, a high-gain amplifier Amp1 and two sets of switches $S1$ and $S2$.

During the calibration mode, the switch set $S1$ is turned OFF, cutting off the RF signal from the PD and the switch set $S2$ is turned ON connecting the amplifier Amp1 outputs to V_{b1} and V_{b2}, which are bulk terminals of $M_{1,2}$ and $M_{3,4}$, respectively, as labelled in Fig. 3. Any unwanted DC voltage difference at nodes V_{O1} and V_{O2} because of mismatches between the main PD circuit and the duplicate PD circuit is amplified by Amp1 and fed back to the bulk terminals of $M_{1,2}$ and $M_{3,4}$ through switch set $S2$ to force the differential voltage V_{O1}–V_{O2} to zero. Once calibration is completed, the switch set $S2$ turns OFF to open the feedback loop and the switch set $S1$ turns ON to connect the amplified input signals to the PD input. Note that the Amp1 output voltage is selected based on Monte Carlo simulations to avoid forward biasing the bulk-to-source p-n junctions for all process corners and all possible operating conditions.

Compared with the offset cancellation method in [17], the calibration method described here is expected to achieve higher sensitivity as the inevitable charges injected by the switches $S2$ appear at the bulks of the PD input transistors instead of their gates as in [17]. Since the gain from bulk to drain of a MOSFET is much lower than the gain from gate to drain, the charge injection effect is reduced. In addition, large capacitors C_b at the bulk terminals reduce the charge rejection problems further without affecting the PD bandwidth.

3.1 *PD calibration accuracy and sensitivity*

The PD calibration accuracy is mostly dependent on the loop gain of the calibration system, which is shown in a simplified form in Fig. 4*b*. In this diagram $V_{in,err}$ represents the input-referred error because of all mismatches in the PD circuit before calibration and $V_{out,\ err}$ represents the output DC error after calibration. Blocks PD and Amp1 represent the same blocks as in Fig. 4*a*. The block A_{TS} represents the equivalent voltage gain between the source-to-bulk voltage and threshold voltage of the PD input transistors $M_{1,2,3,4}$. From (5), this technology and bias-dependent parameter are expressed as

$$A_{TS} = \frac{dV_t}{dV_{SB}} = \frac{1}{2}\frac{\gamma_{th}}{\sqrt{|2\phi_F + V_{SB}|}} \tag{6}$$

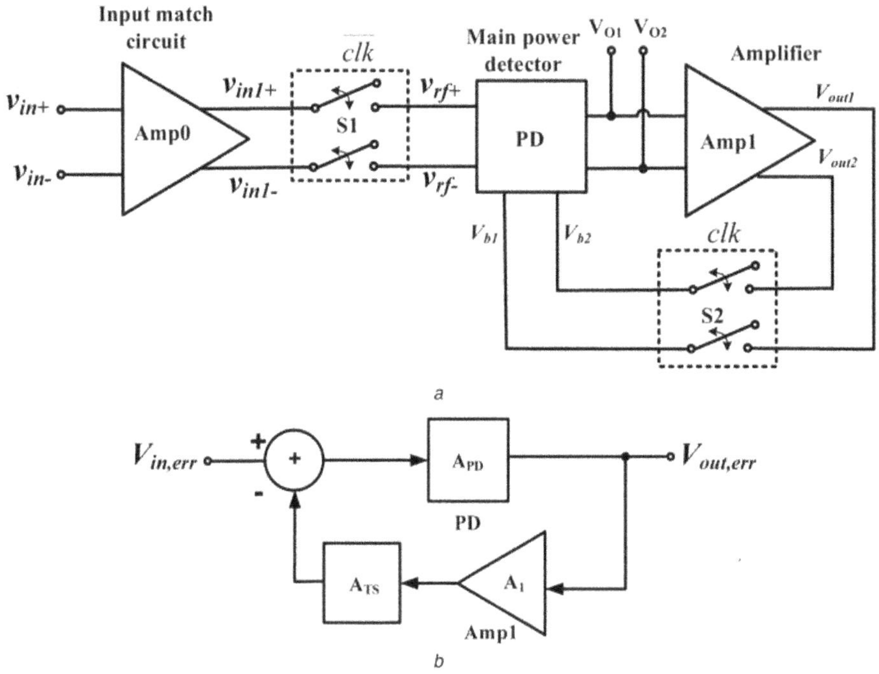

Fig. 4 *PD calibration configuration*
a Calibrated-PD circuit block diagram
b PD calibration block diagram

The closed-loop transfer function of the calibration system is

$$\frac{V_{\text{out, err}}}{V_{\text{in, err}}} = \frac{A_{\text{PD}}}{1 + A_{\text{PD}} A_1 A_{\text{TS}}} \tag{7}$$

where A_{PD} is the open-loop DC voltage gain of the PD, A_1 is the DC voltage gain of Amp1 and $A_{\text{PD}}A_1A_{\text{TS}}$ is the loop gain. Since the loop gain is much larger than 1, the closed-loop gain can be expressed as

$$\frac{V_{\text{out, err}}}{V_{\text{in, err}}} \simeq \frac{1}{A_1 A_{\text{TS}}} \tag{8}$$

When calibration completes, the equivalent error in the differential RF input signal amplitude can be expressed as

$$V_{\text{err}} = \sqrt{\frac{V_{\text{out,err}}}{\xi_p}} = \sqrt{\frac{V_{\text{in,err}}}{\xi_p A_1 A_{\text{TS}}}} \tag{9}$$

where ξ_{p} is the PD detection factor as defined in (4). From (9), V_{err}^2 is inversely proportional to A_{TS}, A_1 and ξ_{p}. Since A_{TS} is mainly determined by the process parameters, A_1 and ξ_{p} are the main parameters that control the calibration accuracy of the PD and should be made large. ξ_{p} can be increased by making load resistor R_{L} large and by widening $M_{1,2,3,4}$ as seen from (4). These however increase the voltage drop across R_{L}, which limits the maximum input power and potentially decreases the detection range. Therefore the calibration accuracy, which determines the PD sensitivity, and the PD detection range need to be traded-off by choosing the proper input transistor sizes and the load resistance R_{L}. According to simulation, for a $V_{\text{in,err}} = 5$ mV before calibration and with $A_1 = 70$ dB, representing the worst-case gain, V_{err}^2 is approximately equivalent to −64 dBm, referenced to100 Ω differential-input impedance, after calibration.

3.2 PD noise analysis

To estimate the effect of noise on PD sensitivity, half of the PD circuit, that is, the main PD circuit in Fig. 3, with its noise sources, as shown in Fig. 5, is considered. $\overline{i_{n1}^2}$, $\overline{i_{n2}^2}$, $\overline{i_{n5}^2}$ and $\overline{i_{n,\text{RL}}^2}$ represent mean-squared thermal noise currents of transistors M_1, M_2, M_5 and load resistor R_{L}, respectively. Simulations show that the thermal noise is dominant and flicker noise is not significant for our circuit and therefore is ignored in the following. The mean-squared thermal noise voltage at the PD half-circuit output V_{O1} can be expressed as

$$\overline{v_{n,\text{O1}}^2} = 4kTR_{\text{L}}\left(2\gamma g_{m1,2}R_{\text{L}} + \gamma g_{m5}R_{\text{L}} + 1\right) \times (f_n - f_s) \tag{10}$$

where $\gamma \simeq 2/3$ is the coefficient of MOSFET channel thermal noise, $g_{m1,\ 2}$ and g_{m5} are transconductances of transistors $M_{1,2}$ and M_5, f_s is related to the duration of the PD detection mode, $f_n \simeq (\pi/2)(1/(2\pi R_{\text{L}}C_{\text{L}}))$ is the noise bandwidth of the $R_{\text{L}} - C_{\text{L}}$

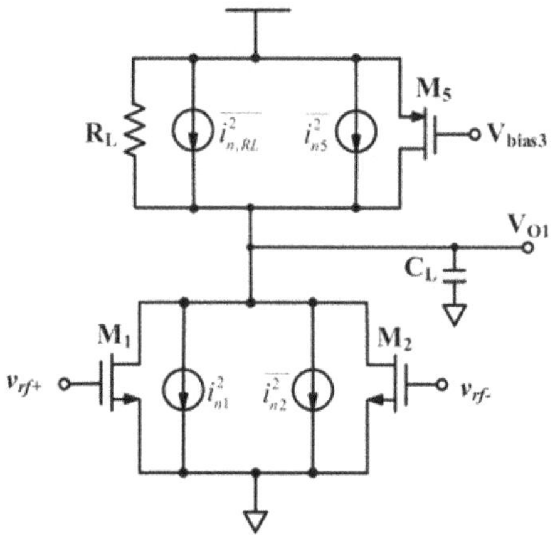

Fig. 5 *Half PD circuit with noise sources*

filter at the PD output, k is Boltzmann's constant and T is the absolute temperature.

To determine the input-equivalent noise power, the output mean-square noise voltage $\overline{v_{n,\mathrm{O1}}^2}$ at V_{O1} is doubled to account for the identical noise contribution of the duplicate PD circuit, identified in Fig. 3, converted to voltage and divided by the PD power detection factor to obtain the PD input-equivalent mean-squared noise voltage $\overline{v_{n,\mathrm{in}}^2}$

$$\overline{v_{n,\mathrm{in}}^2} = \sqrt{2\overline{v_{n,\mathrm{O1}}^2}}/\xi_{\mathrm{p}} = 8\sqrt{\frac{2kT(2\gamma g_{m1-4} + \gamma g_{m5} + R_{\mathrm{L}}^{-1})(f_n - f_s)}{k_n^2\left(\frac{W}{L}\right)_{1-4}^2}} \tag{11}$$

From (11), the mean-squared input-referred noise voltage can be decreased by increasing $(W/L)_{1-4}$ and R_{L} and decreasing g_{m5}.

3.3 Switch charge injection

The switch set $S1$ in Fig. 6 consists of CMOS transmission–gate switches $S1$A–$S1$D. $S1$C and $S1$D connect and disconnect the RF signal to and from the main PD input terminals. The circuit inside the dashed rectangle shown in Fig. 6 is the unit schematic of the transmission–gate switches $S1$A–$S1$D. Same-sized P-channel MOSFET (PMOS) and NMOS transistors reduce charge injection. Small M_{N1} and M_{P1} are selected to reduce feedthrough effects. Small NMOS transistor M_{L1} is a pull-down transistor to avoid floating terminals during the calibration.

To remove the errors caused by the switch charge injection to all PD input terminals during mode transitions, resistors R_{load} are added at the inputs of the reference switches $S1$A and $S1$B, as shown in Fig. 6, to emulate the signal-source resistance for $v_{\mathrm{in1+}}$ and $v_{\mathrm{in1-}}$ and thus provide the same loading to all four inputs to the PD circuit.

The second switch set, $S2$, is a pair of single small NMOS transistors that cause very small charge injection on large C_{b}, thus resulting in only small error voltages on the C_{b}'s. When the differential PD output is measured, the difference because of charge injection becomes even smaller and can be ignored.

3.4 Calibration rate

As discussed above, the mismatch between the PD transistors is calibrated out by adjusting the bulk voltages of $M_{1,2,3,4}$. These voltages are stored on C_{b}'s and are expected to remain constant during the PD detection mode. However, leakage currents, such as from MOSFET source-to-bulk, I_{sb}, drain-to-bulk, I_{db} and deep nwell-to-bulk p–n junctions, I_{nb}, may affect the stored values and decrease PD detection accuracy. The leakage-induced voltage variation on the C_{b}'s, represented by $\Delta V_{C_{\mathrm{b}}}$, depends on leakage currents, C_{b}, and detection sample rate Δt as follows

$$\Delta V_{C_{\mathrm{b}}} = 2(I_{\mathrm{sb}} + I_{\mathrm{db}} + I_{\mathrm{nb}})\frac{\Delta t}{C_{\mathrm{b}}} \tag{12}$$

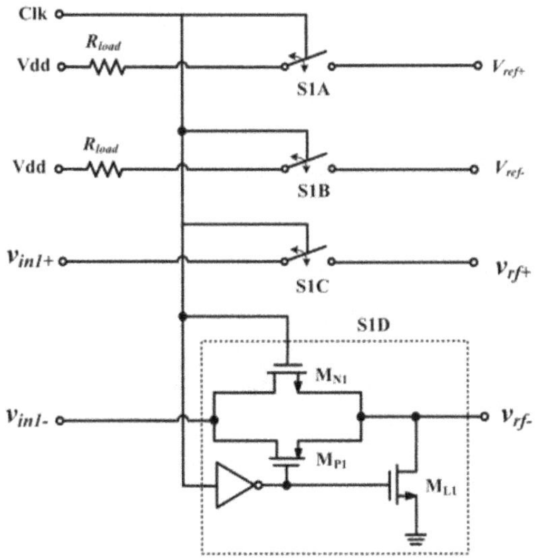

Fig. 6 *Transmission–gate switch set S1*

3.5 PD detection accuracy

The switch charge injection error is relatively small when compared with finite calibration error discussed in Section 3.1 and noise discussed in Section 3.2. Then the minimum RF input signal amplitude squared that the PD can detect, V_{sen}^2, is limited by a combination of V_{err}^2 found from (9) and input-referred noise in (11). From the PD measurement accuracy point of view, V_{sen}^2 appears added to the input signal, which results in the total output voltage $V_{\mathrm{tot}} = V_{\mathrm{O2}} - V_{\mathrm{O1}}$ as

$$V_{\mathrm{tot}} = \underbrace{V_{rf}^2 \xi_{\mathrm{p}}}_{V_{\mathrm{tot},rf}} + \underbrace{V_{\mathrm{sen}}^2 \xi_{\mathrm{p}}}_{V_{\mathrm{tot,sen}}} \tag{13}$$

From (13), the error introduced in the PD measurement can be expressed as

$$\Delta V(\mathrm{dBV}) = 20\,log\left(\frac{V_{\mathrm{tot,sen}}}{V_{\mathrm{tot},rf}}\right) \tag{14}$$

4 Circuits used in the calibrated PD

4.1 Amp1: folded-cascode amplifier with gain boost

To provide high gain, achieve stable closed-loop operation and power-down capability, a fully differential gain-boosted folded-cascode OTA has been implemented as shown in Fig. 7*a*. AN and AP are NMOS-input and PMOS-input gain-boost amplifiers, respectively, shown in Figs. 7*b* and *c*.

The input to the operational transconductance amplifier (OTA) is applied to transistors M_{F1} and M_{F2}, and output voltages V_{out1} and V_{out2} at nodes E and F, which are connected to the bulk terminals V_{b1} and V_{b2} as shown in Fig. 4*a*. Nodes E(F) have the highest output resistance, which provides high gain for the OTA. In addition, this resistance together with capacitors at node E(F), which includes the parasitic capacitances at these nodes and C_{b}, form the dominant pole of the OTA. Once calibration is complete, switch set $S2$ disconnects the C_{b}'s from load nodes E and F, and the main dominant pole location moves to higher frequency, which may cause a stability problem in the folded-cascode amplifier. The amplifier is turned off in this mode to prevent oscillation. The three poles associated with the OTA in the calibration mode can be expressed as follows

$$p_1 = \frac{-1}{R_{\mathrm{out},E}(C_b + C_{gd5} + C_{gd7} + C_{gd11} + C_{gb11})} \tag{15}$$

$$p_2 = \frac{-1}{R_{\mathrm{out},A}(C_{L,A} + C_{\mathrm{AN}})} \tag{16}$$

and

$$p_3 = \frac{-1}{R_{\mathrm{out},B}(C_{L,B} + C_{\mathrm{AP}})} \tag{17}$$

where

$$C_{L,A} = C_{\mathrm{gd9}} + C_{\mathrm{db9}} + C_{\mathrm{db1}} + C_{\mathrm{gb7}} + C_{\mathrm{gs7}} + C_{\mathrm{gd1}} \tag{18}$$

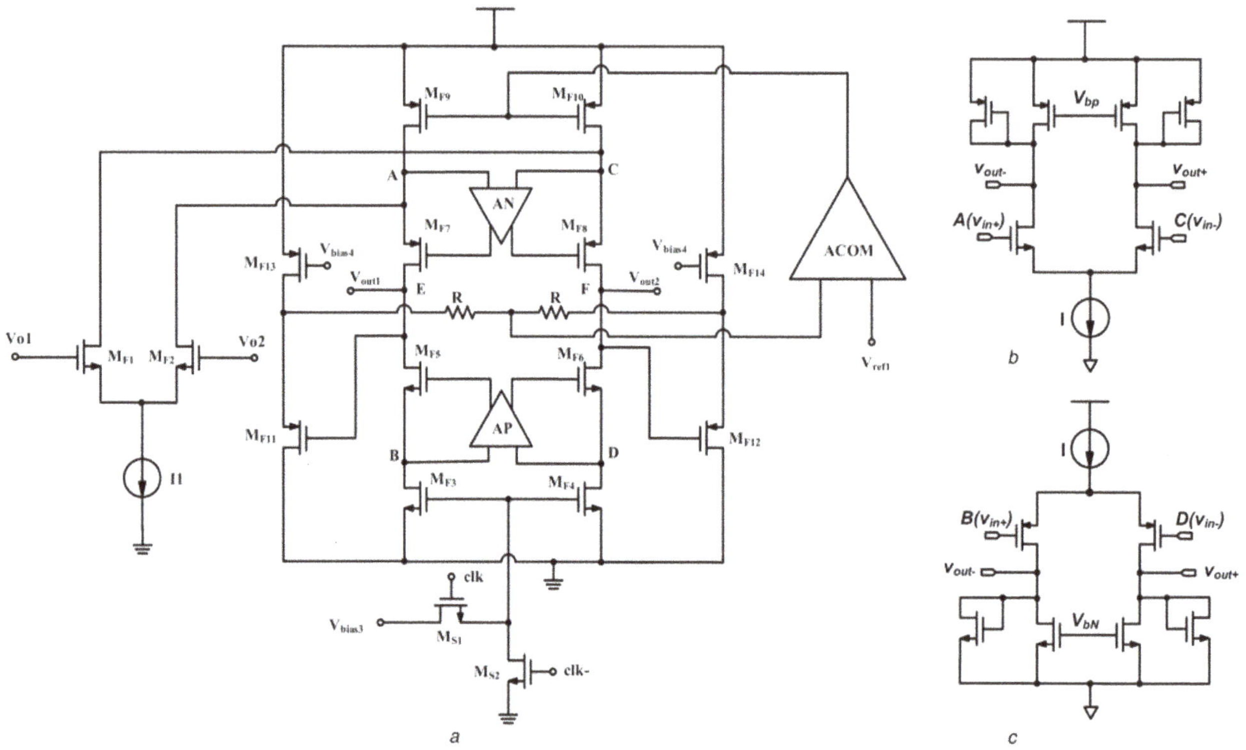

Fig. 7 *Amp1: folded-cascode amplifier with gain boost*
a Schematic of the folded-cascode OTA with gain boosting
b AN amplifier for gain boosting
c AP amplifier for gain boosting

$$C_{L,B} = C_{\text{gd3}} + C_{\text{db3}} + C_{\text{gs5}} + C_{\text{gs5}} \tag{19}$$

C_{AN} is the input capacitance of AN and C_{AP} is the input capacitance of AP, the C_{gs}'s, C_{gd}'s, C_{gb}'s and C_{db}'s are the gate–source, gate–drain, gate–bulk and drain–bulk capacitances of each transistor, respectively. The second dominant pole is p_2, which is at nodes A (C) because of the larger node parasitic capacitance compared with that of nodes B(D) forming a pole at p_3. Gain-boosting amplifiers AN and AP also add extra poles to the system. In this design, amplifiers AN and AP were designed so that the poles at their outputs were placed at high frequencies.

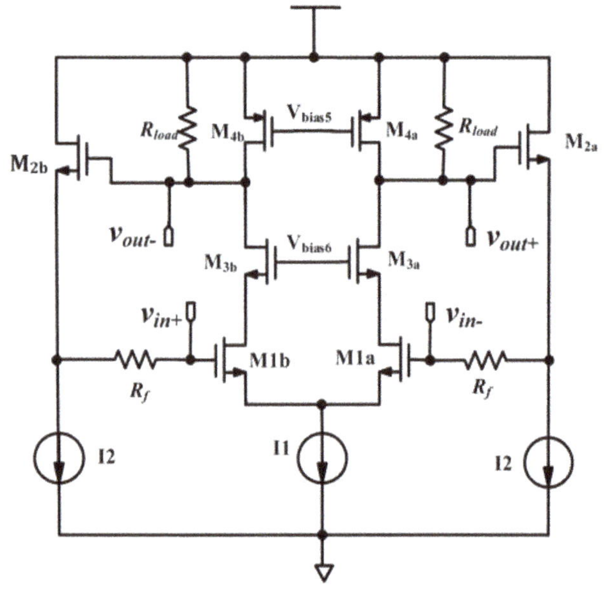

Fig. 8 *Input power-match circuit*

The common-mode feedback (CMFB) circuit, depicted in Fig. 7*a*, is made with two identical resistors *R*, amplifier ACOM (common-mode feedback amplifier) and buffers M_{F11} to M_{F14}, which isolate the CMFB circuit from the main amplifier output and avoid loading from feedback resistors, *R*. PMOS transistors M_{F11} to M_{F14} are chosen for the buffers because of the low common-mode output voltage of the OTA. Transistors M_{S1} and M_{S2} are used to turn ON the OTA load during calibration and turn it OFF during the PD detection to save power consumption and to prevent the folded-cascode amplifier from oscillation.

During the detection cycle, the self-calibrated PD forms a closed-loop negative feedback circuit, which includes the PD circuits, the amplifier Amp1 and the switch set *S*2. As discussed above, there are three poles generated by Amp1. The PD contributes another pole determined by the PD output resistance $R_{\text{out,PD}}$ and output capacitance $C_{\text{out,PD}}$

$$p_0 = \frac{-1}{R_{\text{out, PD}} C_{\text{out, PD}}} \tag{20}$$

For the PD to have high gain to overcome noise when sensing low input power, to provide filtering of the fundamental frequency at the PD output, and to reduce the offset error residual after

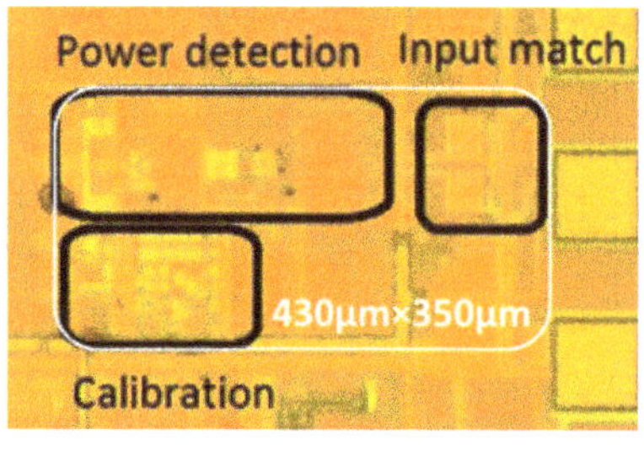

Fig. 9 *PD chip micrograph*

calibration (see Section 3.1), both R_L and C_L are chosen to be large resulting in p_0 to be a low-frequency pole. The other low-frequency pole, p_1 in (15), is located at the output of Amp1 and is formed by the large capacitance C_b, which stores the calibration voltages, and the large output resistance of Amp1. To maintain stability of the detection loop, the pole p_1 is designed to be the dominant pole. Two other Amp1 poles, p_2 in (16) and p_3 in (17), are placed at higher frequencies than pole p_0 to make the closed-loop circuit stable.

4.2 Input match circuit

Power matching of the PD to the 100 Ω differential-input signal source was accomplished with a differential amplifier Amp0 [22]. As shown in Fig. 8, this amplifier is a differential cascode amplifier, consisting of $M_{1a,1b}$ and $M_{3a,3b}$, with active feedback through $M_{2a,2b}$ and R_f for wideband matching and low-noise characteristics. Transistors $M_{4a,4b}$ are added to reduce the voltage drop across load

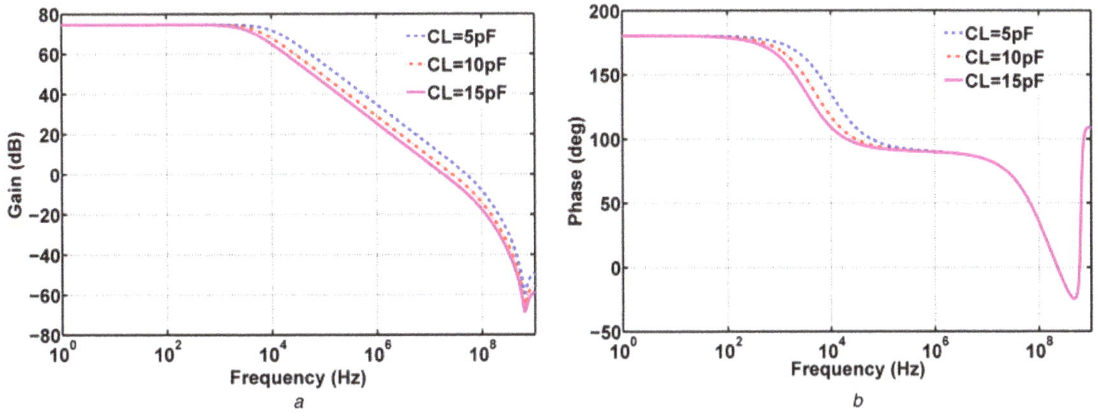

Fig. 10 *Simulated*
a Typical-corner gain
b Typical-corner phase of the folded-cascode gain-boosted amplifier with load capacitances C_b of 5, 10 and 15 pF

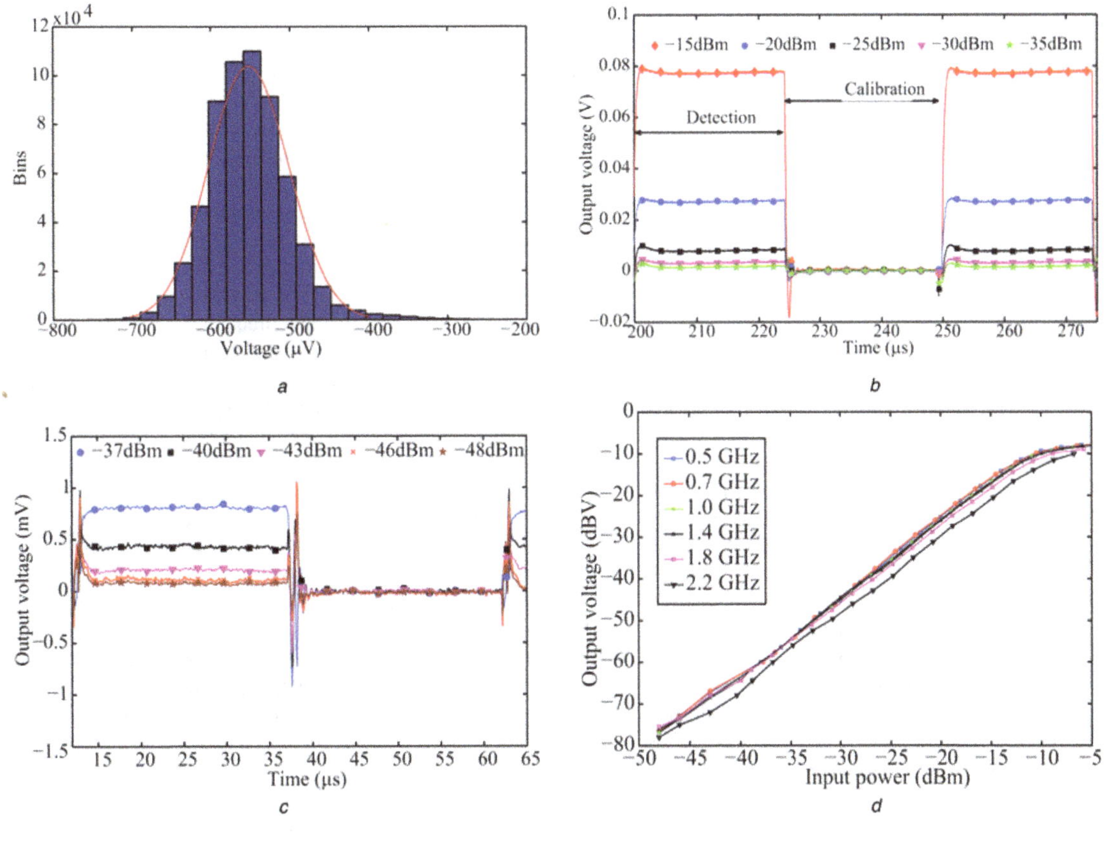

Fig. 11 *Measured*
a Histogram of the PD noise output distribution
b PD output voltage at input power levels of −20, −25, −30 and −35 dBm
c PD output voltages at input power levels of −37, −40, −43, −46 and −48 dBm
d Output voltage against input power at frequencies: 0.5, 0.7, 1.0, 1.4, 1.8 and 2.2 GHz

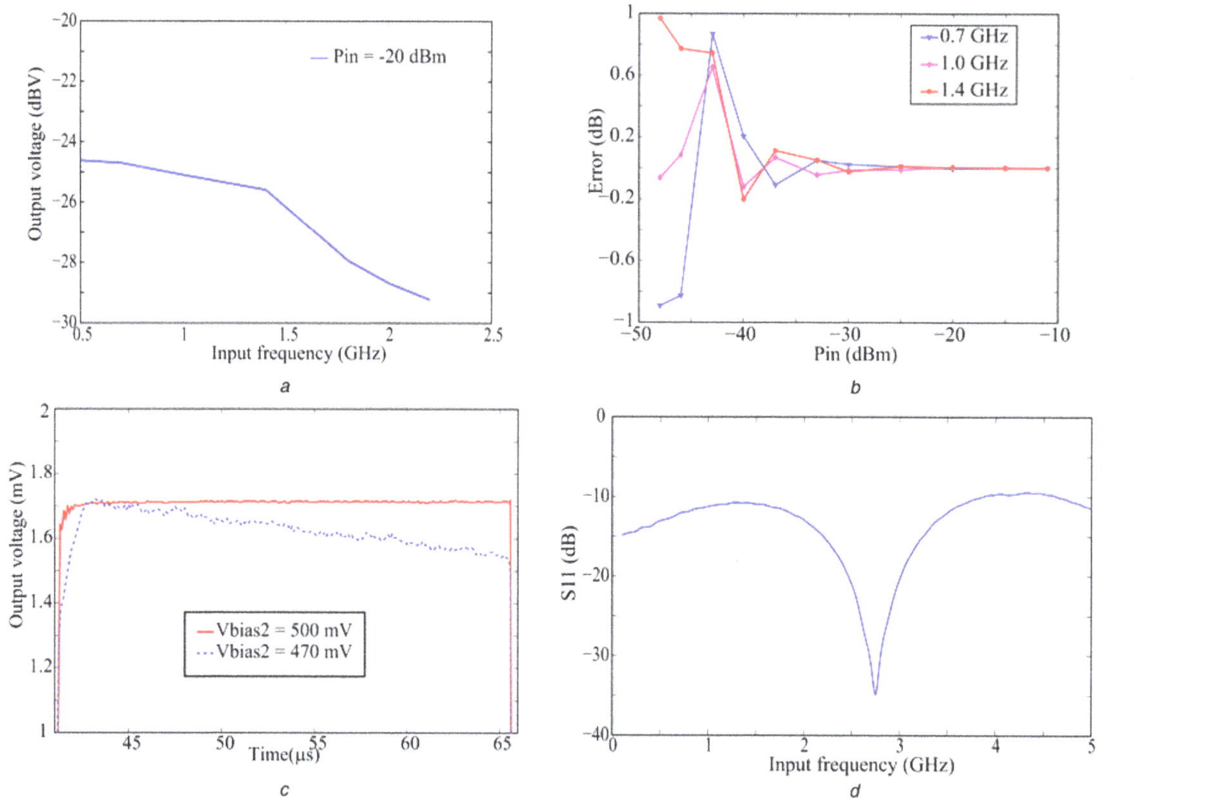

Fig. 12 *Measured*
a PD frequency response measured with −20 dBm signal
b Power detection accuracy at 0.7, 1.0 and 1.4 GHz
c Bulk-leakage effect on the PD output
d Input reflection coefficient

resistors R_{load}. The input match is achieved when

$$g_{m_{2a}} = \left[R_S(1 + g_{m_{1a}} R_{load}) - R_f\right]^{-1} \tag{21}$$

where R_s is the signal-source resistance and g_{m1a} and g_{m2a} are transconductances of $M_{1a,1b}$ and $M_{2a,2b}$, respectively. With a proper selection of g_{m2a} and feedback resistance R_f, 100Ω differential-input power matching is achieved.

5 Simulations and measurement results

The PD was fabricated in TSMC 65 nm CMOS with 1.2 V power supply. The chip micrograph is shown in Fig. 9. The fabricated circuit occupies 430 μm × 350 μm including the input match circuit. The total power consumption is 7 mW. About 1.2 mW is consumed by the PD and the folded-cascode amplifier circuit, and the other 5.8 mW is consumed by the input match circuit, which will not be needed once the PD is embedded in the SKA receiver chain.

Fig. 10 shows the simulated gain of the folded-cascode amplifier Amp1 for three different C_b's. This amplifier provides 75 dB of DC gain as shown in Fig. 10*a*. The minimum phase margin is 60° with C_b of 5 pF, and 75° with C_b of 15 pF, which is the value used in the final design for storage of the calibration voltage. The second pole location is independent of C_b and is at 98 MHz. The total power consumption of the amplifier circuit is 0.61 mW.

Fig. 11*a* shows the histogram of the PD output noise obtained from measurement of the differential output voltage of the PD when the input is turned off and calibration is disabled. The measurements show that the noise exhibits an approximate Gaussian distribution with standard deviation of 50.7 μV and a mean of 510 μV. Since the calibration was turned off for this measurement, the mean value, that is, the DC offset, is not zero. From these measurements, the estimated input-equivalent noise power is −48.3 dBm, which is close to a simulation result of −49.3 dBm. Note that the simulated input-match circuit's noise figure is ~4 dB and its gain is 6 dB, therefore because of large noise of the PD the input-match circuit has negligible effect on the noise performance of the complete circuit.

Figs. 11*b* and *c* show the measured PD differential output voltage during calibration and detection modes for different input power levels with the input signal frequency set to 1.4 GHz. The minimum detected power was approximately −48 dBm. The PD output settles in <5 μs after the detection mode is turned on. Similarly, calibration completes in <5 μs. Fig. 11*d* shows the measured output voltage against the input power at different frequencies. The PD output voltage (dBV) shows a linear relationship with input power from −48 to −11 dBm at all frequencies. Note that the maximum level is limited by the input-referred $P_{1\,dB} = -11$ dBm of the input-match circuit. The 3 dB bandwidth of the PD is ~1.8 GHz as seen from the PD frequency response in Fig. 12*a*.

Fig. 12*b* shows the PD detection accuracy with different input power levels at 0.7, 1.0 and 1.4 GHz, which covers the SKA midband frequency range. The maximum detection error is 0.95 dB at input power of −48 dBm and this error decreases when input power increases. The effect of bulk-leakage currents on the PD output was also measured and shown in Fig. 12*c*. It is observed that when the source-terminal voltage of the PD input transistors, V_{bias2}, is set properly, the discharge of C_b is reduced and the PD output voltage is kept constant with time. Fig. 12*d* shows the PD input reflection coefficient, which is <−9.8 dB from 100 MHz to 5 GHz. The summary of the performance of this proposed PD circuit and other recently published PD circuits is shown in Table 1.

Table 1 Performance summary and comparison

Parameters	This work	[18]	[13]	[20]	[17]	[23]
process	65 nm CMOS	65 nm CMOS	0.13 μm CMOS	0.13 μm CMOS	0.18 μm CMOS	40 nm CMOS
sensitivity, dBm	−48	−25	−35	−33	−39	−10.5
detection range, dB	37	25	20	43	20	32.5
operating frequency, GHz	0.5–1.8	N/A	0.125–1.4	16	3.1–10.6	5
linearity error for specified input range	± 0.95 dB for 37 dB ± 0.5 dB for 29 dB	N/A	±0.5 dB for 18 dB	±1 dB for 43 dB	±2.9 dB for 20 dB	±0.6 dB for 32.5 dB
power consumption, mW	1.2	N/A	0.18	35.2	10.8	0.349
circuit area, mm^2	0.036	N/A	0.013	0.75	0.36	0.009
calibrated	yes	no	no	no	yes	no
measured	yes	no	yes	yes	yes	yes
inputs/outputs	D/D[a]	D/S[a]	S/D[a]	S/S[a]	D/D[a]	D/S[a]

[a]D: differential and S: single-ended.

6 Conclusions

This paper presented a high-sensitivity differential CMOS analogue self-calibrated PD. The proposed circuit is targeted for use in the SKA receiver. The circuit topology and detailed theoretical analysis of circuit operation are described in this paper. This PD operates over an input power range from −48 to −11 dBm with output voltage offset <0.95 dB and input-referred $P_{1\ \mathrm{dB}}$ of −11 dBm, determined by the input match circuit. The 3 dB bandwidth of the PD is about 1.8 GHz, which covers the SKA midband frequency range from 0.7 to 1.4 GHz. The power consumption is only 1.2 mW for the PD itself.

7 Acknowledgments

This work was supported by the University of Calgary, the Natural Sciences and Engineering Research Council of Canada, the NRC Herzberg Astronomy and Astrophysics, CMC Microsystems, the Alberta Ingenuity Fund and the NSERC's Strategic Research Opportunity program.

8 References

[1] Zhang S., Madic J., Bretchko P., Mokoro J., Shumovich R., McMorrow R.: 'A novel power-amplifier module for quad-band wireless handset applications', *IEEE Trans. Microw. Theory Tech.*, 2003, **51**, (11), pp. 2203–2210

[2] Wu C.-P., Tsao H.-W.: 'A 110-MHz 84-dB CMOS programmable gain amplifier with integrated RSSI function', *IEEE J. Solid-State Circuits*, 2005, **40**, (6), pp. 1249–1258

[3] Wu G., Belostotski L., Haslett J.: 'A broadband automatic gain control amplifier for the square kilometer array'. Eighth IEEE Int. NEWCAS Conf. (NEWCAS), Montreal, Canada, June 2010, pp. 153–156

[4] Dewdney P., Hall P., Schilizzi R., Lazio T.: 'The square kilometre array'. Proc. of the IEEE, August 2009, vol. 97, pp. 1482–1496

[5] Imbriale W.: 'The square kilometre array: engineering opportunities'. IEEE Int. Conf. on Wireless Information Technology and Systems (ICWITS), 2010, p. 1

[6] Navaratne D., Belostotski L.: 'Wideband CMOS amplification stage for a direct-sampling square kilometre array receiver', *IEEE Trans. Microw. Theory Tech.*, 2012, **60**, (10), pp. 3179–3188

[7] Wu G., Belostotski L., Haslett J.: 'A broadband variable gain amplifier for the square kilometer array'. 2013 IEEE Int. Symp. on Circuits and Systems (ISCAS), Beijing, China, June 2013, pp. 2267–2270

[8] Jones D.: 'Technology challenges for the square kilometer array', *IEEE Aerosp. Electron. Syst. Mag.*, 2013, **28**, (2), pp. 18–23

[9] Xu Y., Belostotski L., Haslett J.W.: 'A 65-nm CMOS 10-GS/s 4 bit background-calibrated non-interleaved flash ADC for radio astronomy', *IEEE Trans. Very Large Scale Integr. (VLSI) Syst.*, 2014, doi:10.1109/TVLSI.2013.2291563

[10] Zhang T., Eisenstadt W., Fox R., Yin Q.: 'Bipolar microwave rms power detectors', *IEEE J. Solid-State Circuits*, 2006, **41**, (9), pp. 2188–2192

[11] Milanovic V., Gaitan M., Bowen E., Tea N.H., Zaghloul M.: 'Thermoelectric power sensor for microwave applications by commercial CMOS fabrication', *IEEE Electron Device Lett.*, 1997, **18**, (9), pp. 450–452

[12] Yin Q., Eisenstadt W., Fox R., Zhang T.: 'A translinear RMS detector for embedded test of RF ICs', *IEEE Trans. Instrum. Meas.*, 2005, **54**, (5), pp. 1708–1714

[13] Zhou Y., Chia M.Y.W.: 'A low-power ultra-wideband CMOS true RMS power detector', *IEEE Trans. Microw. Theory Tech.*, 2008, **56**, (5), pp. 1052–1058

[14] Meyer R.: 'Low-power monolithic RF peak detector analysis', *IEEE J. Solid-State Circuits*, 1995, **30**, (1), pp. 65–67

[15] Milanovic V., Gaitan M., Marshall J., Zaghloul M.: 'CMOS foundry implementation of schottky diodes for RF detection', *IEEE Trans. Electron Devices*, 1996, **43**, (12), pp. 2210–2214

[16] Ferrari G., Fumagalli L., Sampietro M., Prati E., Fanciulli M.: 'CMOS fully compatible microwave detector based on MOSFET operating in resistive regime', *IEEE Microw. Wirel. Compon. Lett.*, 2005, **15**, (7), pp. 445–447

[17] Townsend K., Haslett J.: 'A wideband power detection system optimized for the UWB spectrum', *IEEE J. Solid-State Circuits*, 2009, **44**, (2), pp. 371–381

[18] Gorisse J., Cathelin A., Kaiser A., Kerherve E.: 'A 60 GHz 65 nm CMOS RMS power detector for antenna impedance mismatch detection'. Proc. of ESSCIRC, Athens, Greece, September 2009, pp. 172–175

[19] Li C., Gong F., Wang P.: 'A low-power ultrawideband CMOS power detector with an embedded amplifier', *IEEE Trans. Instrum. Meas.*, 2010, **59**, (12), pp. 3270–3278

[20] Kim K., Kwon Y.: 'A broadband logarithmic power detector in 0.13 μm CMOS', *IEEE Microw. Wirel. Compon. Lett.*, 2013, **23**, (9), pp. 498–500

[21] Gao Y., Zheng Y., Diao S., *ET AL.*: 'Low-power ultrawideband wireless telemetry transceiver for medical sensor applications', *IEEE Trans. Biomed. Eng.*, 2011, **58**, pp. 768–772

[22] Borremans J., Wambacq P., Soens C., Rolain Y., Kuijk M.: 'Low-area active-feedback low-noise amplifier design in scaled digital CMOS', *IEEE J. Solid-State Circuits*, 2008, **43**, (11), pp. 2422–2433

[23] Francois B., Raynaert P.: 'A transformer-coupled true-rms power detector in 40 nm CMOS'. ISSCC Digest Technical Papers, February 2014, pp. 62–63

13

Implementing voice over Internet protocol in mobile *ad hoc* network – analysing its features regarding efficiency, reliability and security

Naveed Ahmed Sheikh[1], Ashfaq Ahmad Malik[1], Athar Mahboob[2], Khairun Nisa[3]

[1]*PN Engineering College, National University of Science & Technology, Karachi, Pakistan*
[2]*Department of Electrical Engineering, DHA Suffa University, Karachi, Pakistan*
[3]*Department of Computer Science and Engineering, University of Engineering & Technology, Lahore, Pakistan*
E-mail: aamalik2009@yahoo.com; ashfaqahmadmalik@gmail.com

Abstract: Providing secure and efficient real-time voice communication in mobile *ad hoc* network (MANET) environment is a challenging problem. Voice over Internet protocol (VoIP) has originally been developed over the past two decades for infrastructure-based networks. There are strict timing constraints for acceptable quality VoIP services, in addition to registration and discovery issues in VoIP end-points. In MANETs, *ad hoc* nature of networks and multi-hop wireless environment with significant packet loss and delays present formidable challenges to the implementation. Providing a secure real-time VoIP service on MANET is the main design objective of this paper. The authors have successfully developed a prototype system that establishes reliable and efficient VoIP communication and provides an extremely flexible method for voice communication in MANETs. The authors' cooperative mesh-based MANET implementation can be used for rapidly deployable VoIP communication with survivable and efficient dynamic networking using open source software.

1 Introduction

Our objective in this paper was to develop a prototype system to establish reliable and efficient voice over Internet protocol (VoIP) communication and provide an extremely flexible method for VoIP in mobile *ad hoc* networks (MANETs) using Smartphone and tablets running the Android operating system. We successfully established cooperative mesh-based MANET with optimised link state routing (OLSR) as the routing protocol for rapidly deployable VoIP communication. The system is survivable and efficient supporting dynamic networking using open source software (OSS) modules such as modified wireless fidelity (WiFi) Tethering [1] and PTTDroid [2] VoIP application.

After this short introduction Section 2 discusses the background information. Section 3 is about solution implementation and describes the configuration of the application on Android and personal computer (PC) platforms. Section 4 covers security aspects considered for the proposed application. Section 5 discusses performance-based results gathered on pre-set criteria. Finally, we conclude and discuss future work in Section 6.

2 Background information

2.1 MANETs

MANETs are peer-to-peer self-organising networks and do not require a fixed infrastructure. The nodes of a MANET have to self-configure in some arbitrary manner in order to perform their function in the absence of a fixed infrastructure. The areas with depleted fixed infrastructures can be a place where a natural disaster has occurred such as an earthquake, flood or where a chaotic situation has been created by humans such as in a battlefield. The MANET nodes can directly communicate with each other because either they are in the required radio range of each other or otherwise multi-hop routing can be used to establish connectivity. Well-known multi-hop routing protocols include OLSR, a proactive routing protocol and *ad hoc* on-demand distance vector, a reactive routing protocol. Since the MANET nodes may be in continuous motion and may move into and out of the radio range of each other, frequent breakage of data links can occur, resulting in high vulnerability of wireless link between nodes. The topology of a MANET is highly dynamic, therefore routing information also changes. There is also a bandwidth constraint in such a wireless network because it lacks access points and high-speed wired links. Furthermore, since the current routing protocols do not focus much on the security aspects, MANETs are also more vulnerable to security threats as compared with traditional wired networks [3].

2.2 Survey of MANETs implementations on Android

Ad hoc networking support on Linux-based laptop and desktop PCs has matured in recent years [4]. However, it is still an open research problem for Android devices reason being that Android Open Source Project has not formally supported *ad hoc* network implementation in Android Kernel [5]. Primarily, Android OS is designed to work in infrastructure mode or managed network scenario. In a managed network, the client node is connected to an access point (AP) after authentication. It utilises the wpa_supplicant command line utility for configuration as is done in standard Linux as well. *Ad hoc* networks on the other hand do not use AP devices for centralised management and each device has the capability to do the routing for others. The versions of Android older than the Ice Cream Sandwich (ICS) only supported managed networking. The direct-WiFi utility in ICS also does not cater to full *ad hoc* support. Hence, one has to resort configuring wireless chip through use of iwconfig command supplying the parameters of wireless interface. The Linux Kernel must also support wireless extension (WEXT) application programming interface (API) [6]. Next, we review some of existing approaches to enabling *ad hoc* networking in Android.

The Smartphone *ad hoc* network (SPAN) project [7] has conducted significant work in this area and has developed few applications such as MANET Manager, MANET Voice Chat and MANET Visualiser. Stoker's SPAN project used Mueller's WiFi Wireless Tethering [8] initially. Its functions include tethering and executing the OLSR Daemon. Its graphical user interface (GUI) is very user friendly. However, it only supports WEXT Kernel resulting in applicability to a limited number of Android devices. Samsung Galaxy Tab 10.1, S II Epic Touch 4G SPH-D710, Nexus SCH-I515, S III GT-I9300 I9300UBDLJ1, S-IIIGT-I9300, I9300XXBLH1 and ASUS Transformer Prime TF201, Nexus-7 are the devices supported by SPAN's MANET Manager [6].

We have also successfully tested SPAN's MANET Manager application with limited functionalities on XPOD Bubblegum Tablet (installed with ICS 4.0.4) and XPOD Tabloid running Jelly Beans (JB 4.1.1). Details are provided in the implementation section. However, it did not run successfully on all Smartphone devices such as Q-Mobile A8 or A950 installed with JB 4.1.1.

Commotion [9] is another project demonstrating work on mesh networking for Android-based Smartphone, tablets, laptops, PCs, cellular networks and routers. The characteristic features of their implementation include tethering and executing OLSR protocol. It has a user friendly GUI. However, several shortcomings of their implementation hinder it from wide adoption. These include failure of proper execution of tethering application. Consequently, OLSR protocol also fails to execute.

There are a few other projects also that have worked on MANETs or mesh networks on Android platform such as Thinktube [4], Barnacle [10], Open Garden [11], WiFi Tethering by Opengarden [12], WiFi *ad hoc* Enabler [13], Serval [14] etc. In our paper, we have mainly improved the WiFi Wireless Tether [8] application to work reliably in *ad hoc* mode along with OLSR. Details are provided in the implementation section.

2.3 VoIP in MANET

The VoIP technology has been a hot issue in the Information Technology industry for more than ten years now. Providing a secure real-time VoIP service on the MANET is the main design objective of this research. It is perceived to be a difficult task because of restrictions in device resources, adverse properties of the wireless channel, dynamic topology and the lack of central administration. In addition, the flexibility of the VoIP system and the convergence of voice and data networks bring in additional security risks. All devices in an Internet protocol (IP) network become potential active or passive adversaries. It can be seen that the underlying IP data network facility that a VoIP system relies on complicates the security assurance requirement.

3 Solution implementation

3.1 Implementing MANET Manager

As already discussed in Section 1, Stoker's MANET Manager has contributed significantly to implementation of *ad hoc* network on Android devices. We were able to localise two low-end Android tablets available as commercial off the shelf (COTS) which support WEXT and successfully installed MANET Manager application available on Play Store [5]. This was managed by lot of testing and trials. Snapshot of MANET Manager running in *ad hoc* mode is shown in Fig. 1.

XPOD Bubblegum and Tabloid Tablets were assigned IP addresses as 10.10.10.20/24 and 10.10.10.60/24. However, the tablets did not connect to each other in *ad hoc* mode. A special arrangement was made as shown in Fig. 2.

Fig. 1 *Running MANET Manager snapshot*

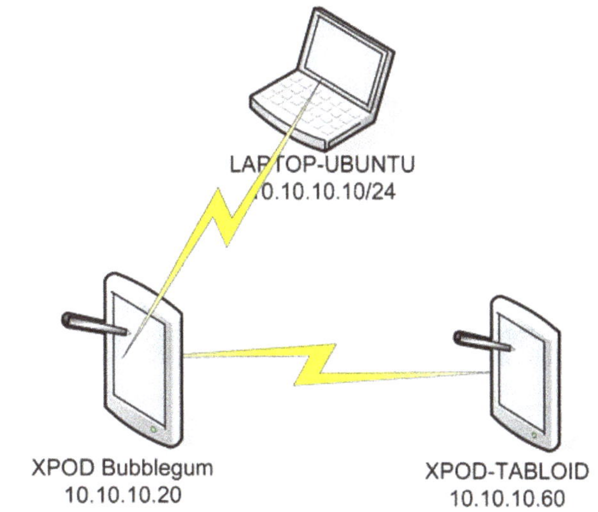

Fig. 2 *Ad hoc network configuration*

Following configuration was set on laptop running with Ubuntu-12.04:

- sudo ifconfig wlan0 10.10.10.10 netmask 255.255.255.0
- sudo iwconfig wlan0 mode *ad hoc*
- sudo iwconfig wlan0 essid 'test'
- sudo iwconfig wlan0 channel 4
- sudo iwconfig wlan0 key off

Owing to some built-in shortcomings of the Android JB 4.1.1 for XPOD Tabloid, the *ad hoc* node with essid 'test' could not be detected on the Tabloid device. However, it was displayed as (*) test on XPOD Bubblegum Tablet which was connected in *ad hoc* mode with Ubuntu laptop. Following routing information was also updated on XPOD tablets as shown in Fig. 3.

The Stoker's MANET Visualiser application created following link diagram as shown in Fig. 4.

Using MANET Manager's built-in messenger, hello chat messages were also exchanged on both tablets as shown in Fig. 5.

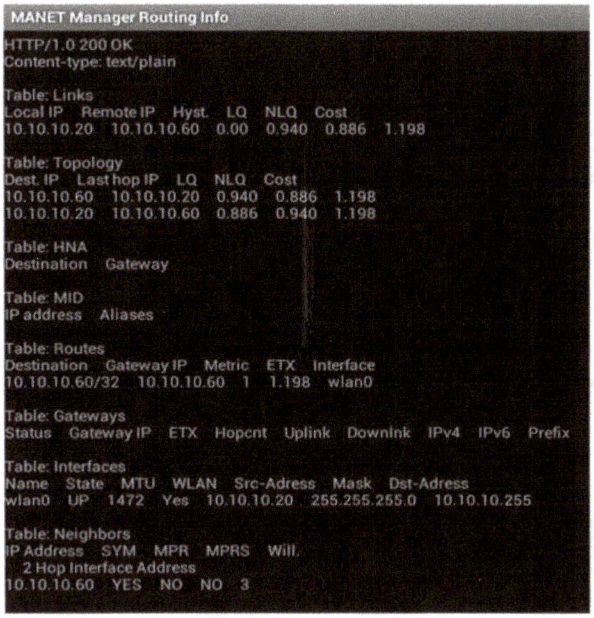

MANET Manager Routing Info

HTTP/1.0 200 OK
Content-type: text/plain

Table: Links
Local IP Remote IP Hyst. LQ NLQ Cost
10.10.10.20 10.10.10.60 0.00 0.940 0.886 1.198

Table: Topology
Dest. IP Last hop IP LQ NLQ Cost
10.10.10.60 10.10.10.20 0.940 0.886 1.198
10.10.10.20 10.10.10.60 0.886 0.940 1.198

Table: HNA
Destination Gateway

Table: MID
IP address Aliases

Table: Routes
Destination Gateway IP Metric ETX Interface
10.10.10.60/32 10.10.10.60 1 1.198 wlan0

Table: Gateways
Status Gateway IP ETX Hopcnt Uplink Downlnk IPv4 IPv6 Prefix

Table: Interfaces
Name State MTU WLAN Src-Adress Mask Dst-Adress
wlan0 UP 1472 Yes 10.10.10.20 255.255.255.0 10.10.10.255

Table: Neighbors
IP Address SYM MPR MPRS Will.
2 Hop Interface Address
10.10.10.60 YES NO NO 3

Fig. 3 *Routing information on MANET Manager*

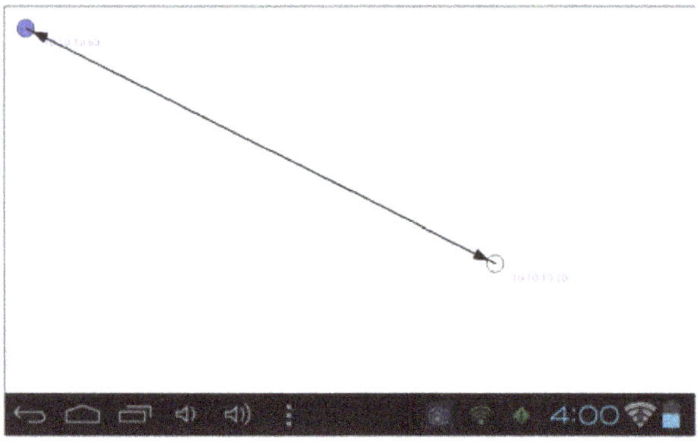

Fig. 4 *Link diagram by MANET visualiser*

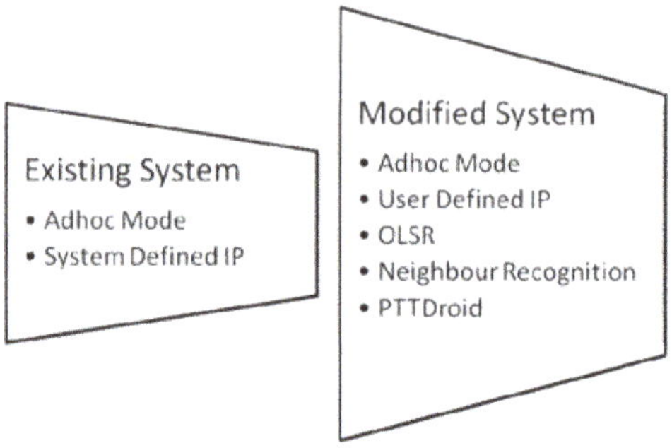

Fig. 6 *Modified WiFi ad hoc system features*

The VoIP communication between these tablet devices configured in *ad hoc* mode could not be satisfactorily established despite several attempts using MANET Voice and PTTDroid VoIP applications. This may be attributed to non-fulfilment of hardware support for MANET Manager installed. This forced us to explore other methods to establish *ad hoc* networks to run VoIP successfully.

3.2 Implementation using WiFi Wireless Tether app

(1) *Preparation:* The WiFi Wireless Tether application [8] mentioned earlier was tried. We describe the experience with WiFi Tether app below. The existing WiFi Tethering application was tried as a basis of our mesh network for VoIP communication system. Original WiFi Tether application does not have the functionality of defining an IP address while executing the application. An *ad hoc* node is assigned a fixed IP address. This restricted the system in becoming a pure MANET. It also lacked the feature to track topology information and neighbour recognition. The enhanced system thus developed and implemented covers all these aspects along with VoIP functionality. The overall improvements are shown in Fig. 6. The implementation has been subdivided into multiple sections which are discussed in the subsequent sub-sections.

(2) *User defined IP:* The existing system is modified to accept a user defined IP. On the user interface, a section is provided for this purpose as shown in Fig. 7 by 'Change IP' option. A text box is added that takes an IP address given by the user as input. This feature makes the network dynamic by allowing any two nodes to connect with each other. As a result, user does not have to depend on dynamic host control protocol (DHCP) service to assign an IP address to the nodes connecting in the network.

(3) *OLSR:* When the start button is pressed in tethering application, a file olsr.conf is fetched from memory containing OLSR configuration. This configuration drives the information sharing process of topology. OLSR is a proactive routing protocol and is required for multi-point relaying (MPR). MPR helps in increasing the diameter of MANET network by allowing two or more nodes to connect through single or multiple bridges.

(4) *Neighbour recognition:* It is essential to know about neighbours in order to communicate with them. This feature helps in recognising the name and IP address of neighbours. The function picks information generated by OLSR for display purpose.

(5) *Configuration of system:* The configuration steps are different depending on whether the device uses a full-blown Linux-based system or an embedded Linux system. Each step of the configuration needs to be executed in a methodical way to achieve a reliable setup of the *ad hoc* network. Configurations are done to build a connection between the devices to form a MANET, and hence be able to perform VoIP communication. If both devices are of the same type, they follow the same setup procedure.

(6) *Setup on PC for real-time monitoring:* Fig. 8 shows different steps involved in installation of OLSR, its configuration, inserting required plugins and mapping software for *ad hoc* nodes.

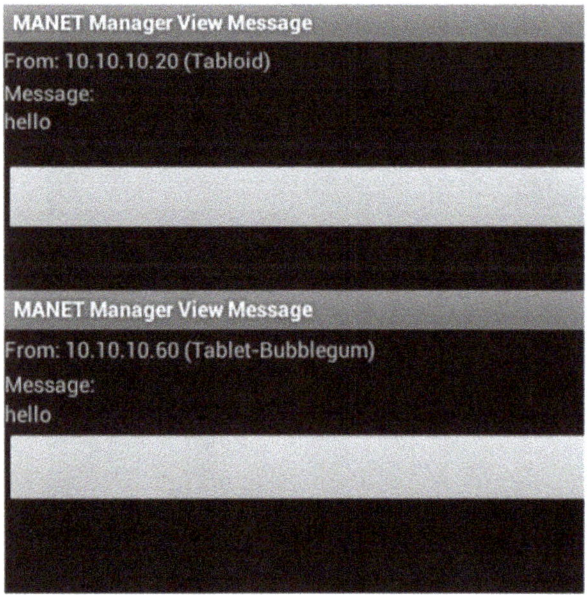

Fig. 5 *Exchange of hello messages*

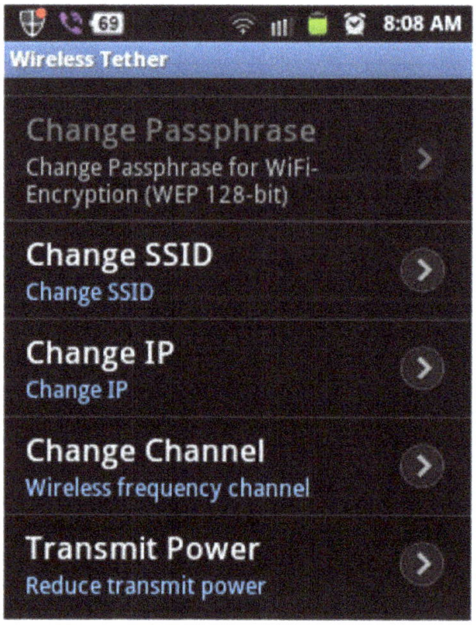

Fig. 7 *Static IP option*

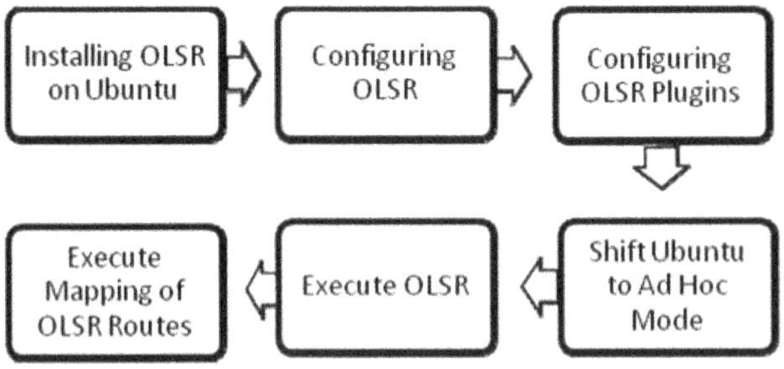

Fig. 8 *PC setup*

To install OLSR [15], a set of pre-requisite software tools are required. These comprise of flex, bison, libc6 and git-core. These can be installed using standard software installation procedure.

Once OLSR is installed with its pre-requisites, it is configured. The OLSR configuration file is set to optimise its response time for detecting nodes entering and leaving the system. There are five plugins of OLSR that are inserted, namely: TXT INFO PLUGIN, HTTP INFO PLUGIN, DOT DRAW PLUGIN, NAME SERVICE PLUGIN and OLSR SECURE PLUGIN. These plugins are used to observe information provided by OLSR in a user friendly way.

Ubuntu 12.04 is joined to *ad hoc* network using the default network manager. It is a user interface that interacts with WiFi protected access (WPA) supplicant file for smooth interaction between hardware and software layers. An IP address is given to the device to be recognised on the network. Channel is selected to join a network operating on a specific frequency. Service set identifier is also set to give the wireless network a recognisable name. OLSR is then executed. Multiple packages are required to make the plugin work that draws the network topology. These packages contain commands that are given for visual graphics. The graphical view of the network topology is generated in real time. Fig. 9 shows an example of the topology view that is generated when different nodes are connected in the network.

7. *Setting up custom WiFi Tether and VoIP on Android:* The steps involved in setting up of Custom WiFi Tether and PTTDroid VoIP application are highlighted in Fig. 10.

There are multiple methods for rooting Android phones depending on device [16]. However, there is a universal method as well. We used the universal method to root the phone. Android terminal emulator and busybox are installed from Google Play Store. Android terminal emulator helps in giving a command line interface

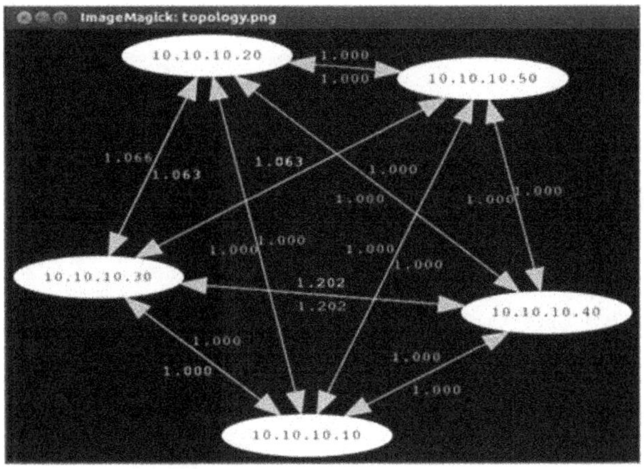

Fig. 9 *Topology overview*

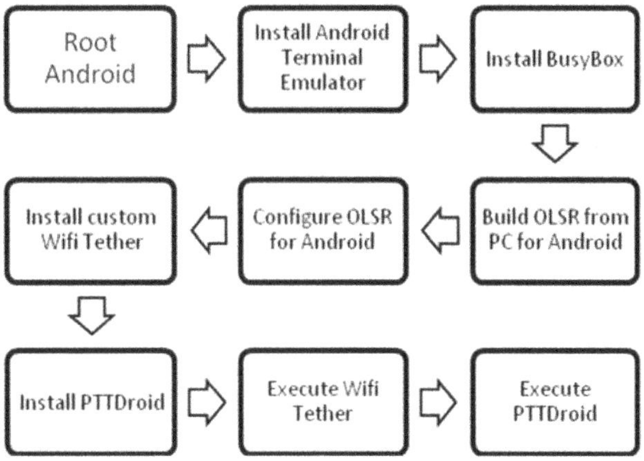

Fig. 10 *Android setup steps*

(CLI) on Android. It is the same thing as using a CLI on a Linux system. Busybox helps in the execution of the basic commands of Linux on Android devices.

OLSR is then built on PC for Android compatibility using a cross-compiler. A number of different utilities and tools are required to build the OLSRd for Android. These include bison, flex, libc6, Android-software development kit (SDK), Android-native development kit (NDK), git-core etc.

Both the Android SDK and NDK are also required. These are extracted and installed in /opt folder to avoid build failures. NDK base file location is updated in the Makefile to reflect the actual location of extracted development toolkits. The following commands are executed to generate Android compatible OLSRd:

```
#make OS = Android DEBUG = 0
NDK_BASE = /opt/Android-ndk
#sudo make OS = Android DEBUG = 0 install_all
```

The WiFi Tether application is configured to copy the latest olsrd.conf to the system/etc folder every time it starts, making it easy to change configuration at any stage, even if root exploration is not possible.

The files generated while creating Android compatible OLSRd are copied on the Android phone. Interfaces can be verified through Android Terminal Emulator by executing netcfg command. This command lists the interface that is responsible for WiFi connectivity. They can be modified in the OLSRd configuration file to suit the requirement of connecting to an operational wireless interface. The custom WiFi Tether application and PTTDroid application are then installed.

After installation, the icon for the WiFi Tether is made available on the Application Drawer Panel and is executed by first tapping over the app icon. After successful implementation of *ad hoc* network and launch the OLSRd, we again move to the application drawer and look for the PTTDroid application icon and execute by tapping over it. To establish a voice connection, the IP address is supplied in the configuration option of the application and communication is commenced by tapping over the microphone icon and holding it for the time the user needs the voice to be sent over to the target. As soon as the user lifts his/her finger from the microphone icon on the screen, the application switches out of the sending voice data mode, although the device remains to be in the listening mode that is, receive data at all times.

4 Security aspects

Multiple options were explored to secure the VoIP system. The objective is to avoid unauthorised access to the MANET. The CSipSimple VoIP application [17] could not be used. It has a feature in which it calls on the Android network manager which is consulted to check the wireless interface state. As the system is functional in *ad hoc* mode and unfortunately because of Android's lack of native *ad hoc* mode supports the network manager indicates to the software that WiFi state is off, the software fails to execute indicating the WiFi state. For security of VoIP during transmission, we encrypted the data using a custom encryption software. Owing to heavy processing load and limited processing power that stream ciphering operation required, PTTDroid Application halted leaving it unresponsive and unstable, and hence became unusable for the system. As another approach to improve encryption performance, IPSec tunnelling was implemented and packets were analysed with the help of the Wirehshark in promiscuous mode. It was discovered that the channel was indeed being established with the help of Internet Security Association and Key Management Protocol negotiation. However, the tunnels were not activated and data were not transferred as the nodes remained stuck in negotiating the parameters for IPSec connectivity. Although the methods listed above were all valid approaches towards securing the VoIP system, every method had some deficiency.

We applied the mechanism for securing the OLSR protocol as implemented in [18] as an extension to the OLSR source code provided from UniK [15]. The mechanism is based on signing each OLSR control packet with a digital signature for authenticating this message. In addition, the mechanism provides a timestamp exchange process. The timestamps are used to prevent replay attacks on the OLSR routing protocol. This security mechanism does not need synchronised time. It only secures the routing messages and not the user traffic being routed in the MANET.

A pre-shared key file is placed at a known location in all the nodes. If it matches, that is, verified, the packets are accepted and OLSR sending node joins the network. If the signature cannot be verified, the message is discarded and the node sending the packet is prevented from joining the system. The same method is used as well for the intrusion detection. If someone is trying to join the network without a key, as it does not contain the digital signature and therefore it is rejected by the OLSR Daemon. The sender IP address is also displayed on such action in the running OLSR Daemon. This can be viewed in the terminal emulator of the Android GUI.

5 VoIP quality performance results

5.1 Parameters

In this section, we provide VoIP performance results using our implemented system. The results were collected for two different scenarios. Results are collected for each of the parameters mentioned below along with the tools used for performance measurement:

(1) *Ping:* It is used for measuring the round trip time (RTT) taken by the packets from source to destination and way back in the network. Total time taken is measured in milli-seconds (ms). The command used for measuring it is as follows:

```
# ping -c60 10.10.10.40
```

(2) *Throughput and bandwidth:* Throughput is the average rate of successful message delivery over a communication channel and is generally measured in bits per second (bps). This is measured with the help of 'iperf'; a simple application installed in Android which calculates the amount of data transferred and the optimum bandwidth achievable. The 'iperf' command used for it is as follows: < colcnt = 1 >

```
# iperf -s -i 1 (on server or side)
# iperf -c 10.10.10.40 -i 1 -t 60 (on client side)
```

Table 1 Jitter values range

Qualities	Ranges, ms	Remarks
good	0–20	acceptable
average	20–30	marginally acceptable
poor	above 30	not acceptable

Table 2 Packet loss percentage range

Qualities	Ranges, %	Remarks
good	0–1	acceptable
average	1–2.9	marginally acceptable
poor	above 3	not acceptable

(3) *Jitter:* Jitter is the variance in measuring successive ping tests. Zero jitter means the results were exactly the same every time and anything above zero is the amount by which they varied. Like the other quality measurements, a lower jitter value is better. While some jitter should be expected over the Internet, having it be a small fraction of the ping result is ideal. Table 1 shows the values for jitter [19].

Jitter was also measured using 'iperf' for Android. The following commands are used:

```
# iperf -s -u -i 1 (on server or side)
# iperf -c 10.10.10.40 -u -i 1 -t 60 (on client side)
```

(4) *Signal strength and signal-to-noise ratio (SNR):* This is measured with the help of a simple app named WiFi analyser. The values given are in dBm, that is, it is the measure of the absolute power received by the node at a given point. SNR is the ratio of the received RF signal, to that of background noise. SNR of greater than zero signifies that the received signal power is greater than the background noise.

(5) *Packet loss:* If you have anything less than complete success in transmitting and receiving 'packets' of data, then you are experiencing this problem with your Internet connection. It can mean much slower download and upload speeds, poor quality VoIP audio, pauses with streaming media etc. Packet loss is a metric where anything greater than 0% should cause concern. Packet loss range of values have been categorised in Table 2 [19].

(6) *Mean opinion score (MOS):* MOS is a test used for telephony networks to obtain user's views for quality of network. It is a subjective measurement. We have used following criteria of MOS as per Table 3 [20] for rating quality of VoIP on our MANET implementation and an average score is calculated through a survey of 20 people.

5.2 Scenario-1 (maximum distance 500 ft)

In this scenario, performance values are collected for the node-to-node connectivity using line of sight approach, that is, no barriers. An illustration of scenario is shown in Fig. 11. It gives

Table 3 Sound quality–MOS chart

MOS	Qualities	Impairments
5	excellent	undetectable
4	good	detectable but not annoying
3	fair	slightly annoying
2	poor	annoying
1	bad	very annoying

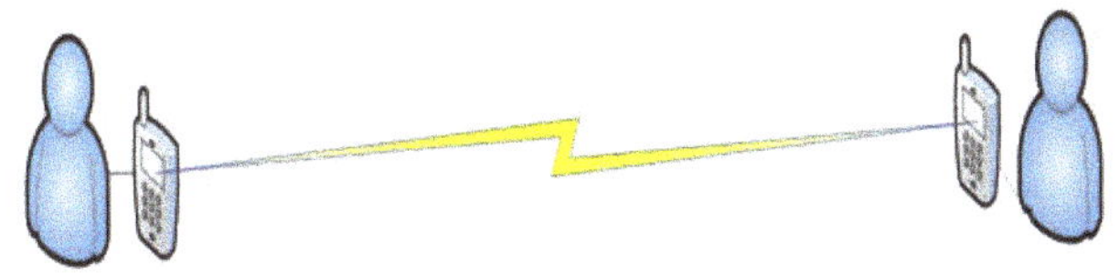

Fig. 11 *Node–node scenario*

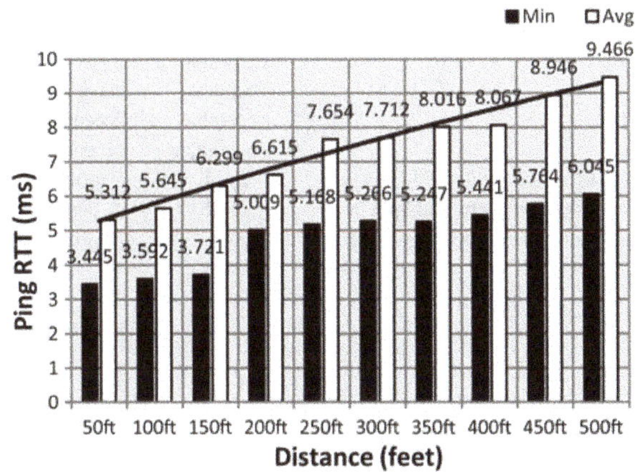

Fig. 12 *Ping RTT against distance*

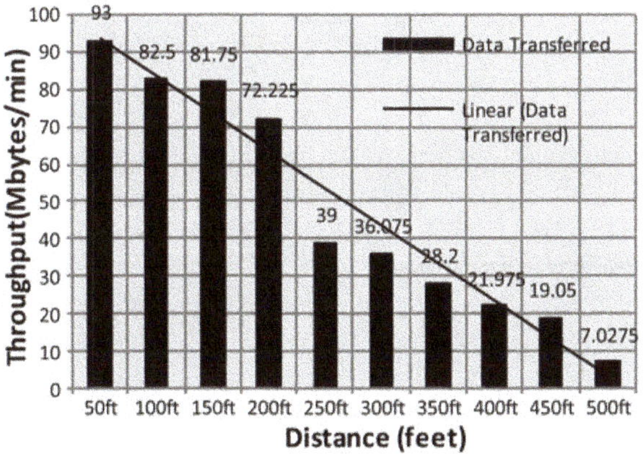

Fig. 13 *Throughput against distance*

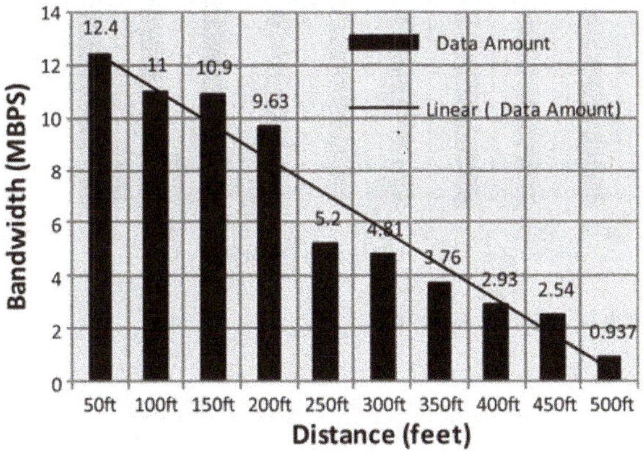

Fig. 14 *Bandwidth against distance*

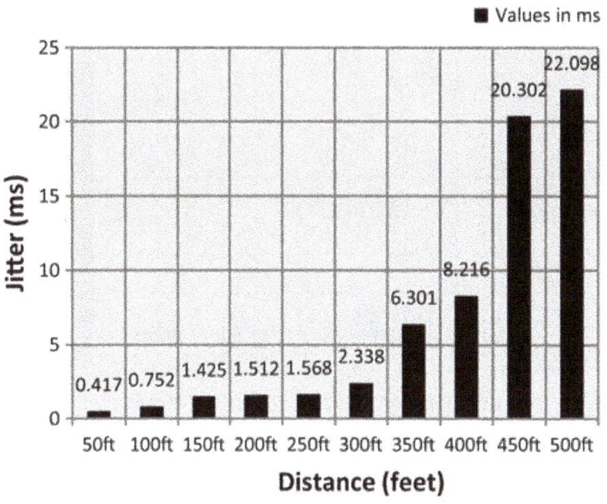

Fig. 15 *Jitter against distance*

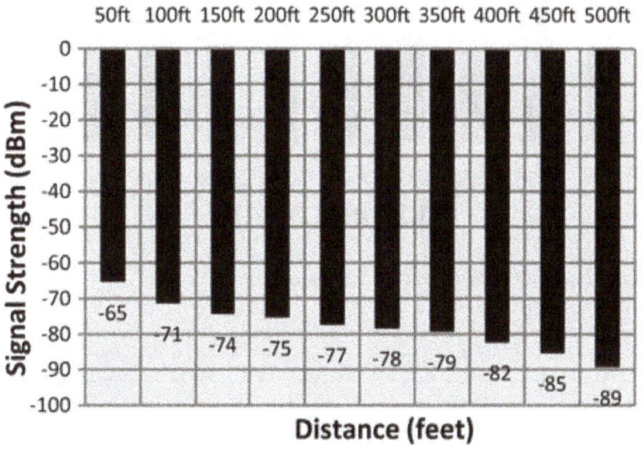

Fig. 16 *Signal strength against distance*

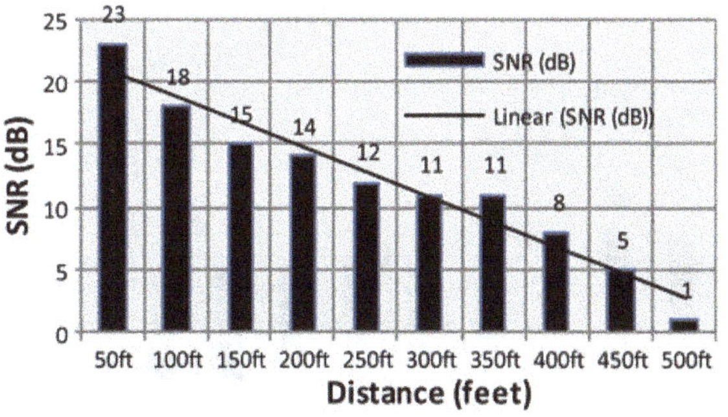

Fig. 17 *SNR against distance*

the idea of the way these tests are conducted in the real-world environment.

The following graphs in Figs. 12–21 illustrate the network performance giving the idea of how the *ad hoc* network performed for the above-mentioned parameters.

5.3 Scenerio-2 (maximum distance 800 ft)

In this case, values are collected for the node–bridge–node scenario using line of sight approach that is, no barriers between node–bridge but at the same time node–node are out of sight. In short, nodes are connected via a bridge. A case scenario illustration is

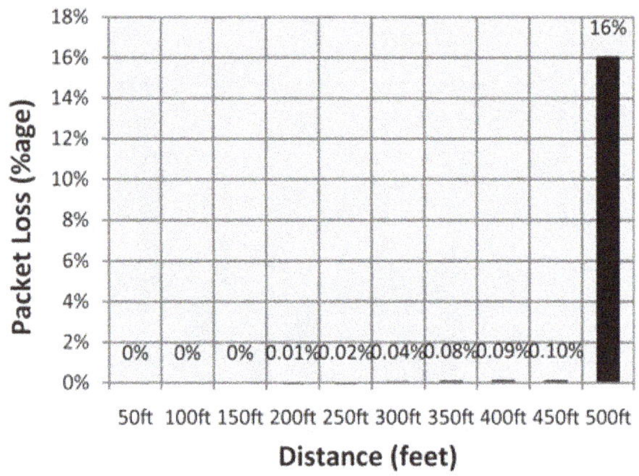

Fig. 18 *Packet loss (percentage against distance)*

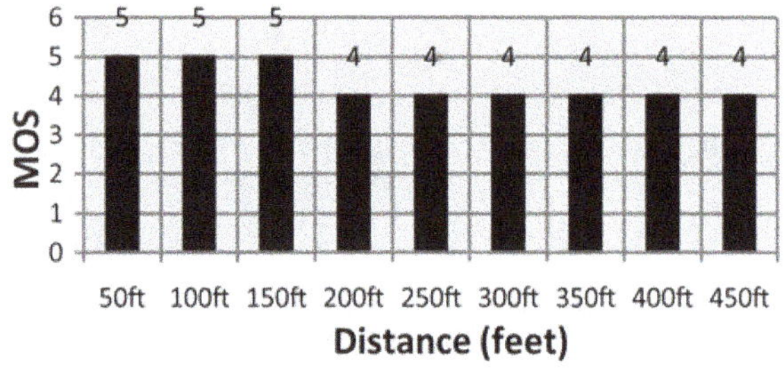

Fig. 19 *Sound quality with Codec at 3.95 kbps*

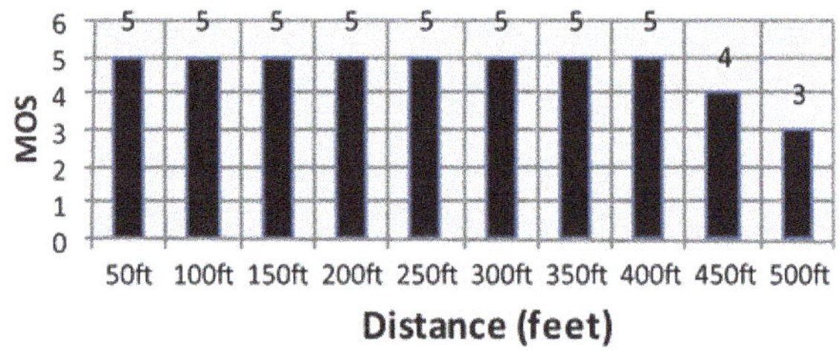

Fig. 20 *Sound quality with Codec at 24 kbps*

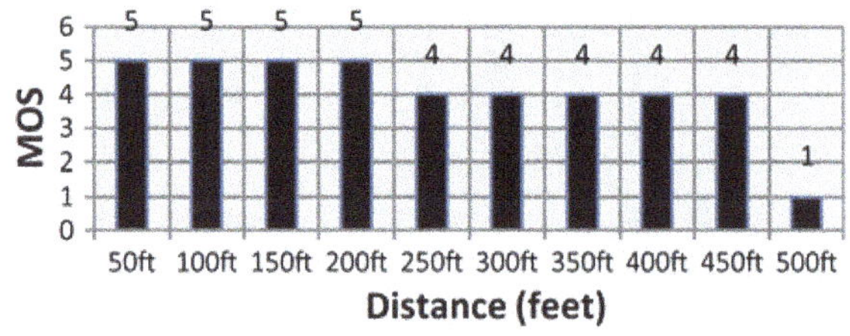

Fig. 21 *Sound quality without Codec*

shown in Fig. 22 giving the idea of the way these tests are conducted in real-world environment.

The results are collected and measured on the same pattern as the first test case scenario. In this case scenario, a Smartphone is used as bridge between two nodes that acts as relay for communication. The bridge is active but does not take part in the communication nor can it hear any communication between the two end nodes. The maximum distance between the node–bridge is 400 ft, whereas the distance between the two end nodes is collectively 800 ft. The entire tests have been taken starting from 200/200 ft setting and ended up at 400/400 ft. The results are plotted in graphs as shown in Figs. 23–30.

5.4 Summary of results

The overall results are summarised for both the scenarios as under:

Features	Scenario 1	Scenario 2
ping	acceptable	acceptable
jitter	acceptable	acceptable
signal strength	acceptable	NA
SNR	acceptable	NA
bandwidth	acceptable	acceptable
throughput	acceptable	acceptable
packet loss	acceptable	acceptable
MOS-quality of sound (with 3.95 kbps)	good	good
MOS-quality of sound (with 24 kbps)	good	good
MOS of quality of sound (without Codec)	good	average

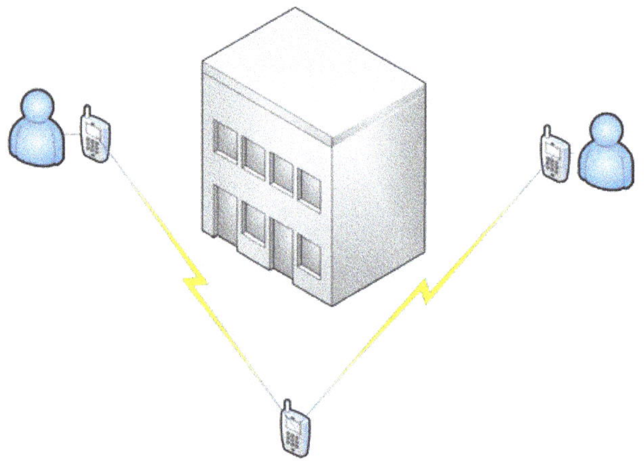

Fig. 22 *Node–bridge–node scenario*

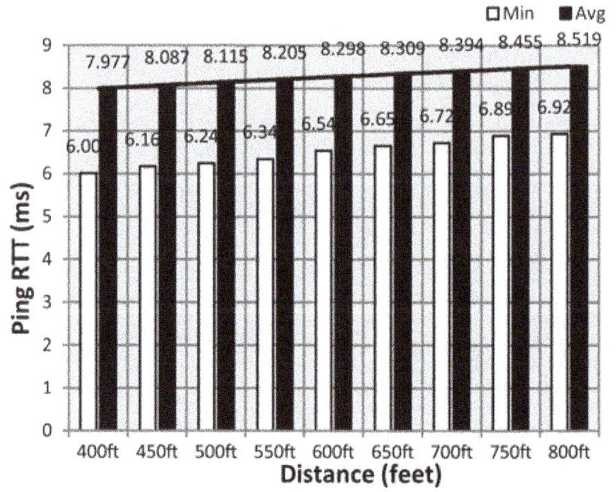

Fig. 23 *Ping RTT against distance*

Results provided in this section indicate that an acceptable quality for VoIP service can be established up to distances of 800 ft in the MANET, even when packets are relayed by an intermediate node and using the COTS Android Smartphone devices.

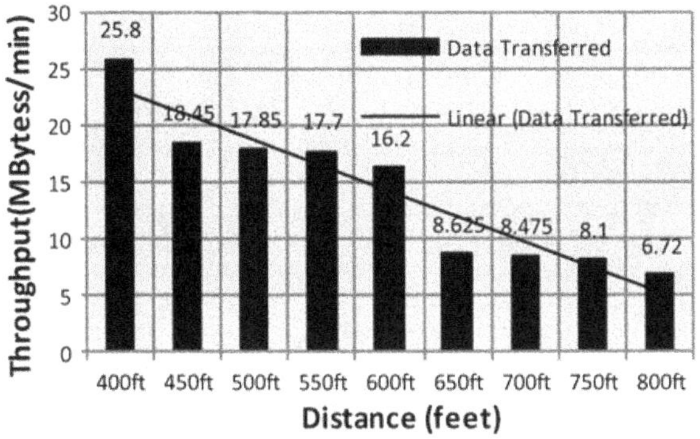

Fig. 24 *Throughput against distance*

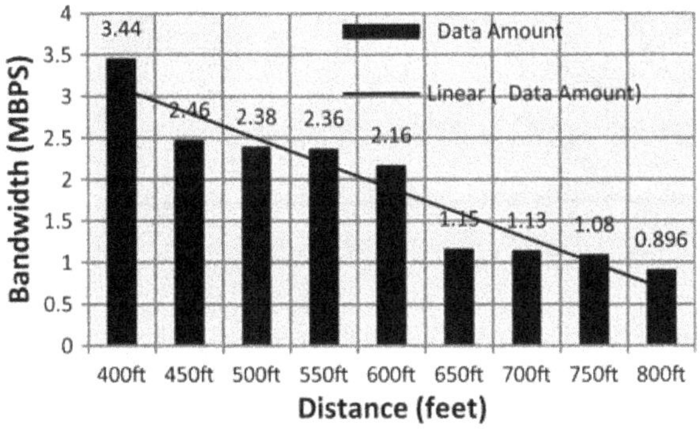

Fig. 25 *Bandwidth against distance*

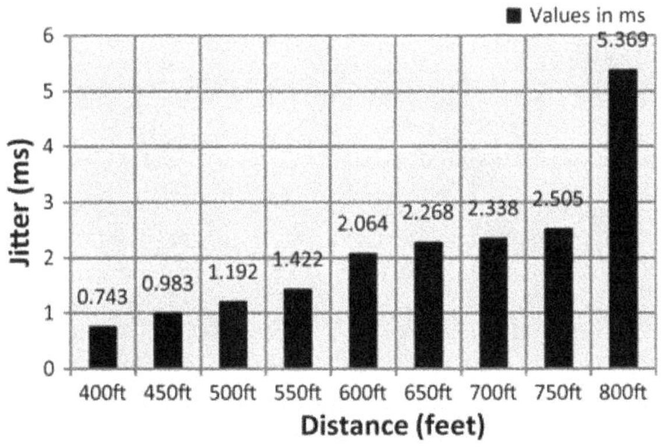

Fig. 26 *Jitter against distance*

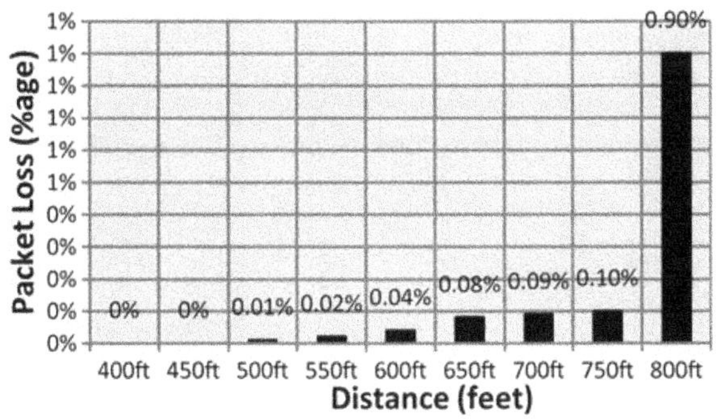

Fig. 27 *Packet loss against distance*

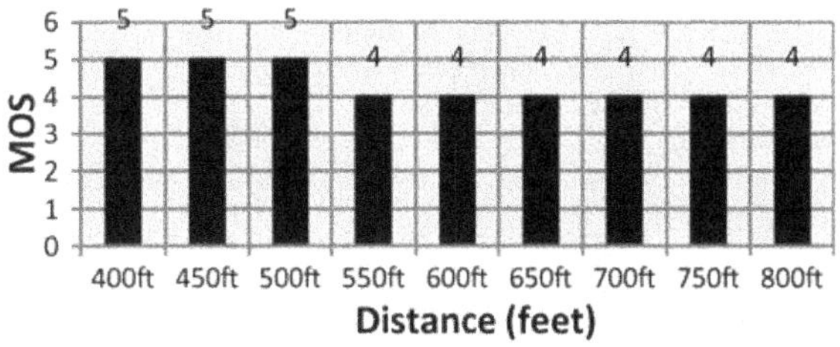

Fig. 28 *Sound quality with Codec at 3.95 kbps*

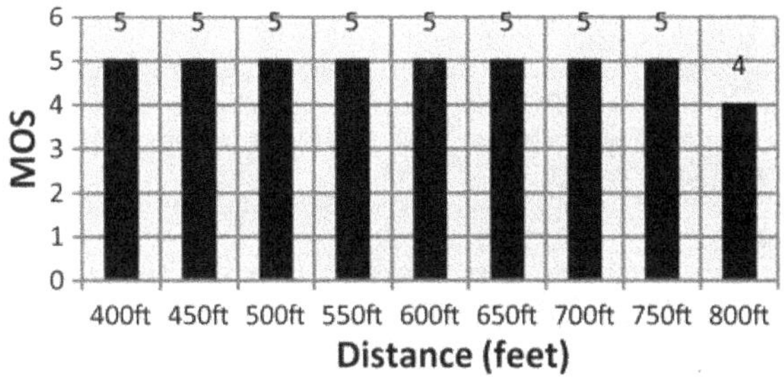

Fig. 29 *Sound quality with Codec at 24 kbps*

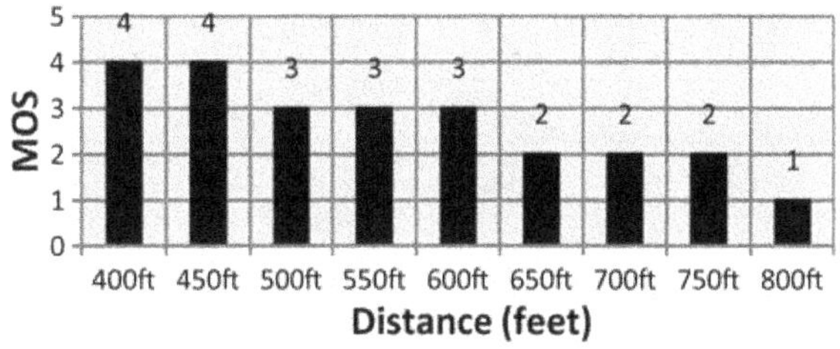

Fig. 30 *Sound quality without Codec used*

6 Conclusion and future works

6.1 Conclusion

We have successfully developed a prototype system which has established secure VoIP communication and also provides an extremely flexible method for VoIP in MANETs. Hence, we can conclude that we were able to establish a cooperative mesh-based mobile *ad hoc* network for rapidly deployable VoIP communication with survivable, efficient, dynamic and secure networking.

6.2 Future works

The customised WiFi Tether application can be worked out further to make it more flexible by doing research and implementing dynamic IP addressing, encryption of packets, providing full duplex mode of communication and by enhancing GUI for enhancing user experience.

We are exploring further ways to improve on the Commotion, MANET Manager and other similar projects to have reliable and generic solutions in MANETs or mesh networks. We are also endeavouring for IPSec implementation in our Android *ad hoc* network solution to have end-to-end security.

7 References

[1] Google web page. Available at http://www.code.google.com/p/Android-wifi-tether/downloads/list, visited on 28 September 2013

[2] Google web page. Available at http://www.code.google.com/p/pttdroid/, visited on 28 September 2013

[3] Malik A.A., Khan M.A., Mahboob A., Haider F.: 'Secure geo-sharing in MANETs'. Proc. IEEE Int. Conf. Collaboration Technologies & Systems CTS-2012, 2012, pp. 59–64

[4] Chang L.-H., Sung C.-H., Chiu S.-Y., Lin Y.-W.: 'Design and realization of ad-hoc VoIP with embedded p-SIP server', *J. Syst. Softw.*, 2010, **83**, (12), pp. 2536–2555

[5] Thinktube Inc: 'Ad-Hoc (IBSS) mode support for Android 4.2.2' Page. Available at http://www.thinktube.com/component/content/article/19-technicalinformation/46-Android-wifi-ibss, visited on 01 November 2013

[6] Thomas J., Robble J., Modly N.: 'Off grid communications with android meshing the mobile world'. 2012 IEEE Conf. Technologies for Homeland Security (HST), 13–15 November 2012, pp. 401–405, doi: 10.1109/THS.2012.6459882

[7] Stoker's MANET Manager. Available at https://www.play.google.com/store/apps/details?id=org.span&hl=en

[8] Android WiFi Tether. Available at http://www.code.google.com/p/Android-wifi-tether/, web page visited 20 September 2013

[9] Commotion project home page. Available at https://www.commotionwireless.net/

[10] Barnacle WiFi Tether web page. Available at http://www.szym.net/barnacle/, visited on 25 September 2013

[11] Opengarden Project web page. Available at http://wwwopengarden.com/, visited on 26 September 2013

[12] WiFi Tether web page. Available at https://www.play.google.com/store/apps/details?id=og.Android.tether, visited on 26 September 2013

[13] WiFi Ad Hoc enabler for Android web page. Available at http://www.arenddeboer.com/wifi-ad-hoc-enabler-for-Android, visited on 25 September 2013

[14] The Serval Project. Available at http://www.servalproject.org

[15] OLSRd an Adhoc Mesh Routing Protocol Home Page. Available at http://www.olsr.org/, visited on 01 November 2013

[16] Raja H.Q.: Addictivetips Home Page. Available at http://www.addictivetips.com/mobile/how-to-root-your-Android-phone-device/, visited on 08 November 2013

[17] Google web page. Available at http://www.code.google.com/p/csipsimple/, visited on 01 November 2013

[18] Hafslund A., Tønnesen A., Rotvik R.B., Andersson J., Kure O.: 'Secure extension to the OLSR protocol'. Published in Proc. OLSR Interop and Workshop, 2004

[19] Telecompute Voice Over Internet Protocol (VoIP) Page. Available at http://www.telecompute.com/voip.asp, visited on 23 December 2013

[20] Wikipedia Mean Opinion Score Page. Available at http://www.en.wikipedia.org/wiki/Mean_opinion_score, visited on 23 December 2013

Design of a full-band polariser used in WR-22 standard waveguide for satellite communications

Soon-mi Hwang, Kwan-hun Lee

Reliability & Failure Analysis Center, Korea Electronics Technology Institute, Sungnam-Si, Republic of Korea
E-mail: asfara@keti.re.kr

Abstract: This paper studies design of a full-band waveguide polariser using iris for satellite communications. Operation theory and design parameters of a full-band polariser are introduced, and a systematic design method has been proposed. The performance of the polariser is analysed using the well-known commercial electromagnetic simulation software high-frequency structure simulator. Using the proposed method, a full-band polariser operating in WR-22 waveguide band (33–50 GHz) is designed, fabricated and tested. Measurements of the fabricated polariser show that the phase difference is <10° as a reference point by 90°, the axial ratio is <1.3 dB, insertion loss is <0.1 dB and return loss is >25 dB in the required overall band.

1 Introduction

Radio waves used in satellite communication are usually circularly polarised to avoid the Faraday rotation caused by the ionosphere [1]. To generate a linearly polarised wave in reflector antenna feeds, a waveguide-type polarisation transformer, commonly called polariser, is generally employed [2, 3]. Of many different types of polariser, iris polariser in the square waveguide finds widespread applications because of the simplicity in structure, design and fabrication. Moreover, recently multi-band antenna feeds for satellite communication are used as the number of frequency bands increases [4, 5]. Iris polariser has such desirable properties as wide or multi-band operation, compactness and simple interfacing requirements [6, 7].

The design of a square waveguide iris polariser boils down to finding an optimum set of irises placed at a proper interval that gives low axial ratio and small reflection coefficient over a required frequency range. Although many papers have been published on the subject of an iris polariser, the design of an iris polariser operating over the full bandwidth of a rectangular waveguide has not been presented yet. This paper studies design of a full-band waveguide polariser using iris for satellite communications. Operation theory and design parameters of a full-band polariser are introduced, and a systematic design method has been proposed. The performance of the polariser is analysed using the well-known commercial electromagnetic simulation software high-frequency structure simulator (HFSS). Using the proposed method, full-band polariser operating in WR-22 waveguide band (33–50 GHz) is designed, fabricated and tested.

2 Operation theory

The iris polariser consists of a number of thin metallic fins arranged on two opposite walls of a square waveguide. A pair of fins placed on two opposite walls is called an iris. Fig. 1 illustrates the structure of an iris polariser.

When the iris of the same interval is arranged inside the waveguides, the iris is represented by Fig. 2

$$\beta' l = \cos^{-1}[\cos\beta l - B/Y_0 \sin\beta l] \tag{1}$$

$$\frac{Y_0'}{Y_0} = \left[1 - \frac{B^2}{Y_0^2} + 2\frac{B}{Y_0}\cot\beta l\right]^{1/2} \tag{2}$$

Between the empty waveguide and the waveguide with iris, the phase difference ($\Delta\theta$) is expressed as follows

$$\Delta\theta = \beta' l - \beta l \tag{3}$$

If $B/Y_0 > 0$, the phase difference is positive and the iris acts as a capacitor. If $B/Y_0 < 0$, the phase difference is negative and the iris acts as an inductor. The iris acts as a shunt capacitor in the

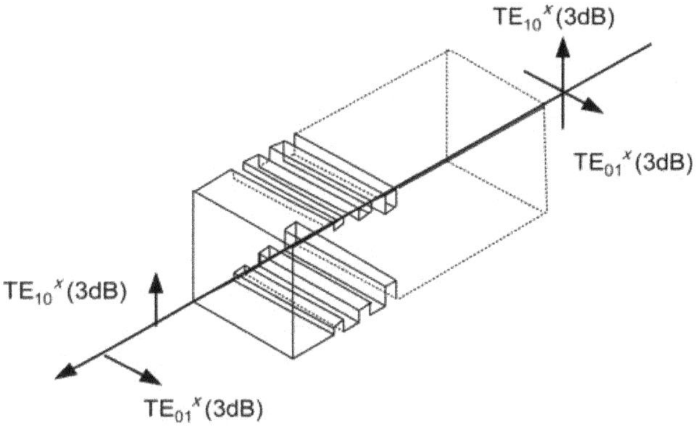

Fig. 1 *Structure of the iris polariser*

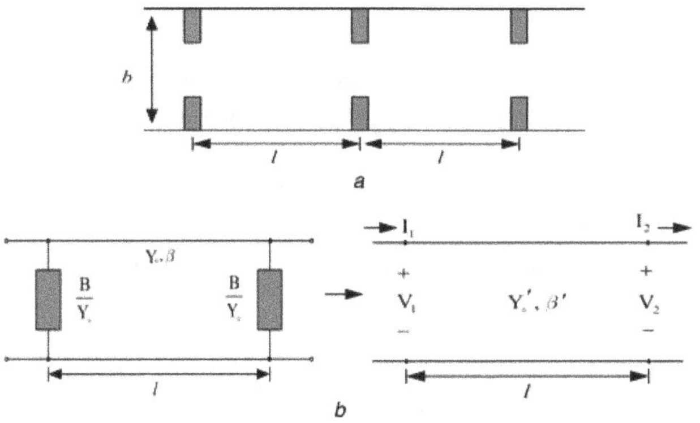

Fig. 2 *Periodically loaded transmission lines and its equivalent circuit*
a Inside iris
b Equivalent circuit (β, β': phase propagation constants, l: length, B/Y_0: normalised iris susceptance and Y_0, Y_0': admittance) [8]

transmission line for the mode with its electric field normal to the iris. When the electric field is parallel to the iris, the iris can be represented as a series inductor [9]. In the former case, the capacitive loading increases the phase propagation constant, whereas in the latter case the phase propagation constant is decreased. When a linearly polarised wave with its electric field at 45° to the iris is incident on the input plane of the iris polariser, it is converted to a circularly polarised wave after propagating through the polariser. In other words, the electric field applied in 45° angle direction can be split into the vertical mode (TE^{x}_{10}) and the horizontal mode. After passing through a polariser, the horizontal mode phase becomes smaller by 90° than the vertical mode phase, so that a right-handed circular polarisation signal is generated. If the circuit has been fully matched, voltage standing wave ratio and phase difference are as follows

$$\mathrm{VSWR}_{\mathrm{MAX.}} = \left[\frac{Y_0'}{Y_0}\right]^2 \text{ or } \left[\frac{Y_0}{Y_0'}\right]^2 \tag{4.1}$$

$$\Delta\theta = 180^\circ - 2\beta l \tag{4.2}$$

3 Design parameters

The design parameters of the iris polariser are the square waveguide size, the iris thickness, the iris interval, the iris number and the iris height. Fig. 3 illustrates the design parameters of the iris polariser.

Square waveguide size (a) is determined so that all channel signals are passed and the higher-mode signal is isolated. Waveguide size is determined by using the two equations below

$$a = (0.8 - 1.2)\lambda_m \tag{5.1}$$

$$2a = \lambda_{c'} = \frac{c}{f_{c'}}(f_{c'} = (0.7{:}1.2)f_c) \tag{5.2}$$

where $f_{c'}$ is a centre frequency and λ_m is a wavelength of centre frequency.

Iris thickness (t) is one element to control the phase characteristic of the polariser. Since phase difference is increased with iris thickness, a thin iris is useful for phase control. Generally, we use the below value

$$t = (0.05 - -0.1)\lambda_m \tag{6}$$

Iris interval (e) brings increase of phase difference when there are many irises. Generally iris interval is twice the iris thickness

$$e \geq 2t \tag{7}$$

Iris height (d) is a parameter which gives big influence at phase change and reflect characteristic. The phase difference increases when iris height is high. Generally, we use below range value [10]

$$d = (0.01 - 0.1)\lambda_m \tag{8}$$

In the case of iris number (N), generally we use 12–24. In the case of full band, if it is possible we select iris number to be $\Delta\phi = 90°$ in the required all frequency range.

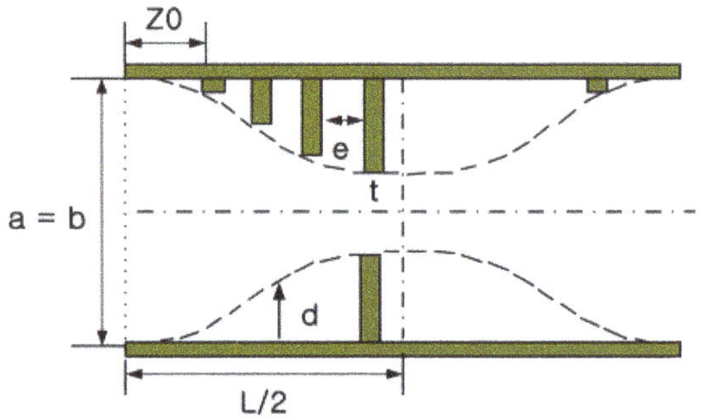

Fig. 3 *Design parameters of the iris polariser*

4 Design method

The focus of the design is to develop design procedures for a polariser having an excellent performance in the entire operating frequency range of a standard rectangular waveguide. The design of a full-band square waveguide iris polariser starts with the choice of the waveguide size. For a full-band operation, the waveguide size is made larger than that of the standard rectangular waveguide. Next, spacing between irises, iris thickness, iris height and the number of irises are optimised to satisfy axial ratio specifications. The iris height is properly tapered to realise a low reflection. Guidelines for selecting dimensional parameters of a full-band square waveguide iris polariser are derived by analysing the performance of the polariser against changes in polariser dimensions. If investigated through a commercial software, HFSS, the design process of a waveguide iris polariser used in the overall band of the WR-22 standard waveguide (33–50 GHz) is as follows.

4.1 Determine the size of square waveguide

Determine the size of waveguide using (5). After setting $f_{c'}$ to 23.1 GHz, which is 0.7 times of 33 GHz, the lowest frequency of operational frequency band, the output value of 'a' becomes about 6.49 mm. The reason to set $f_{c'}$ to the value of 0.7 times of the lowest frequency of operational frequency band is that this point is less affected by attenuation, which gets more serious as it is approaching closer to cutoff frequency, and to reduce the waveguide length, which is lengthened towards infinite as it is approaching closer to cutoff frequency. Vary the size of waveguide to more or less 6.49 mm, which is obtained theoretically. At that time, if higher mode is not occurring in overall band, check that the phase difference between these two modes is close to 90° and find the most appropriate size of waveguide. This process is conducted through computer simulation. Select the thinnest value within the possible limit, because the thinner the iris thickness the shorter the time of computer simulation and the size of computer memory.

4.2 Determine the iris thickness

Determine the iris thickness within the range suggested in (6). The thinner the iris thickness, the easier it will be to adjust overall reflection characteristic and phase characteristic when the number of irises is increasing.

4.3 Determine the iris height

Determine the number of irises as 5, because it should be more than 5 in order to give a taper in iris. Set the space between irises with two times the iris thickness and adjust the iris height so as to equalise the phase difference of transmission coefficients in the minimum frequency and the maximum frequency of the operational frequency band [11]. Determine the iris height using (8).

4.4 Adjust the space between irises

Make a fine tuning on the space between irises so as to equalise the phase difference of transmission coefficients in the minimum frequency and the maximum frequency of the operational frequency band.

4.5 Make the phase difference of 90°

Multiply the number of irises by $90°/\Delta\phi$ and make $\Delta\phi = 90°$. For example, if $|\Delta\phi| = 15°$, $90°/15° = 6°$, then the total number of iris becomes $5 \times 6 = 30$. When calculating the axial ratio, $\mathrm{TE}_x = [A_x, \phi_x]$, $\mathrm{TE}_y = [A_y, \phi_y]$, it is determined by (9) according to the sizes and phases of the two modes

$$A_x = A_y \tag{9.1}$$

$$\left|\phi_x - \phi_y\right| = 90° \pm \Delta \tag{9.2}$$

where Δ is a value to satisfy axial ratio ≤ 1.5 dB and its value is 10.

4.6 Change the iris height into the taper to improve the reflection characteristic

Taper is used to improve the reflection characteristic. The kinds of tapers are linear taper, sine taper and exponential taper. Among these, the exponential taper makes good return loss characteristic [12]. Since the reflection characteristic in a waveguide is largely impacted by the height of the front iris, the purpose of including a taper is also to reduce the height of the front iris. If the overall iris volume is not changed in that moment, there should be theologically no change in the transmission phase difference. To enhance the reflection characteristic, the lower the height of the iris is the better. However, it should be normally more than 0.2 mm in consideration of machining. Fig. 4 illustrates the vertical structure of the polariser according to the number of irises

$$\text{linear taper: } d(z) = d_1 + \frac{d_o - d_1}{L} z \tag{10.1}$$

$$\text{sine taper: } d(z) = d_1 + (d_o - d_1)\left[\frac{1 - C}{L} z + C \sin^2 \frac{\pi z}{2L}\right] \tag{10.2}$$

$$C = \text{taper coefficient},\ 0 < C < 1$$

$$\text{exponential taper:} d(z) = d_1 \exp\left[\ln\left(\frac{d_o}{d_1}\right)\frac{z}{L}\right] \tag{10.3}$$

4.7 Make the transmission phase difference to exactly 90°

Giving a taper in the iris height actually brings a slight change to the transmission phase difference. Theoretically speaking, giving a taper should not bring any change in the total iris volume inside the waveguide, but the design frequency is so high that a change in phase difference occurs even by a minute change in volume because of an extremely small value change in the design dimension. As the last step, make a fine tuning on the iris height and interval so as to make the transmission phase difference exactly 90° in the required band.

5 Design and measurement result

Using the proposed method, a full-band polariser operating in the WR-22 waveguide band (33–50 GHz) is designed, fabricated and tested. The design of the polariser is analysed using the well-known commercial electromagnetic simulation software HFSS. The number of irises in finally designed iris polariser is 23 and the waveguide size is 7.5 mm, the iris thickness is 0.5 mm, the iris interval is

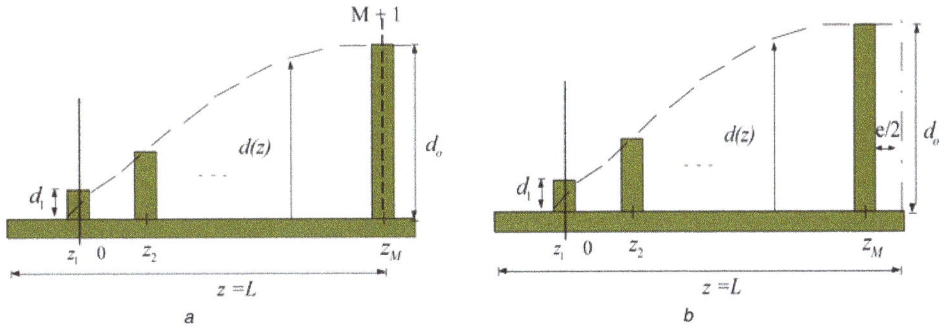

Fig. 4 *Vertical structure of the polariser according to the number of iris*
a Even number (2*M*)
b Odd number (2*M* + 1)

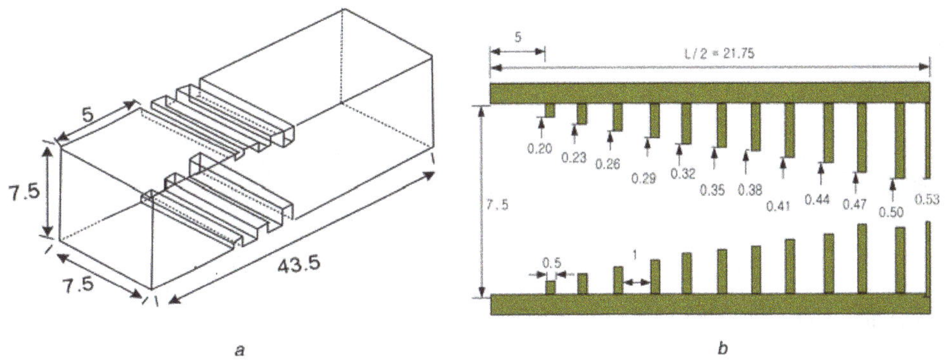

Fig. 5 *Design dimensions of the WR-22 standard waveguide polariser*
a External
b Inside (unit: millimetres)

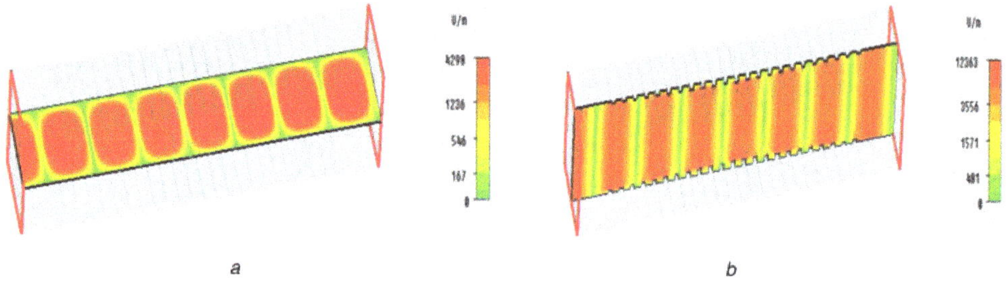

Fig. 6 *Electric-field distribution of the polariser*
a TE_{10} mode
b TE_{01} mode

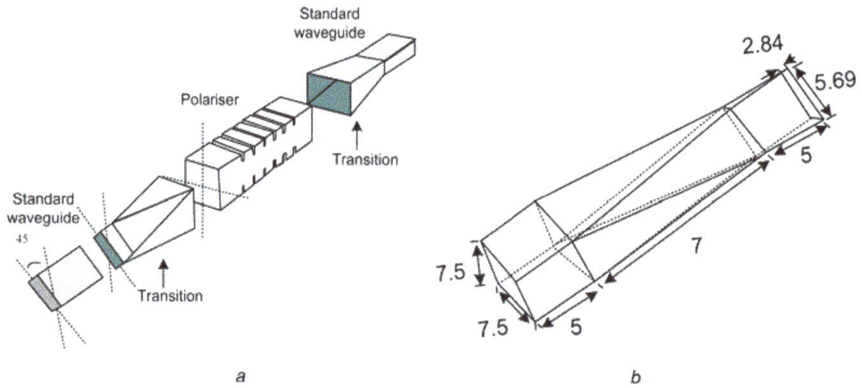

Fig. 7 *Design result of the transition*
a Position of the transition
b Design dimensions (unit: millimetres)

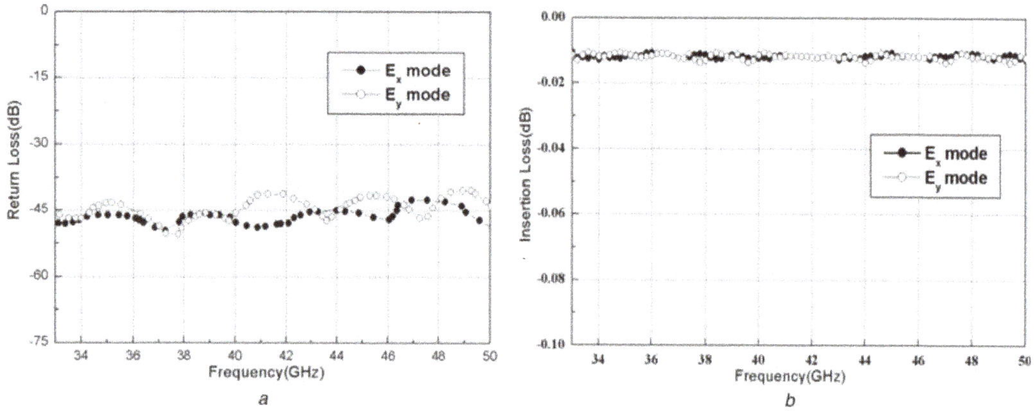

Fig. 8 *Simulation result of the transition*
a Return loss
b Insertion loss

1.0 mm and the total length of the polariser is 43.5 mm. For the iris height, an exponential taper is used. Fig. 5 illustrates the design dimensions and Fig. 6 illustrates the electric-field distribution.

Since the port of the polariser is not a standard size, the transition device is required to connect polariser and standard waveguide for measurements. The transition is designed to have low loss over a required frequency range. Fig. 7 illustrates the design result and Fig. 8 illustrates the simulation result.

The designed polariser was produced with milling processing. Since it was impossible to make an all-in-one processing, it was divided into four parts for production. Two transitions were produced for measurement. Fig. 9 illustrates the fabricated polariser and transition.

Fig. 9 *Fabricated the WR-22 standard waveguide polariser and transition*

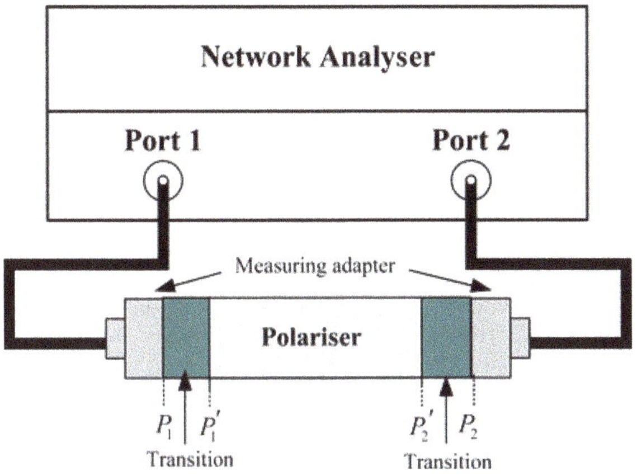

Fig. 10 *Measurement structure of the polariser*

Two-port network analyser was used to measure characterisation of the polariser. For the accurate measurements, the calibration of the network analyser was done at the end (P1′ and P2′) of polariser using TRL(THRU/REFLECT/LINE) method. The TRL calibration method is widely used in two-port microwave circuit measurement [13]. The TRL calibration, a full two-port calibration method using three standards: THRU, REFLECT and LINE, is often employed [14]. In this paper, we used short load as the REFLECT, waveguide of quarter times length of the centre wavelength as the LINE, and the THRU was used connecting directly to each other waveguide. Fig. 10 illustrates measurement structure of the polariser and Fig. 11 illustrates the results of the design and measurement of the WR-22 standard waveguide polariser.

In the case of simulation results, the return loss is <30 dB and the insertion loss is close to zero in overall band. The transmission phase difference is (87°–97°), and the axial ratio is within 1.1 dB in the required overall band. The result of measurement, the return loss is <25 dB and the insertion loss is an excellent characteristic that was <0.1 dB in the required overall band. The transmission phase difference is (80°–100°) and axial ratio is within 1.3 dB in the required overall band.

6 Conclusion

This paper studies design of a full-band waveguide polariser using iris for satellite communications. Operation theory and design parameters of a full-band polariser are introduced, and a systematic design method has been proposed. Using the proposed method, the full-band polariser operating in the WR-22 waveguide band is designed, fabricated and tested. Measurements of the fabricated polariser shows that the phase difference is <10° as a reference point by 90°, the axial ratio is <1.3 dB, the insertion loss is <0.1 dB and the return loss is >25 dB in the required overall band. Agreements between simulated and measured performances are good. The

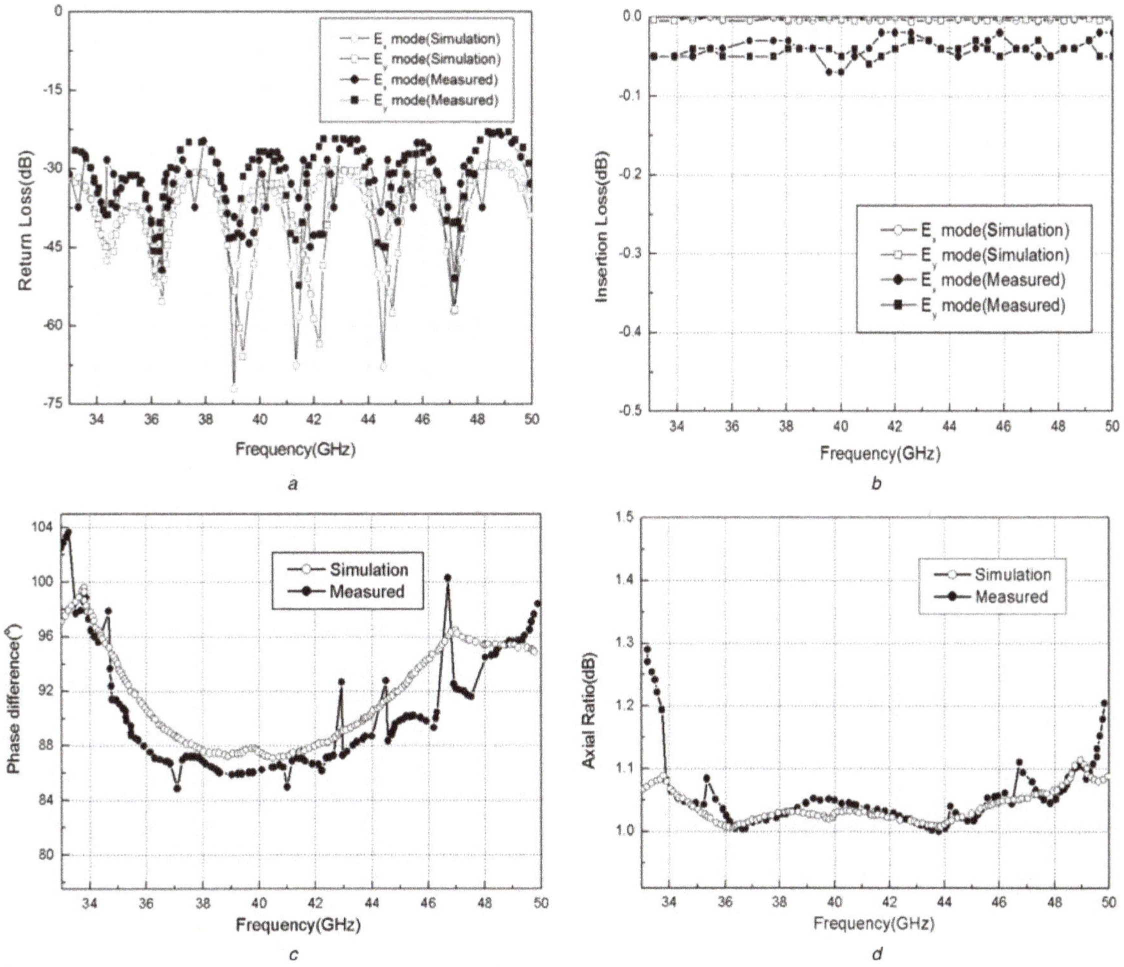

Fig. 11 *Results of the design and measurement of the WR-22 standard waveguide polariser*
a Return loss
b Insertion loss
c Phase difference
d Axial ratio

method presented in this paper can be used effectively in design of a polariser operating at the full-band of any rectangular waveguide.

7 References

[1] Dondl P.: 'Standardization of satellite component of the UMTS', *IEEE Pers. Commun.*, 1995, **2**, (5), pp. 68–74

[2] Ghedia L.: 'Developments in mobile satellite communications'. IEE Workshop on Microwave and Millimetre Wave Communications the Wireless Revolution, 29 November 1995, pp. 4/1–4/5

[3] Kitsuregawa T.: 'Advanced technology in satellite communication antennas' (Boston Artech House, 1990), pp. 363–366

[4] Brown K.W., Prata Jr., A.: 'A design procedure for classical offset dual reflector antennas with circular apertures', *IEEE Trans. Antennas Propag.*, 1994, **42**, (8), pp. 1145–1153

[5] Chu T.S.: 'Polarization properties of offset dual reflector antennas', *IEEE Trans. Antennas Propag.*, 1991, **39**, (12), pp. 1753–1756

[6] Ege T., McAndrew P.: 'Analysis of stepped septum polarizers', *Electron. Lett.*, 1985, **21**, (24), pp. 1166–1168

[7] Chambelin P.: 'Design and optimization of dual and wide band polarizer for low cost Ka band applications'. IEEE Antennas and Propagation Society Int. Symp., 9–14 July 2006, pp. 1595–1598

[8] Patzelt H., Arndt F.: 'Double plane steps in rectangular waveguides and their application for transformers, irises, and filters', *IEEE Trans. Microw. Theory Tech.*, 1982, **30**, (5), pp. 771–776

[9] Simmons A.J.: 'Phase shift by periodic loading of waveguide and its application to broad-band circular polarization', *IRE Trans. Microw. Theory Tech.*, 1955, **3**, (6), pp. 18–21

[10] Chen M.H., Tsandoulas G.N.: 'A wide-band square waveguide array polarizer', *IEEE Trans. Antennas Propag.*, 1973, **21**, (3), pp. 389–391

[11] Rebollar J.M., de Frutos J.: 'Dual-band compact square waveguide corrugated polarizer'. IEEE Antennas and Propagation Society Int. Symp., 11–16 July 1999, pp. 962–965

[12] Tucholke U.: 'Field theory design of square waveguide iris polarizers', *IEEE Trans. Microw. Theory Tech.*, 1986, **34**, (1), pp. 156–160

[13] Engen G.F., Hoer C.A.: 'Thru–reflect–line: an improved technique for calibrating the dual six-port automatic network analyzer', *IEEE Trans. Microw. Theory Tech.*, 1979, **27**, (12), pp. 987–993

[14] Kim Y.S., Lee C.S.: 'An improved error correction method of TRL removing manufacturing inaccuracy'. Antennas and Propagation Society Int. Symp., 20–25 June 2004, pp. 3625–3628

15

Framework for optical millimetre-wave generation based on tandem single side-band modulation

Maryam Niknamfar, Mehdi Shadaram

Department of Electrical and Computer Engineering, University of Texas at San Antonio, San Antonio, TX, USA
E-mail: mniknamfar@gmail.com

Abstract: A novel scheme for optical millimetre-wave (mm-wave) generation based on tandem sub-carrier multiplexing in hybrid communication systems is suggested. The method is analysed mathematically for optimum settings. The tandem single side-band (TSSB) method is tolerant to fibre chromatic dispersion. However, the TSSB technique is prone to critical interference of unwanted harmonics which may result in poor data transmission bit error rate (BER). Based on the proposed mathematical derivation for TSSB modulation, a framework is provided to enhance the mm-wave generation system performance. Up to 2 Gb/s data transmission over multiple 60 GHz sub-bands is considered. The model includes a single-mode fibre with chromatic dispersion factor of 16 ps/(nm km). Two cascaded dual electrode Mach–Zehnder modulators are used to generate TSSB 60 GHz sub-bands. Three oscillators are required to generate two tandem mm-wave signals. A proper selection of these three frequencies is vital to guarantee a successful data transmission. Mathematical analysis is carried out to verify the proposed scheme and a summary of appropriate frequency sets is presented and discussed. The BER curves are obtained. The results verify that the TSSB-based mm-wave generation framework proposed in this study guarantees data transmission with an acceptable BER.

1 Introduction

Wireless transmission in the lower microwave band is congested by applications such as wireless fidelity, global system of mobile etc. Thus, unlicensed 60 GHz frequency band (57–64 GHz) and 70–94 GHz band have been considered in the past few years [1–4]. Since propagation loss is the biggest problem for millimetre-wave (mm-wave) wireless communication, this technology can be used for short distances. In addition, geographical consideration is crucial for antenna base stations (BSs) installment. Owing to large number of required BSs and the high throughput of each BS, deployment of an optical fibre backbone is necessary. This network provides a broadband link between central office (CO) and BSs. The hybrid network can decrease the complexity and cost of the BSs by moving the routing, switching and processing functionalities to the CO. This way the cost of equipment can be shared among antenna BSs.

Optically mm-wave frequency (mmf) up-conversion obtained more attention because of electronic components' limited frequency response [5]. Three optical intensity modulation methods are used for the generation of mm-wave signals over optical fibre (i.e. direct intensity modulation, external modulation and remote heterodyning). However, external modulation is considered more often because of its wide bandwidth, small power consumption and large saturation power for input signals. However, electro-absorption modulator has small saturation input power. Thus, only a Mach–Zehnder modulator (MZM) can be used for multi-channel up-conversion [6]. Through different biasing points for MZM, different types of optical mm-wave signals can be generated. These types include optical double side band (ODSB), optical single side band (OSSB) and optical carrier suppression (OCS) [7].

The fibre dispersion effect on optically transmitted signals is critical to be controlled specifically for a long fibre link. To eliminate this impairment, OCS and OSSB techniques can be leveraged [8–11].

ODSB cannot be applied for long distances of fibre. The reason is that experienced phase shifts of light beams are different for various frequencies (i.e. upper and lower ODSB side bands) travelling through a dispersive fibre. Thus, this phase difference affects the detected radio-frequency (RF) power at the photodetector (PD). For the OSSB, there is only one side-band and consequently no RF power degradation will happen. Optical tandem single side-band (TSSB) method is suited to increase optical spectral efficiency. The TSSB carriers are easier to be filtered than those of the OSSB technique [12]. This method can also tolerate the fibre chromatic dispersion.

2 TSSB modulation

In TSSB method, two different RF modulated signals can be transmitted on each side band of the same optical wavelength. In comparison with OSSB format, two different RF modulated sub-carriers are spaced twice apart from each other [12].

2.1 Theoretical analysis

The TSSB signal can be obtained using one dual electrode MZM [12]. Considering the dual electrode MZM, the applied electrical signals to the upper and lower MZM electrodes are as follows

$$V_{\text{RF-upper}} = V_{\text{rf}_2}(t) + V_{\text{rf}_1}(t + \pi/2); \quad V_{\text{DC-upper}} = V_\pi/2 \quad (1)$$

$$V_{\text{RF-lower}} = V_{\text{rf}_1}(t) + V_{\text{rf}_2}(t + \pi/2); \quad V_{\text{DC-lower}} = 0 \quad (2)$$

where $V_{\text{rf}_1}(t)$ and $V_{\text{rf}_2}(t)$ are two different RF signals which can be further elaborated as given in the following

$$V_{\text{rf}_1}(t) = V_{\text{ac}_1}\cos(w_{\text{rf}_1}t); \quad m_1 = V_{\text{ac}_1}/V_\pi \quad (3)$$

$$V_{\text{rf}_2}(t) = V_{\text{ac}_2}\cos(w_{\text{rf}_2}t); \quad m_2 = V_{\text{ac}_2}/V_\pi \quad (4)$$

where V_{ac_1} and V_{ac_2} are RF signals amplitudes. The angular frequencies of two RF signals are w_{rf1} and w_{rf2}. The MZM switching voltage is V_π. Parameters m_1 and m_2 are defined as applied RF normalised amplitudes. Assume a continuous wave (CW) laser with output optical field of $E_{\text{in}}e^{jw_ct}$ is modulated by the applied electric signals. The E_{in} and the w_c represent laser optical field and optical frequency, respectively. The output optical field $E_{\text{out}}(t)$ (assuming equal 50/50 split ratio) for the MZM is represented in (5) [13].

Considering (5), $E_{\text{out}}(t)$ for TSSB is obtained using (1)–(4).

Consequently, the power spectral density of MZM output is derived for the positive half of the frequency domain ($f > 0$) as in (6) where J_n is the nth-order Bessel function. In (6), the first term shows optical carrier and the second and third terms represent two RF sub-carriers. The other terms (term number 4 and above) are unwanted harmonics which must be considered in case the TSSB method is utilised for mm-wave generation. Based on (6), and using MATLAB® software, a flexible framework is proposed in this paper. Thus, the three oscillator frequencies should be chosen considering some constraints to prevent critical system performance degradation because of the unwanted harmonics' interference.

2.2 Framework for mm-wave generation

For generation of mm-wave signal, optical or non-optical techniques can be utilised. However, limited frequency response of electrical devices makes electrical methods inefficient. Numerous optical techniques that are based on external modulation have been suggested [14–16]. The mm-wave generation framework proposed in this paper is based on mathematical analysis of the TSSB modulation. Two cascaded MZMs are applied in this paper to generate mm-wave signals. The first MZM is biased at the quadrature point to generate the optical tandem spectra of two different modulated RF signals. The cascaded MZM is biased at its null point in order to up-convert the RF frequencies to 60 GHz range. Fig. 1 represents the block diagram of the suggested method for optical mm-wave generation.

The power spectral density of the first MZM output is the same as (6). All frequencies in (6) will be shifted in both directions (right and left) by f_3 after the second MZM. The f_1 and f_2 are two different sub-carrier frequencies applied to the first MZM, whereas f_3 represents the oscillating signal frequency applied to the second MZM. For this method to operate successfully, a proper selection of the three frequencies is vital. This requirement is studied through a mathematical algorithm based on (6). This algorithm is implemented using MATLAB® software, and the results are summarised in Table 1.

Table 1 represents those sets of frequencies (12 GHz $\leq f_1$, f_2 and $f_3 \leq$ 25 GHz) which can be used for the proposed system to generate 57 GHz $\leq$ mmfs $\leq$ 64 GHz without interference while $|$'mmf'$_1$–'mmf'$_2| \geq 3$ GHz. In addition, the three frequencies f_1, f_2 and f_3 are considered so that the unwanted harmonics appear at least 3 GHz (in section *A*) or 2 GHz (in section *B*) away from the mm-wave generating frequencies ($-f_1$–f_3, f_3, $-f_3$, f_2+f_3). If frequencies f_1, f_2 and f_3 are not chosen properly, interference between unwanted harmonics and mm-wave generating signals will occur which will drastically degrade the system performance. This degradation is discussed more in Section 3.

3 System model and results

3.1 System model

Fig. 1 is implemented as the simulation model of the link in order to generate optical mm-wave signals. First, a pseudo random binary sequence of order 9 is used to generate data with a rate of 1 Gb/s for each channel. RF signals (which are generated by two oscillators with frequencies of f_1 and f_2) are modulated by non-return to zero smoothed data pulses. Then, the two RF modulated signals are optically modulated by two MZMs in series. The first MZM (i.e. MZM_1) is biased at the quadrature point ($V_\pi/2 = 8.2/2$ V) and the second MZM (i.e. MZM_2) is biased at the null point ($V_\pi = 8.2$ V). A sinusoidal signal with the frequency of f_3 is applied to MZM_2 electrodes with 180° phase difference to up-convert RF to the 60 GHz band. Both dual electrode MZMs have extinction ratios of 30 dB. Two different sets of frequencies are examined to verify the model performance. The frequency set #1 of $f_1 = 30$ GHz, $f_2 = 27$ GHz and $f_3 = 15$ GHz, which is not included in Table 1, is chosen randomly. In addition, from Table 1, the frequency set #2

$$E_{\text{out}} = \text{Real}\left\{[E_{\text{in}}\, e^{jw_c t}/2][\exp(j(\pi V_{\text{RF-upper}}(t)/V_\pi + \pi V_{\text{DC-upper}}/V_\pi)) + \exp(\pm j(\pi V_{\text{RF-lower}}(t)/V_\pi + \pi V_{\text{DC-lower}}/V_\pi))]\right\} \tag{5}$$

$$\begin{aligned}
S_E(f) = {} & (E_{\text{in}}^2/32)\Big[J_0^2(m_1\pi)J_0^2(m_2\pi)\delta(f-f_c) + 2J_0^2(m_2\pi)J_1^2(m_1\pi)\delta\big(f-f_c+f_{\text{rf}_1}\big) \\
& + 2J_1^2(m_2\pi)J_0^2(m_1\pi)\delta\big(f-f_c-f_{\text{rf}_2}\big) + J_1^2(m_2\pi)J_1^2(m_1\pi)\delta\big(f-f_c+\big(f_{\text{rf}_2}-f_{\text{rf}_1}\big)\big) \\
& + J_1^2(m_2\pi)J_1^2(m_1\pi)\delta\big(f-f_c-\big(f_{\text{rf}_2}-f_{\text{rf}_1}\big)\big) + J_1^2(m_2\pi)J_1^2(m_1\pi)\delta\big(f-f_c+\big(f_{\text{rf}_2}+f_{\text{rf}_1}\big)\big) \\
& + J_1^2(m_2\pi)J_1^2(m_1\pi)\delta\big(f-f_c-\big(f_{\text{rf}_2}+f_{\text{rf}_1}\big)\big) + 2J_1^2(m_1\pi)J_2^2(m_2\pi)\delta\big(f-f_c+\big(2f_{\text{rf}_2}-f_{\text{rf}_1}\big)\big) \\
& + 2J_1^2(m_1\pi)J_2^2(m_2\pi)\delta\big(f-f_c-\big(2f_{\text{rf}_2}+f_{\text{rf}_1}\big)\big) + 2J_2^2(m_1\pi)J_1^2(m_2\pi)\delta\big(f-f_c+\big(2f_{\text{rf}_1}+f_{\text{rf}_2}\big)\big) \\
& + 2J_2^2(m_1\pi)J_1^2(m_2\pi)\delta\big(f-f_c-\big(2f_{\text{rf}_1}-f_{\text{rf}_2}\big)\big) + J_0^2(m_2\pi)J_2^2(m_1\pi)\delta\big(f-f_c+2f_{\text{rf}_1}\big) \\
& + J_0^2(m_2\pi)J_2^2(m_1\pi)\delta\big(f-f_c-2f_{\text{rf}_1}\big) + J_2^2(m_2\pi)J_0^2(m_1\pi)\delta\big(f-f_c+2f_{\text{rf}_2}\big) \\
& + J_0^2(m_1\pi)J_2^2(m_2\pi)\delta\big(f-f_c-2f_{\text{rf}_2}\big) + J_2^2(m_1\pi)J_2^2(m_2\pi)\delta\big(f-f_c+2\big(f_{\text{rf}_2}+f_{\text{rf}_1}\big)\big) \\
& + J_2^2(m_2\pi)J_2^2(m_1\pi)\delta\big(f-f_c+2\big(f_{\text{rf}_2}-f_{\text{rf}_1}\big)\big) + J_2^2(m_2\pi)J_2^2(m_1\pi)\delta\big(f-f_c-2\big(f_{\text{rf}_2}-f_{\text{rf}_1}\big)\big) \\
& + J_2^2(m_2\pi)J_2^2(m_1\pi)\delta\big(f-f_c-2\big(f_{\text{rf}_2}+f_{\text{rf}_1}\big)\big) + 2J_0^2(m_2\pi)J_3^2(m_1\pi)\delta\big(f-f_c-3f_{\text{rf}_1}\big) \\
& + 2J_0^2(m_1\pi)J_3^2(m_2\pi)\delta\big(f-f_c+3f_{\text{rf}_2}\big) + J_3^2(m_1\pi)J_1^2(m_2\pi)\delta\big(f-f_c+\big(3f_{\text{rf}_1}+f_{\text{rf}_2}\big)\big) \\
& + J_1^2(m_2\pi)J_3^2(m_1\pi)\delta\big(f-f_c-\big(3f_{\text{rf}_1}-f_{\text{rf}_2}\big)\big) + J_1^2(m_2\pi)J_3^2(m_1\pi)\delta\big(f-f_c+\big(3f_{\text{rf}_1}-f_{\text{rf}_2}\big)\big) \\
& + J_3^2(m_1\pi)J_1^2(m_2\pi)\delta\big(f-f_c-\big(3f_{\text{rf}_1}+f_{\text{rf}_2}\big)\big) + J_3^2(m_2\pi)J_1^2(m_1\pi)\delta\big(f-f_c+\big(3f_{\text{rf}_2}+f_{\text{rf}_1}\big)\big) \\
& + J_1^2(m_1\pi)J_3^2(m_2\pi)\delta\big(f-f_c-\big(3f_{\text{rf}_2}-f_{\text{rf}_1}\big)\big) + J_3^2(m_2\pi)J_1^2(m_1\pi)\delta\big(f-f_c+\big(3f_{\text{rf}_2}-f_{\text{rf}_1}\big)\big) \\
& + J_3^2(m_2\pi)J_1^2(m_1\pi)\delta\big(f-f_c-\big(3f_{\text{rf}_2}+f_{\text{rf}_1}\big)\big) + \ldots\Big]
\end{aligned} \tag{6}$$

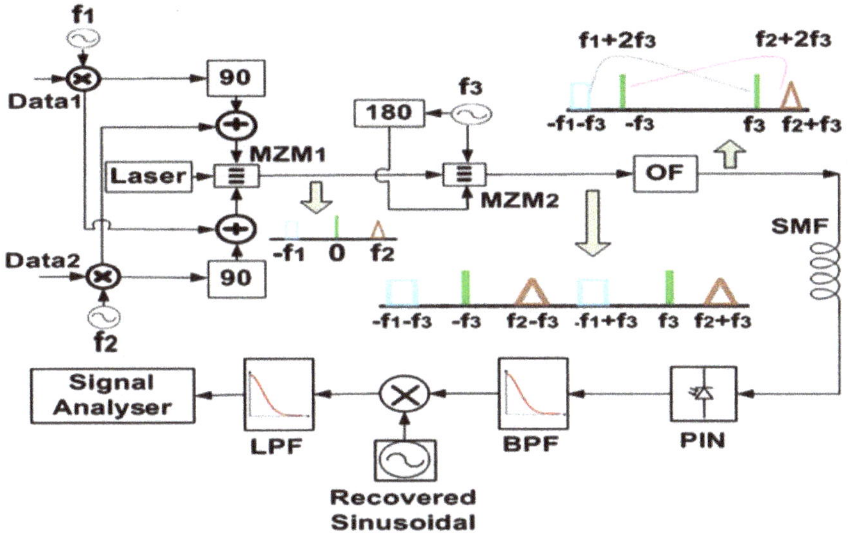

Fig. 1 *Block diagram of the suggested method for optical mm-wave generation*

that consists of $f_1 = 25$ GHz, $f_2 = 22$ GHz and $f_3 = 18$ GHz is examined. A CW laser which emits a beam with line-width of 7 MHz and power of 1 mW feeds the first MZM. The emission wavelength of the laser is 1553 nm. The MZM_2 output is filtered by a band stop filter to remove the unwanted signals. Transmission is done over 400 km of single-mode fibre with the dispersion factor of 16 ps/(nm km). Attenuation of the fibre is not considered in the first step in order to focus only on the chromatic dispersion effect of the fibre. A high-speed PD with responsivity of 0.5 A/W is applied. Dark current and thermal noise of the PD are 1 pA and 5 $pA/(Hz)^{0.5}$, respectively; shut noise is also included. After the PD, the 60 GHz range RF modulated channels are separated using band pass filters. Then, they are mixed with their corresponding synchronised carriers in order to detect the data. After the low-pass filter at the receiver side of the link, an analyser is used to evaluate the link performance.

The model is also examined when a fibre attenuation of 0.1 dB/km is included and the data rate is increased to 2 Gb/s. To compensate for the attenuation effect, an optical amplifier is added to the set-up. The link performance is still acceptable for the frequency set #2, and for the fibre length of about 250 km. In the next section, how to choose the model oscillators' frequencies to guarantee the system performance is discussed in more detail.

Table 1 Frequency sets that guarantee the system performance based on mathematical analysis

A			At the PD	
f_1, GHz	f_2, GHz	f_3, GHz	mmf_1, GHz	mmf_2, GHz
21	24	18	57	60
25	22	18	61	58
25	22	19	63	60
18	15	21	60	57
16	19	22	60	63
15	12	24	63	60
B			At the PD	
21	25	18	57	61
24	20	19	62	58
25	20	19	63	58
19	23	20	59	63
21	17	20	61	57
24	18	20	64	58
24	19	20	64	59
15	20	21	57	62
20	16	21	62	58
22	16	21	64	58
22	17	21	64	59
22	18	21	64	60
18	14	22	62	58
20	14	22	64	58
16	12	23	62	58
18	13	23	64	59

3.2 *Results and discussion*

First, consider the frequency set #1 of $f_1 = 30$ GHz, $f_2 = 27$ GHz and $f_3 = 15$ GHz. It can be seen that the unwanted harmonics $-2f_1 + f_3$ and $f_1 + f_2 - f_3$ interfere with mm-wave generating frequencies at $-f_1 - f_3 = -30 - 15 = -45$ GHz and $f_2 + f_3 = 27 + 15 = 42$ GHz, respectively. In addition, shifted version of optical carrier harmonic at -15 GHz is interfered by shifted version of $-f_1/f_3$ to the right $(-f_1 + f_3)$. These interferences prevent an acceptable mm-wave signal generation at the PD. The generated mm-wave electrical spectra at the PD for this frequency set is obtained using simulation software VPI® and is shown in Fig. 2*a*. As expected from mathematical analysis in (6), unwanted harmonics interfere with the RF frequencies and result in the performance degradation. The signal distortion is recognisable from Fig. 2*a* for channel frequency of $f_2 = 27$ GHz. This channel generates 57 GHz at the PD. The

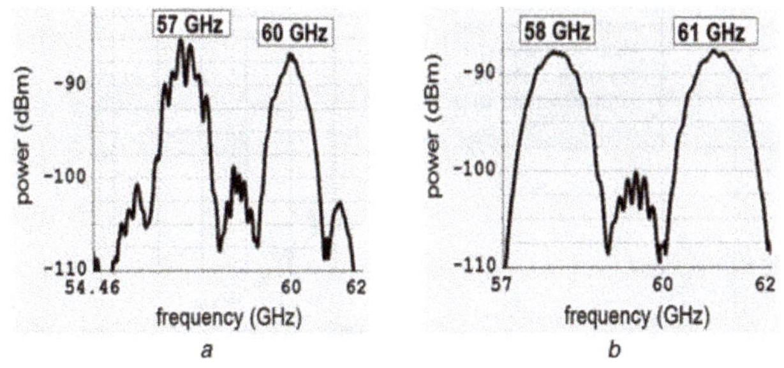

Fig. 2 *Generated mm-wave electrical spectra at the PD*
a For frequency set #1
b For frequency set #2

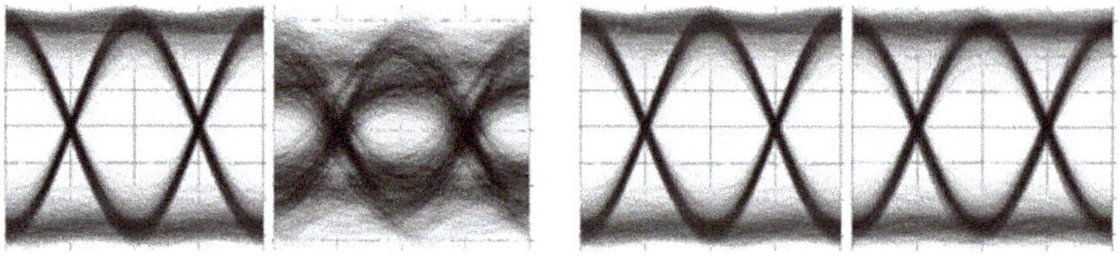

a *b*

Fig. 3 *Eye diagrams of the detected mm-wave signals*
a For frequency set #1: $f = 60$ GHz (left) and $f = 57$ GHz (right)
b For frequency set #2: $f = 61$ GHz (left) and $f = 58$ GHz (right)

normalised RF amplitude is set to 0.15 ($m_1 = m_2 = 0.15$). To guarantee the system performance, the frequency set needs to be chosen from the proposed Table 1. For instance, from Table 1, frequency set #2 including $f_1 = 25$ GHz, $f_2 = 22$ GHz and $f_3 = 18$ GHz is examined. Fig. 2*b* shows acceptable spectra at the PD for this set. Figs. 3*a* and *b* represent eye diagrams for set #1 and set #2, respectively. An acceptable bit error rate (BER) performance is obtained for both channels of frequency set #2 (for mmf = 58 GHz the BER is 1.08×10^{-17} and for mmf = 61 GHz the BER is 1.62×10^{-16}). However, for the set #1 the BER of the detected 57 GHz channel is 0.005776, which is not applicable for data transmission. Therefore, the frequency set #1 is not appropriate for the system. Although, as the eye diagram shows an acceptable performance for mmf = 60 GHz, we need to consider both channels' performances of the frequency set. For Figs. 2 and 3, fibre length is 400 km while only chromatic dispersion of 16 ps/(nm km) is included (attenuation effect in this step is neglected to focus on the dispersion effect) and the data rate is 1 Gb/s.

For the second examination of the system, the data rate is increased to 2 Gb/s. A fibre length of 220 km with dispersion of 16 ps/(nm km) and attenuation of 0.1 dB/km is included. Fig. 4 represents BER curves for both frequency sets against normalised RF amplitude. It is apparent from this figure that channel frequency selection, regardless of the normalised RF power adjustment, has a critical impact on the BER level of the received data. The BER curves (in green and black) for the frequency set #2 are better for any normalised RF amplitude adjustment. The 30 GHz channel BER curve (in blue) does not show an acceptable BER level for any normalised RF amplitude value.

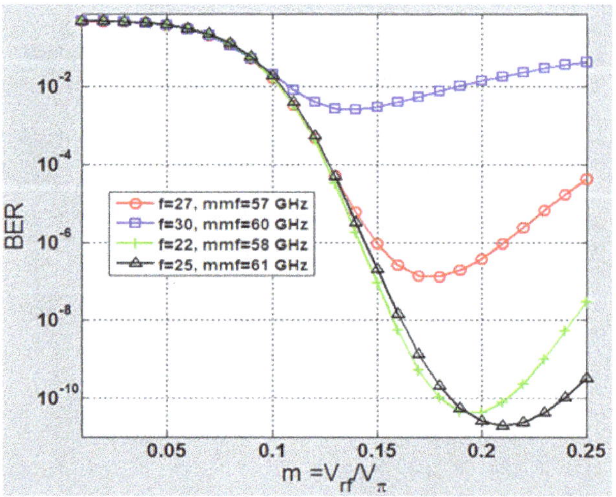

Fig. 4 *BER of both frequency sets obtained at the receiver analyser against normalised RF amplitude 'm'*

4 Conclusion

The novel scheme for optical mm-wave generation of RF modulated signals is suggested based on TSSB modulation method. Equation derivation is done for mathematical analysis of the proposed scheme. To guarantee performance of this method, proper selection of frequencies of the RF oscillators is required. Several appropriate frequency sets that can be used for the proposed scheme are obtained and presented based on the mathematical framework. The results and graphs verify that the unwanted harmonics must be considered carefully in order to avoid interference when generating mm-wave signals. These unwanted harmonics can degrade the system performance drastically. However, a proper RF frequency set selection for the system guarantees its high performance.

5 Acknowledgment

This work is partially supported by an NSF grant (HRD-0932339).

6 References

[1] Chang K., Chowdhury A., Jia Z., ET AL.: 'Key technologies of WDM-PON for future converged optical broadband access networks', *J. Opt. Commun. Netw.*, 2009, **1**, (4), pp. C35–C50
[2] Seeds A.J., Williams K.J.: 'Microwave photonics', *J. Lightwave Technol.*, 2006, **24**, (12), pp. 4628–4641
[3] Chowdhury A., Chien H.-C., Hsueh Y.-T., Chang G.-K.: 'Advanced system technologies and field demonstration for in-building optical-wireless network with integrated broadband services', *J. Lightwave Technol.*, 2009, **27**, (22), pp. 1920–1927
[4] Jia Z., Yu J., Ellinas G., Chang G.-K.: 'Key enabling technologies for optical-wireless networks: optical millimeter-wave generation, wavelength reuse, and architecture', *J. Lightwave Technol.*, 2007, **25**, pp. 3452–3471
[5] Lim C., Nirmalathas A., Bakaul M., ET AL.: 'Fiber-wireless networks and subsystem technologies', *J. Lightwave Technol*, 2010, **28**, (4), pp. 390–405
[6] Yu J., Jia Z., Chang G.K.: 'All-optical mixer based on cross-absorption modulation in electroabsorption modulator', *IEEE Photonics Technol. Lett.*, 2005, **17**, (11), pp. 2421–2423
[7] Yu J., Jia Z., Yi L., Su Y., Chang G.-K., Wang T.: 'Optical millimeter-wave generation or up-conversion using external modulators', *IEEE Photonics Technol. Lett.*, 2006, **18**, (13), pp. 265–267
[8] Sieben M., Conradi J., Dodds D.E.: 'Optical single sideband transmission at 10 Gb/s using only electrical dispersion compensation', *J. Lightwave Technol*, 1999, **17**, (10), pp. 1742–1749
[9] Smith G.H., Novak D., Ahmed Z.: 'Overcoming chromatic-dispersion effects in fiber-wireless systems incorporating external modulators', *IEEE Trans. Microw. Theory Tech.*, 1997, **45**, (8), pp. 1410–1415
[10] Griffin R.A., Lane P.M., Reilly J.J.O.: 'Dispersion-tolerant subcarrier data modulation of optical millimetre-wave signals', *Electron. Lett.*, 1996, **32**, pp. 2258–2260
[11] Niknamfar M., Shadaram M.: 'Optimization of optical single sideband configurations for radio over fiber transmission and multi-type data communication over a DWDM link', *J. Comput. Electr. Eng.*, 2014, **40**, (1), pp. 83–91
[12] Narasimha A., Meng X.J., Wu M.C., Yablonovitch E.: 'Tandem single sideband modulation scheme for doubling spectral efficiency of analogue fibre links', *Electron. Lett.*, 2000, **36**, (13), pp. 1135–1136

[13] SSLckinger E.: 'Broadband circuits for optical fiber communication' (John Wiley & Sons, Inc., New Jersey, USA, 2005)
[14] Chun-Ting L., Chen J., Peng-Chun P., ET AL.: 'Hybrid optical access network integrating fiber-to-the-home and radio-over-fiber systems', *IEEE Photonics Technol. Lett.*, 2007, **19**, (8), pp. 610–612
[15] Yu-Ting H., Zhensheng J., Hung-Chang C., Chowdhury A., Jianjun Y., Gee-Kung C.: 'Multiband 60 GHz wireless over fiber access system with high dispersion tolerance using frequency tripling technique', *J. Lightwave Technol.*, 2011, **29**, (8), pp. 1105–1111
[16] Zizheng C., Jianjun Y., Lin C., Qinglong S.: 'Reversely modulated optical single sideband scheme and its application in a 60 GHz full duplex ROF system', *IEEE Photonics Technol. Lett.*, 2012, **24**, (10), pp. 827–829

16

Energy-efficient network reprogramming scheme with Raptor code by using transmission power control in WSNs

Dongwan Kim, Sunshin An

Department of Electric and Electronics Engineering, Korea University Anam Campus, 145, Anam-ro, Seongbuk-gu, Seoul 136-701, South Korea
E-mail: dongwank@korea.ac.kr

Abstract: In wireless sensor networks, the necessity of reprogramming becomes more and more important for variety of purposes. However, the reprogramming produces a large amount of data and causes large energy consumption and interference. In this Letter, we propose an energy-efficient reprogramming scheme with Raptor code by using transmission power control. By selecting proper relay nodes, relay nodes' transmission power and Raptor code overhead, the proposed scheme minimises energy consumption while guaranteeing reliable transmission. This is verified by comparing it with conventional schemes.

1 Introduction

In general, wireless sensor network (WSN) consists of hundreds of small-sized battery-powered sensor nodes that integrate sensing, computing and communication capabilities. WSNs are employed in various applications such as disaster protection, security surveillance, battle field observation and healthcare infrastructure [1]. In many applications, WSNs are deployed in inaccessible areas. Thus, reprogramming is required to enable new functionalities.

Reprogramming of WSN should be supported not only with 100% accuracy of the disseminated data, but also in an energy-efficiency manner. A Traditional method is to perform manual reprogramming; however, it is costly and impossible since certain nodes cannot be accessed physically. Another conventional reprogramming method is multicast/broadcast schemes [2–5] which perform reprogramming with fixed transmission power. However, this can cause redundant energy consumption and interference. In addition, reprogramming methods with variable transmission power are proposed in [6–8] where transmission power is decided according to the received signal strength indication (RSSI). However, they are unsuitable for applications generating large traffic such as reprogramming because of control overhead of each node. Furthermore, [6–8] lead to unreliable transmission because of reducing transmission power. In this Letter, we propose new reprogramming scheme with Raptor code which optimises transmission power while guaranteeing reliable transmission.

2 Protocol description

The proposed scheme encompasses two phases: topology discovery phase and transmission power control phase. After discovering current topology in the first phase, relay node (RN) and transmission power of RNs are decided in the second phase. After then, sender node transmits reprogramming data by pre-calculating transmission power.

3 Topology discovery phase

In this Letter, WSN consists of beacon nodes (BNs) which have an inherent knowledge of their own position and normal sensor nodes (NSNs). BNs broadcast their location information and NSNs determine their location based on the RSSI and BN's location information using (1)

$$P_i(x, y) = \frac{\sum_{k=1}^{l} \left(\mathrm{WC}_{ik} B_k(x, y)\right)}{\sum_{k=1}^{l} \mathrm{WC}_{ik}} + \frac{\sum_{j=1}^{n} \left(\mathrm{WC}_{ij} P_j'(x, y)\right)}{\sum_{j=1}^{n} \mathrm{WC}_{ij}} \qquad (1)$$

where $P_i(x, y)$ represents the position of node i given by its two-dimensional coordinates. The known position of beacon k is given by $B_k(x, y)$ and the estimated position of node j is given by $P'_j(x, y)$ which is calculated by other BN's location information and other known NSNs location information. l and n indicate, respectively, the number of BNs and estimated number of known nodes that are within the communication range of the considered unknown node. In addition, WC_{im} is the weight cost between the node i and the known position of node m which is defined as

$$\mathrm{WC}_{im} = \frac{\alpha}{\mathrm{LQI}_{im}} \qquad (2)$$

where LQI_{im} is the link quality indicator value between i and m and α is proportional factor between BN and estimated location information node that is determined as follows

$$\alpha = \begin{cases} 1, \quad \text{location information from BN} \\ \min(1 + \text{number of estimation location},\ 5), \\ \quad \text{location information from estimation nodes} \end{cases} \qquad (3)$$

After calculating the location coordinate, each node can be aware of its own position. This information is transmitted to the sink that will be piggybacked in transmitted data. In this Letter, we assume that the reprogramming is performed after a given time of network deployment, thus the sink node recognises the network topology.

4 Transmission power control phase

When the network reprogramming triggers, based on the estimated node location information in topology discovery phase, RNs and their power are determined in transmission power control phase. In the proposed scheme only specific RNs transmit reprogramming data with pre-calculated transmission power.

First of all, the sink node calculates normalised transmission power (NpP) and it is defined as

$$NpP = \frac{P_i}{k} \qquad (4)$$

where P_i is the adjusted transmission power based on node i and adjusted transmission power is calculated by Friis formula with location information. k is the number of nodes within P_i.

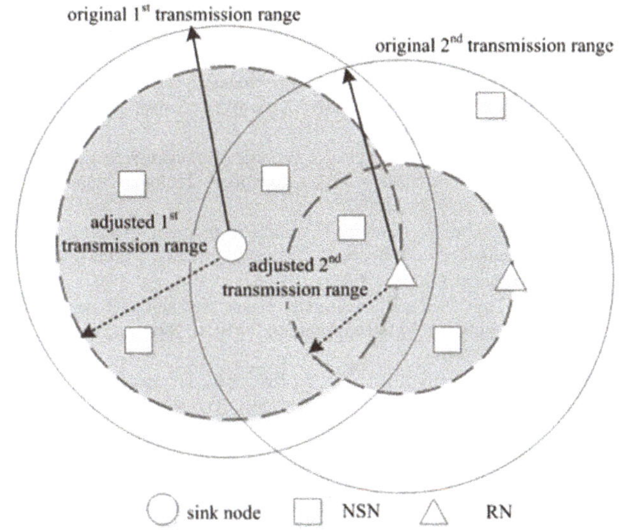

Fig. 1 *Coverage area of proposed scheme*

After then, the sink node selects the RN node that has the minimum *NpP* value. However, if RN is only selected based on *NpP*, the same node will be consecutively selected as RN, which quickly exhausts its energy. To overcome this side effect, the number of cumulative selection is considered when selecting the RN as described by the following formula

$$\text{RN\_ID}(i, l) = \min\left(NpP_1\, m_l,\ \ldots,\ NpP_i\, m_l,\ \ldots,\ NpP_n\, m_l\right) \quad (5)$$

where m_l is the cumulative selecting number as RN at lth reprogramming and its default value is 1.

Fig. 1 shows coverage area of the proposed scheme. The original transmission range indicates the transmission range for a fixed transmission power. The adjusted transmission range represents the optimal transmission range that can cover the node.

Furthermore, we utilise Raptor code, which is a famous application layer forward error correction (AL-FEC), to compensate for unreliable transmission by reducing transmission power. After applying Raptor code, we obtain reliability gain described as

$$P_f(n, P_e) = 0.85 \times 0.567^{n-k-P_e n} \quad (6)$$

where P_f is the expected received symbol erasure probability in AL, n is the output Raptor symbol length, k is the input source symbol length and P_e is the input symbol erasure probability in medium access control (MAC) layer. The relation between n and k is defined as Raptor code overhead and it is described as

$$\varepsilon = \frac{n-k}{k} \quad (7)$$

In addition, P_e is calculated using (8)

$$P_e(l) = 1 - \left(1 - 0.5 \times \operatorname{erfc}\left(\sqrt{\frac{P_r \times W}{N \times f}}\right)\right)^l \quad (8)$$

where P_r is the received power, W is the channel bandwidth, N is the noise power, f is the transmission bitrate, erfc is the complementary error function and l is the symbol length [9].

After determining the first RN_ID, its transmission power, and its Raptor code overhead, sink node determines the second RN's parameter based on the first ones. The calculation for selecting RNs is performed until the final target node is covered by RN node.

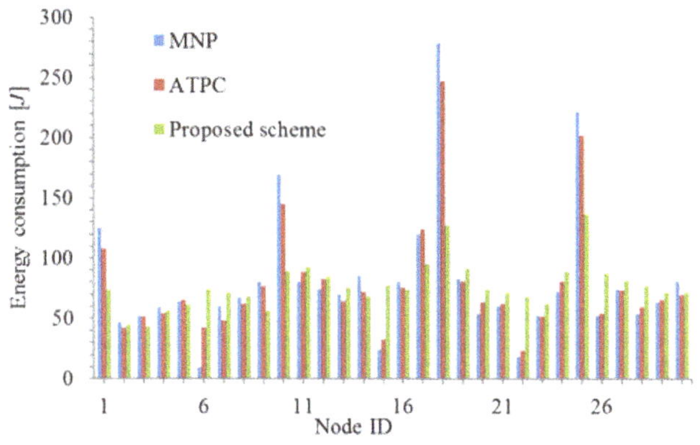

Fig. 2 *Energy consumption comparison*

Finally, the determined parameters are to be transmitted to RNs being piggybacked in data. Eventually, by selecting proper RN, RN's transmission power and Raptor code overhead, the proposed scheme reduces energy consumption while guaranteeing reliable transmission.

5 Simulation results

For the performance evaluation, we use OMNeT++ [10] which is a widely used network simulator and allows the performance evaluation in WSNs. We assume that the new programme size is 30 Kb corresponding to the code size for an operating system such as TinyOS [11].The new programme is divided into 32 pages; each page consists of 30 packets with 32 byte of data payload. The 30 sensor nodes are randomly deployed in a 500 × 500 m field. We assume that the node, positioned at the bottom-left edge, initially floods the new program. In the simulation, the radio characteristics are of CC2240 transceiver [12] and the MAC protocol is based on IEEE 802.15.4 [13]. The proposed scheme is compared with the multi-hop network reprogramming protocol (MNP) [4] and adaptive transmission power control (ATPC) [7]. In the MNP, each source node calculates the number of neighbour requesters that sent the request packets to it. By exchanging messages including its number of neighbour requesters with other sensor nodes, one of the source nodes in their hop range is selected as the sender which has the most neighbour requesters. On the other hand, in the ATPC, every node broadcasts several beacons to measure the RSSI value at diverse transmission powers. Every node then learns the relationship between the RSSI value and the various transmission powers.

Fig. 2 shows the energy consumption of the MNP, ATPC and the proposed scheme. As it can be observed, the proposed scheme has lower energy consumption than the two others. By using both adjusting transmission power and Raptor code, the proposed scheme restrains unnecessary retransmission by transmission errors. The average energy consumption of MNP, ATPC and the proposed scheme are 82.7, 78.6 and 74.3 J respectively. Furthermore, in MNP and ATPC, the node is positioned at the middle of the network; it has a high probability to transfer the received new programme because it has many receiver nodes around its neighbour. However, the proposed scheme selects RN by considering *NpP* and cumulative RN selection number and the proposed scheme has low energy consumption that is distributed equally to all nodes.

6 Conclusion

The proposed scheme aims to design an energy-efficient reprogramming method with Raptor code using transmission power control. Based on *NpP*, cumulative RN selection number, the sink node decides RNs, their transmission power and Raptor code overhead. Additionally, unreliable transmission by adjusting transmission power is compensated by Raptor code. Comparing with MNP and ATPC, we demonstrate the proposed scheme outperforms them both.

7 References

[1] Akyildiz I., Su W., Sankarasubramaniam Y., Cayirci E.: 'A survey on sensor networks', *IEEE Commun. Mag.*, 2002, **40**, (8), pp. 102–114

[2] Jeong J., Kim S., Broad A.: 'Network reprogramming. TinyOS documentation', 2003. Available at http://www.tinyos.net/tinyos-1.x/doc/NetworkReprogramming.pdf

[3] Levis P., Patel N., Shenker S., Culler D.: 'Trickle: a self-regulating algorithm for code propagation and maintenance in wireless sensor networks'. The First Symp. Networked System Design and Implementation, San Francisco, CA, 2004, pp. 15–28

[4] Kulkarni S.S., Wang L.: 'MNP: multihop network reprogramming service for sensor networks'. The 25th IEEE Int. Conf. Distributed Computing Systems (ICDCS), 2005, pp. 7–16

[5] De P., Liu Y., Das S.K.: 'ReMo: an energy efficient reprogramming protocol for mobile sensor networks'. The Sixth annual IEEE Int. Conf. Pervasive Computing and Communications, 2008, pp. 60–69

[6] Kubisch M., Karl H., Wolisz A.: 'Distributed algorithm for transmission power control in wireless sensor networks'. Proc. IEEE Wireless Communications and Networking Conf., New Orleans, LA, 2003, pp. 558–563

[7] Lin S., Zhang J., Zhou G., Gu L., He T., Stankovic J.A.: 'ATPC: adaptive transmission power control for wireless sensor networks'. Proc. Fourth Int. Conf. Embedded Networked Sensor Systems, Boulder, CO, USA, 2006

[8] Jeong J., Culler D., Oh J.: 'Empirical analysis of transmission power control algorithms for wireless sensor networks'. Technical Report, no. UCB/EECS-2005–16, 2007

[9] Lee S., Bhattacharjee B., Banerjee S.: 'Efficient geographic routing in multihop wireless networks'. Proc. ACM MobiHoc, Urbana-Champaign, IL, May 2005

[10] Varga A.: 'The OMNeT++ discrete event simulation system'. European Simulation Multi-Conference (ESM'2001), Prague, Czech Republic, 2001

[11] Han C.C., Rengaswamy R.K., Shea R., Kohler E., Srivastava M.: 'A dynamic operating system for sensor nodes'. The Third Int. Conf. Mobile Systems, Applications, and Service, Washington, SA, 2005

[12] Texas Instruments, CC2420 datasheet. Available at http://www.ti.com/

[13] IEEE 802.15.4. (2006).Wireless medium access control (MAC) and physical layer (PHY) specifications for low-rate wireless personal area networks (WPANs)

17

A 99%-efficiency GaN converter for 6.78 MHz magnetic resonant wireless power transfer system

Yoshiyuki Akuzawa, Yuki Ito, Toshihiro Ezoe, Kiyohide Sakai

Kamakura Office, Mitsubishi Electric Engineering Co. Ltd, 730 Kamimachiya, Kamakura, Kanagawa 247-0065, Japan
E-mail: akuzawa.yoshiyuki@ma.mee.co.jp

Abstract: The authors developed a high-efficiency gallium-nitride (GaN) Class-E converter for a 6.78 MHz magnetic resonant wireless power transfer system. A negative-bias gate driver circuit made it possible to use a depletion mode GaN high-electron-mobility transistor (HEMT), and simplified the converter circuit. As the depletion mode GaN HEMT with very small gate–source capacitance provided almost ideal zero-voltage switching, the authors attained a drain efficiency of 98.8% and a total efficiency of 97.7%, including power consumption of a gate driver circuit, at a power output of 33 W. In addition, the authors demonstrated a 6.78 MHz magnetic resonant wireless power transfer system that consisted of the GaN Class-E converter, a pair of magnetic resonant coils 150 mm in diameter with an air-gap distance of 40 mm, and a full-bridge rectifier using Si Schottky barrier diodes. The system achieved a dc–dc efficiency of 82.8% at a power output of 25 W. The efficiencies of coil coupling and the rectifier were estimated to be ~94 and 90%, respectively.

1 Introduction

Inductively coupled wireless power transfer (WPT) systems have been used for power supply and battery charging of mobile electronics and home appliances. Magnetic resonant coupling (MRC), reported in 2007, has attracted attention in the research community because MRC provides high coupling efficiency even if the transmitting coil and receiving coil are widely separated [1]. An industrial alliance of MRC recently legislated for a practical specification [2]. However, there are drawbacks of low efficiency for a total magnetic resonance WPT system including dc–ac and ac–dc converters, because a high frequency within the industrial, scientific and medical (ISM) band must be selected under radio regulations to meet coil size requirements for mobile equipment [3].

To increase the total efficiency of an MRC WPT system, many kinds of dc–ac converter have been studied as transmitters. Gallium-nitride (GaN) high-electron-mobility transistor (HEMT) fabricated on a silicon substrate is the most promising candidate for high-power and high-frequency transistors. Switching converter topologies using Class-D and Class-E amplifiers have been studied because of their inherent high efficiency (theoretically 100%). As frequency stabilisation using Class-D amplifiers is difficult in the ISM band, Class-E amplifiers are desirable for both high efficiency and precise frequency tuning [4, 5]. *Chen et al.* [5] demonstrated a Class-E converter using an enhancement mode GaN HEMT with a drain efficiency of 93.6% at a power output of 26.8 W, and system efficiency of 73.4% at a frequency of 13.56 MHz. However, there is a large scope to increase the efficiency of Class-E converters and MRC WPT systems in the ISM band.

We have developed a high-efficiency Class-E converter using a depletion mode GaN HEMT with a drain efficiency of 98.8% at a frequency of 6.78 MHz and demonstrated an MRC WPT system with a dc–dc power transfer efficiency of 82.8%, which consisted of a Class-E converter, a pair of transmitting and receiving coils with an air-gap distance of 40 mm and a full-bridge rectifier. To our knowledge, this is the highest efficiency Class-E converter for MRC WPT systems at a frequency of 6.78 MHz reported to date.

2 Experimental results

Fig. 1 shows a schematic circuit diagram of Class-E converter using a depletion mode GaN HEMT. Although enhancement mode GaN HEMT has been studied for switching devices to achieve highly efficient converter, we adopted a depletion mode GaN HEMT (prototype of Mitsubishi Electric Co.) to idealise zero-voltage switching (ZVS) operation because of its very small gate–source capacitance of 34 pF, low drain–source on-resistance of 69 mΩ, and high drain voltage rating of above 200 V. The use of a negative-bias gate driver circuit made it possible to simplify the converter circuit, because a cascade design combined with, for example, an Si metal–oxide semiconductor field effect transistor was not suitable for low-power application of <50 W. A shunt capacitance *C*1 of 445 pF and a choke inductance *L*2 of 1.8 μH were integrated in the vicinity of the GaN HEMT. In the experiment, we designed a load resistance (*RL*) of 31 Ω, which was close to the maximum output impedance of the following demonstration of an MRC WPT system.

Fig. 2 shows waveforms of the negatively biased gate pulses and the GaN drain voltage in Fig. 2*a*, and waveforms of the drain voltage and output voltage at the load resistance *RL* in Fig. 2*b*. The drain voltage waveform shows almost ideal Class-E operation, and the output voltage waveform at load resistance was a sinusoidal wave with very small distortion. However, a small phase difference between the start point of the drain voltage off state and peak point of the negative output voltage was observed because the fall time of the gate pulse voltage was delayed due to the influence of the rippled gate voltage, and the phase difference caused a decrease in the drain efficiency.

The drain efficiency and total efficiency, including the power consumption of the gate driver circuit, are shown in Fig. 3. A drain efficiency >95% was maintained over the wide range of the

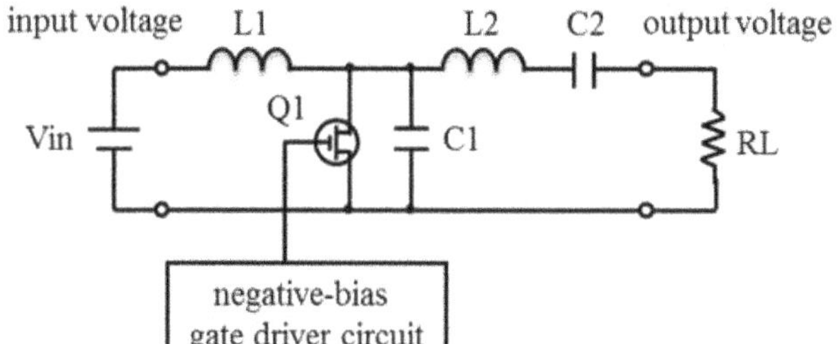

Fig. 1 *Schematic circuit diagram of Class-E converter, where depletion mode GaN HEMT was operated by negative-biased pulses from the gate driver circuit*

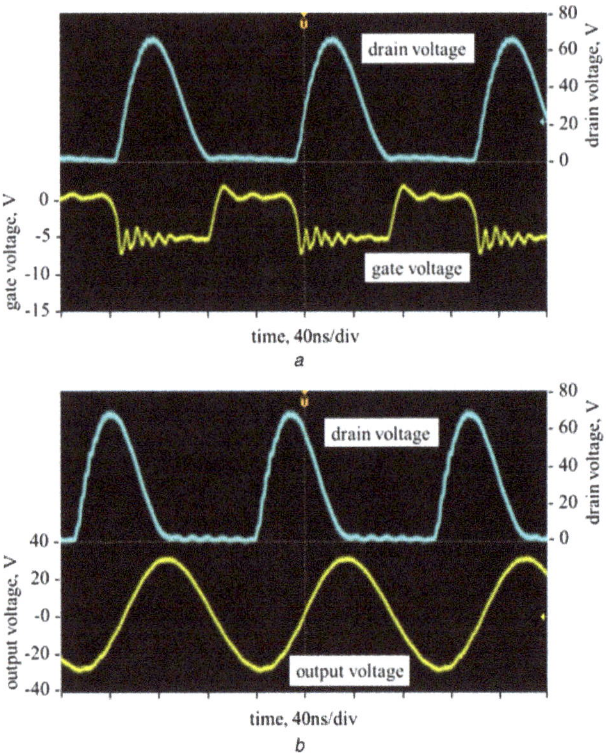

Fig. 2 *Waveforms of Class-E converter as follows*
a Drain voltage and gate voltage
b Drain voltage and output voltage at the load resistance *RL*

power output from 5 to 36 W. The highest drain efficiency of 98.8% and the total efficiency of 97.7% were attained at a power output of 33 W. As the power output was measured with an oscilloscope in the experiment, a certain small error was probably included due to distortion of the output voltage waveforms.

Fig. 4 shows the schematic circuit diagram of a 6.78 MHz MRC WPT system using the GaN Class-E converter. A photograph of the test set is shown in Fig. 5. Assuming desktop battery charging application, we designed a pair of helical coils with a diameter of 150 mm at a coil distance of 40 mm, where the coils were made of Litz wire with a length of 1.9 m and a diameter of 3 mm. Air-gap variable capacitors were adopted at both ends of the coils to precisely tune the resonant frequency and to achieve a monopole frequency response. The series compensation on the primary side with the capacitor

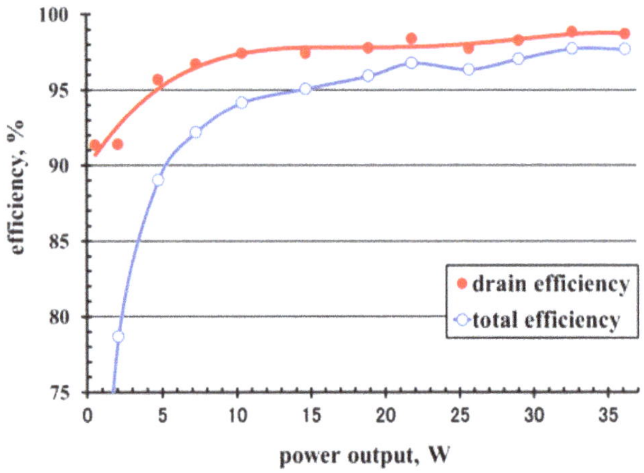

Fig. 3 *Drain efficiency (solid circles) and total efficiency including power consumption of the gate driver circuit (open circles) as a function of power output*

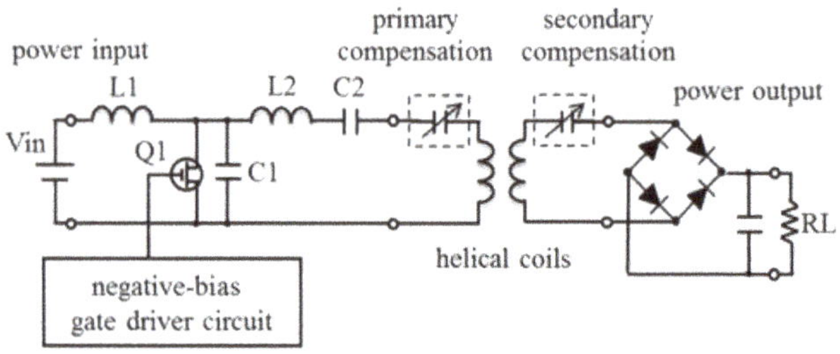

Fig. 4 *Schematic circuit diagram of magnetic resonant WPT system using GaN Class-E converter, with series–series compensating design at both sides of the coils and full-bridge rectifier with Si Schottky barrier diodes*

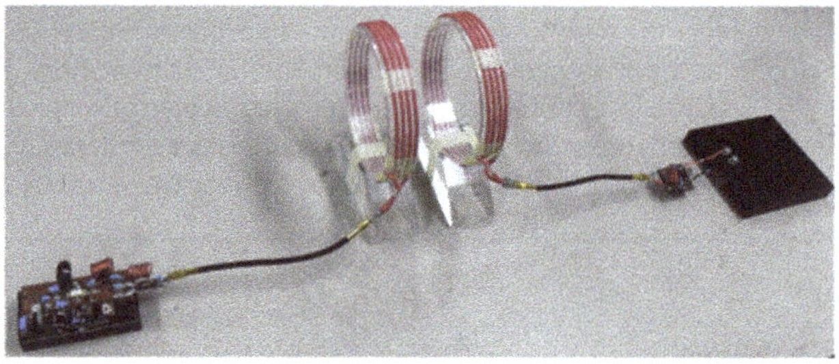

Fig. 5 *Photograph of MRC WPT system*

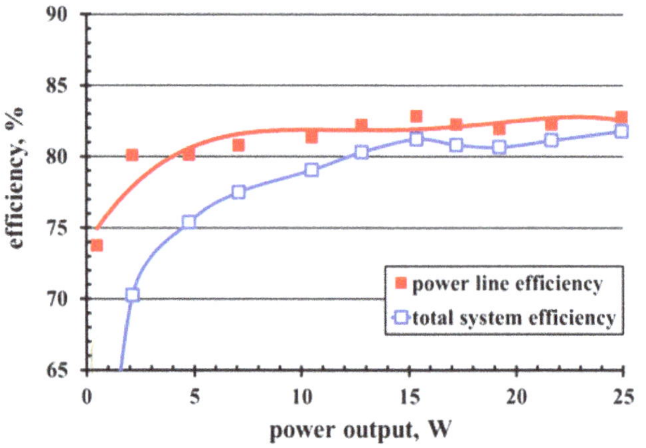

Fig. 6 *Power line efficiency (solid squares) and total system efficiency of WPT system, including power consumption of gate driver circuit (open squares), as a function of power output*

reduced high-voltage stress to the GaN HEMT, and the series compensation on the secondary side was empirically selected to increase the efficiency. To increase the rectifier efficiency, Si Schottky barrier diodes with very small forward voltage of 0.5 V and total capacitance of 30 pF were applied. The variable capacitors were precisely tuned to decrease the power reflection from the rectifier.

A power line efficiency and total system efficiency, including power consumption of the gate driver circuit, are shown in Fig. 6. The system efficiency from the dc input to the dc output was 82.8% at a power output of 25 W, and the corresponding efficiencies of coil coupling and the rectifier were estimated to be ~94 and 90%, respectively. A decrease in power reflection from the rectifier appeared to increase the rectifying efficiency, rather than a secondary higher rectifying mode effect [6].

3 Conclusion

We have developed a high-efficiency GaN Class-E converter and the prototype of an MRC WPT system using the converter at a frequency of 6.78 MHz. As the depletion mode GaN HEMT with very small gate–source capacitance provided an almost ideal ZVS, we attained a drain efficiency of 98.8% and a total efficiency of 97.7%, including power consumption of the gate driver circuit at a power output of 33 W. In addition, the dc–dc efficiency of the total MRC WPT system was 82.8% at a power output of 25 W. The efficiencies of coil coupling and the rectifier were estimated to be ~94 and 90%, respectively.

4 Acknowledgments

We are grateful to A. Inoue, M. Nakayama and J. Yamashita of Mitsubishi Electric Co. for fruitful discussion and the provision of GaN HEMT prototypes.

5 References

[1] Kurs A., Karalis A., Moffatt R., Joannopoulos D.J., Fisher P., Soljacic M.: 'Wireless power transfer via strongly coupled magnetic resonances', *Science*, 2007, **317**, pp. 83–86, doi: 10.1126/science.1143254

[2] Tseng R., Novak B., Shevde S., Grajski K.A.: 'Introduction to the alliance for wireless power loosely-coupled wireless power transfer system specification version 1.0'. Proc. 2013 IEEE Wireless Power Transfer Conf., 2013, pp. 79–83, doi 10.1109/WPT.2013.6556887

[3] Shoki H.: 'Issues and initiatives for practical deployment of wireless power transfer technologies in Japan'. Proc. IEEE, 2013, vol. 101, pp. 1312–1320, doi: 10.1109/JPROC.2013.2248051

[4] Calder R.J., Lee S., Lorenz R.D.: 'Efficient, MHz frequency, resonant converter for sub-meter (30 cm) distance wireless power transfer'. Proc. 2013 IEEE Energy Conversion Congress and Exposition, 2013, pp. 1917–1924, doi: 10.1109/ECCE.2013.6646942

[5] Chen W., Chinga R.A., Yoshida S., Lin J., Chen C., Lo W.: 'A 25.6 W 13.56 MHz wireless power transfer system with a 94% efficiency GaN Class-E power amplifier'. Proc. 2012 IEEE MTT-S Microwave Symp., 2012, pp. 1–3, doi: 10.1109/MWSYM.2012.6258349

[6] Kusaka K., Itoh J.: 'Experimental verification of rectifiers with SiC/GaN for wireless power transfer using a magnetic resonance coupling'. Proc. 2011 IEEE Ninth Int. Conf. Power Electronics and Drive Systems, 2011, pp. 1094–1099, doi: 10.1109/PEDS.2011.6147394

18

Capacitive digital-to-analogue converters with least significant bit down in differential successive approximation register ADCs

Lei Sun[1], *Kong-Pang Pun*[1], *Wai-Tung Ng*[2]

[1]*Department of Electronic Engineering, Chinese University of Hong Kong, Shatin, Hong Kong, N.T., People's Republic of China*
[2]*Department of Electrical & Computer Engineering, University of Toronto, 10 King's College Road, Toronto, Canada*
E-mail: kppun@ee.cuhk.edu.hk

Abstract: This Letter proposes a least significant bit-down switching scheme in the capacitive digital-to-analogue converters (CDACs) of successive approximation register analog-to-digital converter (ADC). Under the same unit capacitor, the chip area and the switching energy are halved without increasing the complexity of logic circuits. Compared with conventional CDAC, when it is applied to one of the most efficient switching schemes, V_{cm}-based structure, it achieves 93% less switching energy and 75% less chip area with the same differential non linearity (DNL)/integral non linearity (INL) performance.

1 Introduction

The capacitive digital-to-analogue converter (CDAC) in the feedback of successive approximation register (SAR) ADC performs the binary searching of the closest digital representation to the sampled input signal during the conversion process. Conventionally, a binary weighted capacitor array (BWA) is used [1]. In differential implementations, symmetrical switching on the positive and negative capacitor arrays is widely adopted to guarantee a constant common-mode (CM) voltage at the inputs of the comparator. It is noted that the CM voltage variation caused by the least significant bit (LSB) switching is the least, an asymmetrical switching on the LSB is proposed for differential DACs in this manuscript. In this scheme, the LSB capacitors only switch down, never up, and thus called as 'LSB down' scheme. This manuscript explains the usages on the ideas of 'LSB down' so that it saves the switching energy and chip area. When it is applied to one of the highest energy efficient V_{cm}-based [2] structure, it achieves even higher efficiency. For the symmetrical switching, any variance on the V_{cm} does not affect the resolution for the differential architecture with good CM rejection. For the LSB transition, the variance and noise on V_{cm} does introduce variances, however, with the least magnitude.

2 Proposed CDACs with LSB down

First, let us look at the basic case with the modification on the LSB switching in the conventional differential BWA DAC as illustrated in Fig. 1 using 2 bit examples. There are two capacitor arrays (positive and negative) in these differential structures. In the proposed scheme, to determine the LSB, only one unit capacitor from either the positive or negative arrays, not from both arrays, switches. This leads to a reduction of the total capacitance by half compared with the conventional approach. As a result, the averaged switching energy is nearly halved.

When LSB-down scheme is applied to one of the best energy efficient CDACs, namely V_{cm}-based [2], the energy efficiency gets doubled. Fig. 2 describes the difference with 3 bits examples. Both switching energy and area are reduced by at least 50%. When it comes into implementation, the switching network, number of cycles and logic circuit remain almost the same as in the original approach, except the tiny modification on switching of the unit capacitor.

To find out the switching energies (E_{up} and E_{dn}) for the 'up' and 'down' transitions in each capacitor array for the differential configuration, the deduction for the second conversion step in Fig. 1 can be followed as [3–5]

$$E_{up}\left[\frac{V_{ref}}{2} \to \frac{3V_{ref}}{4}\right] = V_{ref}\left(\text{net charge from } V_{ref}\right) = V_{ref}\left[\frac{3C_0V_{ref}}{4} - \frac{2C_0V_{ref}}{4}\right] = \frac{C_0V_{ref}^2}{4} \quad (1)$$

$$E_{dn}\left[\frac{V_{ref}}{2} \to \frac{V_{ref}}{4}\right] = V_{ref}\left[\frac{2C_0V_{ref}}{4} + \frac{3C_0V_{ref}}{4}\right] = \frac{5}{4}C_0V_{ref}^2 \quad (2)$$

So

$$E[2] = E_{up}\left[\frac{V_{ref}}{2} \to \frac{3V_{ref}}{4}\right] + E_{dn}\left[\frac{V_{ref}}{2} \to \frac{V_{ref}}{4}\right] = \frac{3}{2}C_0V_{ref}^2 \quad (3)$$

where the index i represents the ith conversion step ($i = 1, 2, \ldots, N$ for N bit example). It is noted that the 'down' transition here is assumed to switch the two capacitors simultaneously. Similarly, steps can be applied to find out the switching energies for all the transitions as indicated in Figs. 1 and 2. The waveform V_P and V_N at the output of differential CDAC is illustrated in Fig. 3.

3 Simulation results on switching energy

The averaging switching energies have been summarised in Table 1 and behaviour simulation plots are shown in Fig. 4. Additionally, a set-to-down [3] DAC is also included for comparison. Since the capacitors in set-to-down are not symmetrically switching during the conversion, it is not valid for applying LSB-down scheme. The digital numbers next to the dotted lines represent the normalised average switching energies for different DACs with the assumption that the input signals are distributed with identical probabilities. The energies are about halved when the LSB-down scheme is applied to the conventional BWA and the V_{cm}-based DACs. Note that when this scheme is applied to the V_{cm}-based, the V_{cm} is $V_{ref}/2$. If there is any fluctuation on V_{cm}, it only affects the resolution on the LSB. Although the input CM voltage for the comparator is varying during the comparison on the LSB, it is not problematic because the variation is the smallest at this moment. On the other hand, the V_{cm}-based structure with LSB down is 50% more efficient than the one without LSB down.

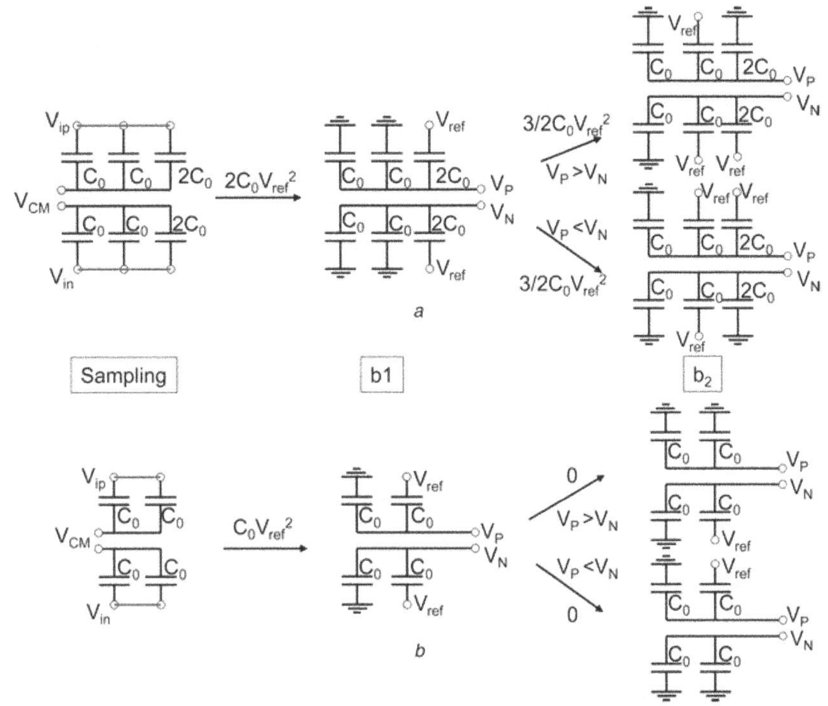

Fig. 1 *Comparison on BWA and BWA employing LSB down (2 bit example); V_P and V_N are outputs of the DAC to the comparator. The energy taken from V_{ref} in each switching step is marked near the corresponding arrow*
a BWA
b BWA with LSB down

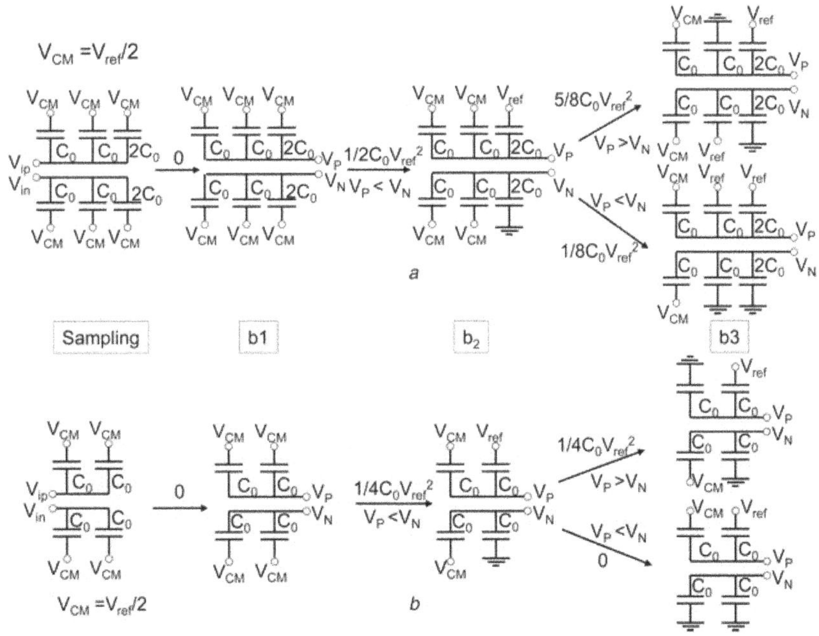

Fig. 2 *Comparison on*
a V_{cm}-based
b V_{cm}-based employing LSB down (3 bits example)

4 INL and DNL performance

Assume that the variation in unit capacitors of the DACs has a Gaussian distribution $[N(0,\ \sigma_0^2)]$, where σ_0 is the standard deviation of unit capacitor matching. The minimum unit capacitor is typically limited by the matching requirement rather than thermal noise requirement if we do not consider the mismatch calibration. Therefore the INL and DNL requirement for an 'N'-bit resolution SAR ADCs can be deduced from

$$V_{\mathrm{DAC}}(n) = \frac{\sum_{i=0}^{N-1}(C_i + \Delta C_i)b_i}{C_{\mathrm{T}}} V_{\mathrm{ref}} \tag{4}$$

$$\mathrm{DNL}(n) = V_{\mathrm{DAC}}(n-1) - V_{\mathrm{DAC}}(n) \tag{5}$$

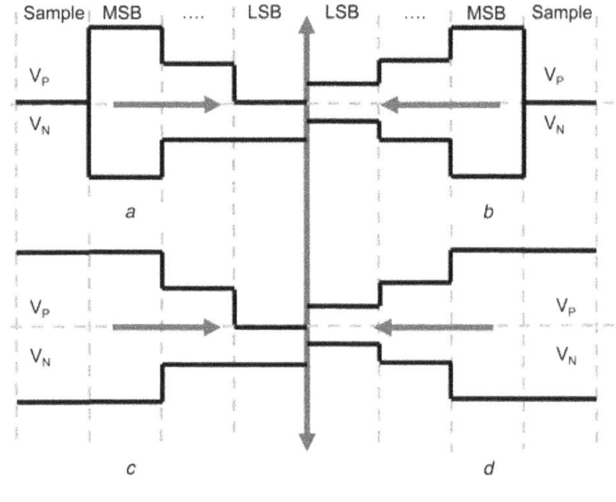

Fig. 3 *Waveform of V_P and V_N on*
a BWA with LSB down
b BWA
c V_{cm}-based with LSB down
d V_{cm}-based (3 bits example)

$$\mathrm{INL}(n) = V_{\mathrm{DAC}}(n) - V_{\mathrm{DAC\_ideal}}(n) \tag{6}$$

where C_T is the total capacitance, n is the corresponding digital codes for the output of DAC, C_i is the capacitance in the CDAC with the mismatch of ΔC_i and V_{ref} is the reference voltage. From the definition on DNL and INL in (5) and (6), the standard deviations of DNL and INL in terms of capacitor mismatch are easily calculated [4, 5]. Note that all the capacitor variations have been referred to the mismatch of the unit capacitor because of the rule of thumb that the mismatch is inversely proportional to the squared of the area [6]. The relationship between the standard deviation of INL and DNL and the mismatch (σ_0/C_0) of unit capacitor has been summarised in Table 1. Explicitly from Table 1, even though the INL/DNL performance becomes worst for the LSB-down schemes, the V_{cm}-based with LSB down can save the switching energy by around 93%, and the area by a quarter for the same INL/DNL performance as the conventional BWA approach. When capacitor mismatch calibration technique [7] is applied, the unit capacitor can be sized only considering the thermal noise requirement. With the same unit capacitance, the LSB-down scheme saved 50% of switching energy and area.

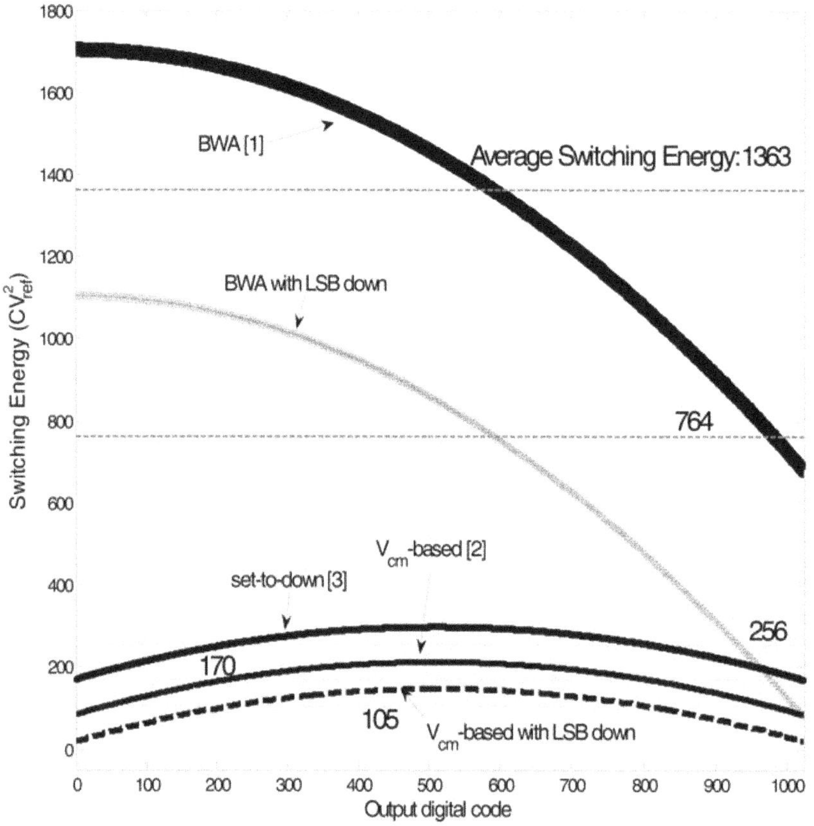

Fig. 4 *Normalised switching energy comparison for 10 bit SAR ADCs employing various CDACs*

Table 1 Comparison of CDACs with differential architectures

	Average power, J	σ_{INL}, LSB	σ_{DNL}, LSB	Total capacitance
BWA [1]	$1.33 \times 2^N C_0 V_{\mathrm{ref}}^2$	$2^{(N/2)-1}(\sigma_0/C_0)$	$2^{(N/2)}(\sigma_0/C_0)$	$2^{N+1}C_0$
BWA with LSB down	$0.75 \times 2^N C_0 V_{\mathrm{ref}}^2$	$\sqrt{2}2^{(N/2)-1}(\sigma_0/C_0)$	$\sqrt{2}2^{(N/2)}(\sigma_0/C_0)$	$2^N C_0$
V_{cm}-based [2]	$0.17 \times 2^N C_0 V_{\mathrm{ref}}^2$	$(\sqrt{2}/2)2^{(N/2)-1}(\sigma_0/C_0)$	$(\sqrt{2}/2)2^{(N/2)}(\sigma_0/C_0)$	$2^N C_0$
V_{cm}-based with LSB down	$0.09 \times 2^N C_0 V_{\mathrm{ref}}^2$	$2^{(N/2)-1}(\sigma_0/C_0)$	$2^{(N/2)}(\sigma_0/C_0)$	$2^{N-1}C_0$

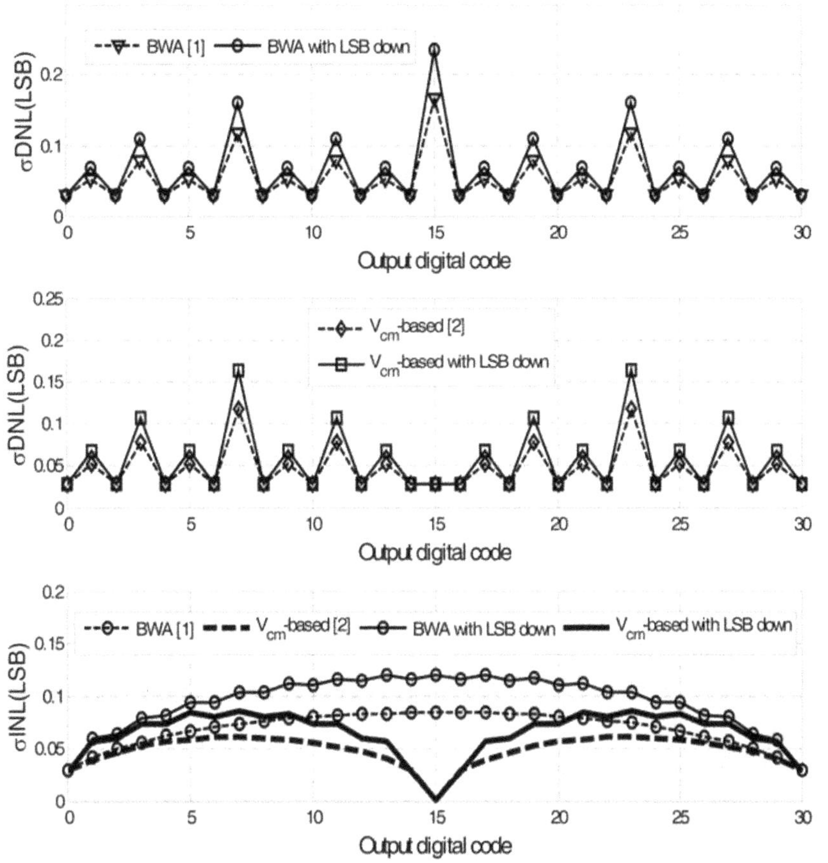

Fig. 5 *DNL and INL of 5 bit SAR ADC with BWA, BWA with LSB down, V_{cm}-based and V_{cm}-based with LSB down DACs5000 Monte Carlo runs were performed with independent identically distributed Gaussian errors in the unit capacitors ($\sigma_0/C_0 = 3\%$)*

Table 1 is a summary comparing the power, INL, DNL and total capacitance of different DAC approaches for an *N*-bit SAR ADC. The results have good achievement with simulations from 5000 Monte Carlo runs as shown in Fig. 5.

5 Conclusions

This manuscript has proposed to make use of asymmetrically switching in the LSB transition in some commonly used CDACs in differential SAR ADCs. When it is applied to V_{cm}-based structure, it achieves 93% energy efficiency with similar INL/DNL performance when compared with the conventional BWA approach. The total capacitance is reduced by a quarter of the BWA without introducing the complexity of logics and switches. It achieves 50% energy saving and area reduction with the same unit capacitor comparing the V_{cm}-based structure without LSB down.

6 Acknowledgment

This work was supported by a grant from the research grants council (RGC) of the Hong Kong SAR, China (project no. CUHK 416113).

7 References

[1] McCreary J., Gray P.: 'All-MOS charge redistribution analog-to-digital conversion techniques', *IEEE J. Solid-State Circuits*, 1975, **SC-10**, (12), pp. 371–379

[2] Zhu Y., Chan C.-H., Chio U.-F., Sin S.-W., U, S.-P., Martins, R.P., Maloberti F.: 'A 10-bit 100-MS/s reference-free SAR ADC in 90 nm CMOS', *IEEE J. Solid-State Circuits*, 2010, **45**, (5), pp. 1111–1121

[3] Liu C.C., Chang S.J., Huang G.Y., Lin Y.Z.: 'A 10-bit 50-MS/s SAR ADC with a monotonic capacitor switching procedure', *IEEE J. Solid-State Circuits*, 2010, **45**, (4), pp. 731–740

[4] Pun K.P., Sun L., Li B.: 'Unit Capacitor array based SAR ADC', *Microelectron. Reliab.*, 2013, **53**, (3), pp. 505–508

[5] Saberi M., Lotfi R., Mafinezhad K., Serdijin W.A.: 'Analysis of power consumption and linearity in capacitive digital-to-analog converters used in successive approximation ADCs', *IEEE Trans. Circuits Syst. I*, 2011, **58**, (8), pp. 1736–1748

[6] Aparicio R., Hajimiri A.: 'Capacity limits and matching properties of integrated capacitors', *IEEE J. Solid-State Circuits*, 2002, **37**, (3), pp. 384–393

[7] Yoshioka M., Ishikawa K., Takayama T., Tsukamoto S.: 'A 10-b 50-MS/s 820-μW SAR ADC with on-chip digital calibration', *IEEE Trans. Biomed. Circuits Syst.*, 2010, **4**, (6), pp. 410–416

Wireless network-on-chip: a survey

Shuai Wang, Tao Jin

State Key Laboratory of Novel Software Technology, Department of Computer Science and Technology, Nanjing University, Nanjing 210046, People's Republic of China
E-mail: swang@nju.edu.cn

Abstract: To alleviate the complex communication problems arising in the network-on-chip (NoC) architectures as the number of on-chip components increases, several novel interconnect infrastructures have been recently proposed to replace the traditional on-chip interconnection systems that are reaching their limits in terms of performance, power and area constraints. Wireless NoC (WiNoC) is among the most promising scalable interconnection architectures for future generation NoCs. In this study, the authors first provide a general description of the WiNoC architecture. Then, they discuss the research problems under five categories: topology, routing, flow control, antenna and reliability. Open research issues for the realisation of the WiNoC are also discussed.

1 Introduction

As the processor architecture enters the multi-/many-core era, the interconnection and communication among the on-chip components are playing major roles in defining the performance, area, power consumption and reliability of the entire chip. The network-on-chip (NoC) architectures have been proposed to replace the conventional global interconnections that will have severe problems because of the scalability and high-bandwidth requirement of the multi-/many-core chips [1–4]. To alleviate the complex communication problems in the NoC architecture, several novel interconnect infrastructures, such as three-dimensional (3D) and photonic NoCs [5, 6], have been recently proposed. Wireless NoC (WiNoC) is among the most promising scalable interconnection architectures for future generation NoCs.

Owing to the invention of different types of on-chip antennas [7–9], the on-chip wireless communication becomes feasible to build the WiNoCs. In the WiNoC architecture, the conventional wired links, for example, the links in a 2D mesh topology, are replaced by the wireless links. In most of today's WiNoC architectures [10–12], the wireless links are commonly used as highways (shortcuts) between long-distance nodes with low transmission latency and less power dissipation. Although the WiNoC designs are newly emerging ideas, some researchers are currently engaged in developing different types of WiNoCs. In this paper, we present a survey of WiNoC designs proposed so far. Our goal is to provide a better understanding of the current research issues in this area and insightful guidance to the future research work.

The rest of the survey is organised as follows. In Section 2, we discuss the topologies of the current WiNoC designs. Section 3 and 4 describe the routing algorithm and flow control in the WiNoC. The on-chip antennas and reliable designs of the WiNoC are presented in Section 5 and 6. 3D-NoC and photonic NoC are discussed and compared in Section 7. Some open research issues are discussed in Section 8. We conclude our paper in Section 9.

2 Topology

The topology of NoCs determines the physical layout as well as the connections between nodes and channels in the network. A topology will have a significant effect on the overall network cost and performance. A topology determines the number of hops a message/packet must traverse as well as the link lengths between hops. Thus, it influences the network latency significantly. As sending data through routers and links incurs energy dispassion, a topology's effect on hop count also directly affects network energy consumption. Furthermore, the topology determines the total number of alternate paths between nodes, which will affect how well the network can distribute traffic and support bandwidth requirements. Traditional NoC topologies are commonly categorised into three groups: (a) direct topologies, including rings, meshes, tori etc., (b) indirect topologies, including butterflies, clos networks, fattrees etc. and (c) irregular topologies, such as the multiprocessor system-on-chip (MPSoC) with a wide variety of heterogeneous intellectual property (IP) blocks.

Most of the traditional NoC topologies are based on the wired links between nodes and evaluated by the degrees of each nodes. Different from the traditional NoC topologies, the WiNoCs exploit the wireless connections between nodes to reduce the network latency, especially for these long-distance nodes. Therefore the WiNoC topology can be totally different from the traditional NoCs'. For example, a WiNoC can be a pure wireless link-based topology, or a variation of the traditional NoC with a hybrid topology including both wired and wireless links. Recent research has proposed some possible solutions for the WiNoC topology and their implementations.

2.1 Wireless-based topology

To provide a scalable, cost-effective and flexible on-chip communication infrastructure, the wireless interconnection was proposed to replace the wired communication. Therefore, in a pure wireless design, all the wired links should be replaced by the wireless links. For example, compared with a fully connected topology in the traditional NoC design, all the links between two nodes should be wireless connected in the WiNoC. However, because of the channel limitation in current/future on-chip wireless communication technology [13], we cannot support all the wireless links of a single node working simultaneously as the number of nodes (processor cores) is increasing to hundreds or even thousands. Therefore the scalability of such fully connected wireless topology is poor.

To implement the WiNoC with the wireless links, recent work [14] proposed a multi-channel WiNoC (McWiNoC) based on the traditional NoC topologies, and can be very flexible according to the transmission range of its wireless/radio frequency (RF) nodes. Fig. 1 shows the topology of the McWiNoC based on a 4×4 2D mesh. The RF node is working as a wireless router (WR) with ultra-wideband (UWB) [15] transceivers and antenna. Each node is associated with a processor tile (P). The wired link in the traditional 2D mesh is replaced by a high-bandwidth RF link.

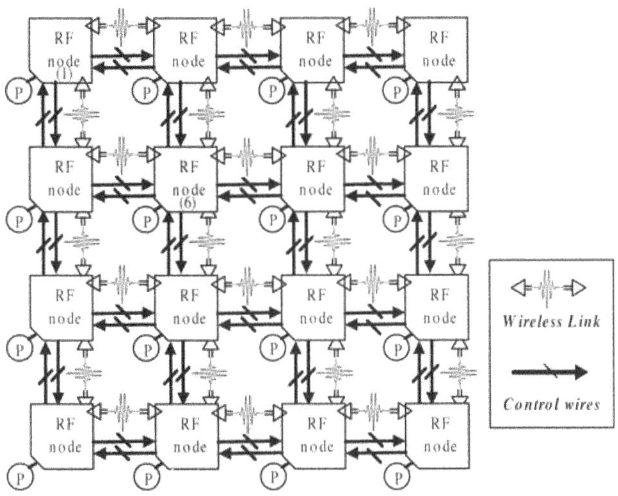

Fig. 1 *Topology of the McWiNoC based on a 4 × 4 2D mesh [14]*

The packets from the processor tile are delivered to their destination through one or multiple hops, similar to a traditional 2D mesh. To achieve the fast and simple channel arbitration, a separated control wires channel is added between two wireless connected nodes in their McWiNoC design.

Different from the traditional NoCs, the McWiNoC also targets at achieving a flexible topology design by changing the RF transmission range of the WR. For example, in Fig. 1, the transmission range (T) is set to the distance (L) between two adjacent RF nodes in the 2D mesh. If the transmission range (T) is increased from L to $\sqrt{2}L$, additional wireless links can be added, for example, between node 1 and node 6, and thus it can arrive at a different topology. If the transmission range (T) can cover any two nodes in the network, the McWiNoC will become fully connected. Therefore there is a tradeoff between the transmission range and hop count. If the transmission range is increased, the average hop count (path length) will be reduced. However, at the same time, the increased transmission range will incur more channel arbitration that will increase the latency and power consumption of the single link transmission. Therefore transmission range needs to be carefully chosen when implementing such McWiNoC design.

2.2 *Hybrid topology*

Instead of using all wireless links to transfer data between different nodes, the hybrid design uses both wired and wireless links to construct the on-chip networks. The hybrid design basically exploits the idea that the simple and reliable short-distance wired link is good for the local traffic within a small group of nodes, whereas the wireless link can significantly improve the performance and power efficiency for the long-distance communication. Note that this kind of hybrid topologies in WiNoCs is different from today's wireless Internet architecture that has the wired backbone for long-distance communication and wireless edges for local connection. Owing to its low design complexity and convenience to be extended from the existing NoC architectures, the hybrid design currently dominates in the WiNoC research. We will discuss the recent research work on the hybrid WiNoC topologies in following sections.

2.2.1 2D mesh-based hybrid topology: In recent work [11], a hybrid WiNoC design based on the existing 2D mesh NoC was proposed for multi-core platforms. The proposed WiNoC is based on a 2D mesh NoC called network-based processor array (NePA) that was introduced in their previous work [4]. The NePA is a little bit different from the conventional mesh design where there is only one bidirectional link between two adjacent nodes. In the NePA, two extra vertical ports are added to help dividing the whole network into two subnetworks: E-subnetwork and W-subnetwork. This partition can reduce the design complexity of the router as well as guarantee deadlock free [4]. To implement the hybrid WiNoC design, some of the conventional routers are replaced by the WRs that have both the wired and wireless links, and can transfer data via both wired and wireless channels. Fig. 2 shows the topology of the hybrid WiNoC design based on a 10 × 10 NePA. Some of the baseline router (BR) is replaced by the WR and the WiNoC is divided into four subnets with one WR in each subnet to provide the long-distance communications between subnets. Frequency division multiple access (FDMA) is used to provide the simultaneous communications between WRs with multiple channels.

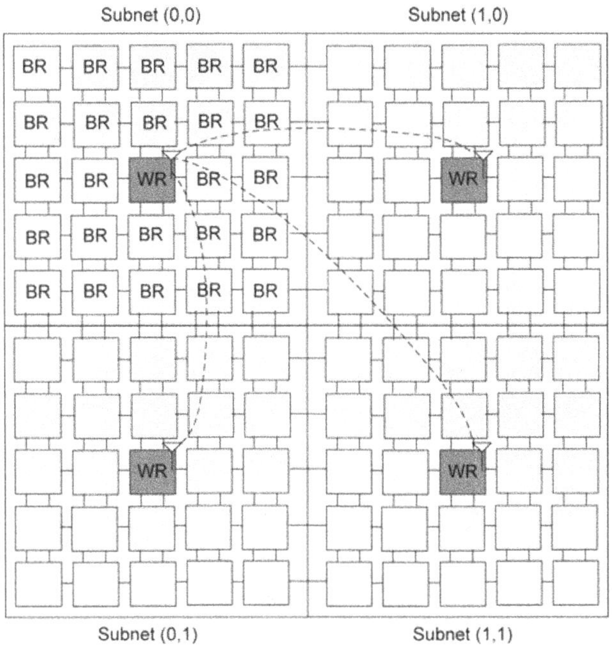

Fig. 2 *Topology of a 10 × 10 NePA-based hybrid WiNoC [11]*

In this topology, the communication between two nodes in different subnets may involve two types of transmissions: the local wired transmission and the long-distance wireless transmission. Therefore, in order to achieve the best performance, a dedicated routing algorithm is required for choosing the minimal path between nodes. Compared with the pure wireless based design, the requirement of multiple channels for simultaneous communications is reduced since the number of WRs has significantly decreased. However, for the scalability of this hybrid WiNoC, the major design issues, especially for a network with hundreds or thousands nodes, are how the WRs are placed and how the whole network is divided/organised.

2.2.2 Multiple tiers hybrid topology: To address the performance and scalability problems in NoCs, recent work [10] proposed a recursive wireless interconnection structure named WCube to deal with scaling limitation when the number of on-chip cores increases to thousands. The proposed WCube is also a hybrid topology with a multi-tier structure, including the wireless backbone and wired edges. The wired edge (CMesh), which is basically a 2D mesh with 4-way concentration (Concentration in on-chip networks is to share the router among different (processor) nodes [16].), is the baseline of the entire hybrid WiNoC. Fig. 3 shows a baseline CMesh with 16 base routers. Since the 4-way concentration is employed in this topology, each base router is attached with four processor cores or L2 caches nodes. Hence, there are total 64 nodes in a baseline CMesh.

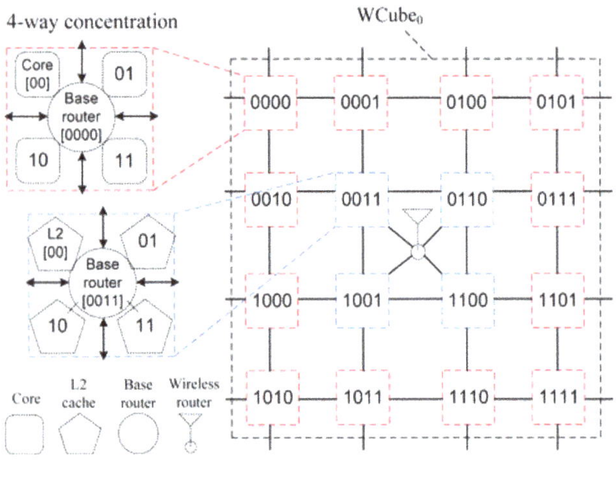

Fig. 3 *Topology of a baseline CMesh for WCube architecture with 16 base routers (64 nodes) [10]*

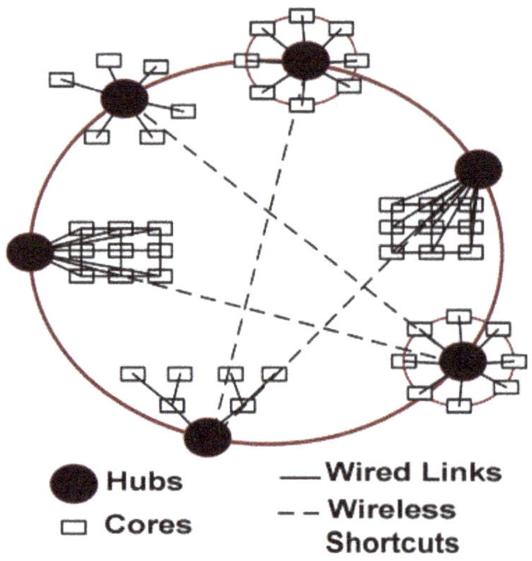

Fig. 4 *Topology of a small-world-based WiNoC [24]*

The wireless backbone is built on the top of the baseline CMesh. A WR is placed in each CMesh for the wireless communication, as shown in Fig. 3. The topology of the wireless backbone can be flexible. By comparing different topologies, such as ring, 2D mesh, fattree, hypercube etc., in terms of the number of WRs, the author claimed that the hypercube is the best choice to form the wireless backbone with 1000 s nodes in an on-chip network. In their study, total of 16 WRs are required to form a 1024-node NoC with the hypercube topology. Therefore they proposed a WCube topology by inheriting some useful properties from the hypercube. The WCube is a recursive multiple-level, 2D structure. The level-0 is the baseline CMesh named WCube$_0$. Then, the each level-i WCube$_i$ is constructed using four level-$(i-1)$ WCube$_{i-1}$s in a 2D mesh structure. To form a 1024-node NoC, a 2-level WCube is sufficient if the WCube is built based on the CMesh (WCube$_0$) shown in Fig. 3. Owing to the recursive structure, the WCube topology demonstrates good scalability compared with other hybrid WiNoC designs.

2.2.3 Small-world-based topology: To reduce the high latency and power dissipation caused by the multi-hop communication, especially for the long-distance communication, recent work [17] proposed to build the NoC by following the principle of small-world graphs [18]. The network topologies with the small-world property have a short average path length (hop count) between any two nodes in the network. The shortest path length in a small-world graph is bounded by a polynomial in $\log(n)$, where n is the number of nodes in the graph [19–21]. Therefore the small-world topology is very attractive to construct the complex networks [22]. For instance, both the Internet and our social network have the small-world property.

Instead of inserting the long-distance wired links to improve the performance of NoCs [17], recent work [23, 24, 12] proposed a WiNoC design based on the small-world property. In the small-world-based WiNoC, the long-distance wired links are replaced by the wireless links to create some shortcuts. These wireless shortcuts can significantly improve the performance of the NoCs compared with some conventional topologies, without the design issues of placing the long wires across the chips. Fig. 4 shows the small-world-based topology of a WiNoC proposed in [12]. The whole network is divided into multiple small groups (subnets). Different from other hybrid topologies discussed above, the topology of each small subnet in the small-world-based WiNoC may vary. It can be any of the conventional NoC topologies, such as mesh, ring, tree etc., and is connected by wirelines. The intrasubnet communication has a relatively short average path length because of its small size. Each subnet is equipped with a wireless hub, which is capable of transmitting and receiving data packets over the wireless channels. Owing to the limit in the number of possible wireless links, the adjacent hubs are still connected by wired links and just some of the distant hubs are connected by the wireless links, as shown in Fig. 4. The intersubnet communication will be achieved through the wired/wireless links, whereas the intrasubnet communication is still via wirelines. Thus, the whole hybrid WiNoC is a hierarchical architecture with subnets at the lower level and a small-world topology of hubs at the upper level.

Since the upper level of the network is constructed with hubs connected by both wired and wireless links, the placement of the wireless link is of critical importance to the network performance. In [12], a simulated annealing (SA) [25] based optimisation technique was employed to insert the wireless links under given resource constraints. For example, in a network of 16 hubs with 6 wireless links, the optimal wireless link arrangement can be found by using the SA techniques.

2.3 *Irregular topology*

Irregular NoC topologies are usually used for MPSoCs with heterogeneous IP blocks. However, the irregular topologies in WiNoC have been seldom studied so far. In [26], a low-cost and

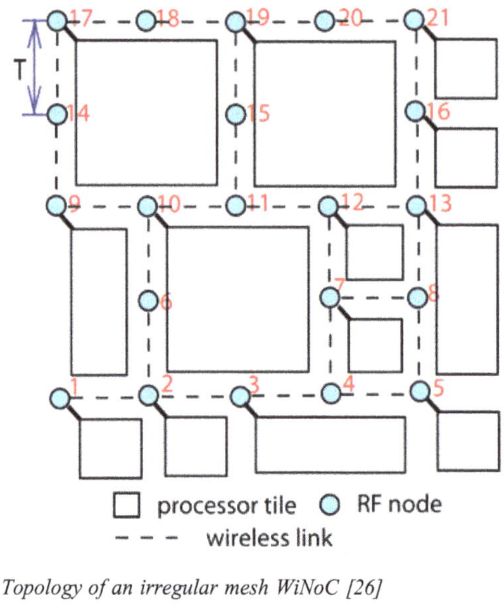

Fig. 5 *Topology of an irregular mesh WiNoC [26]*

high-efficient distributed minimal table based routing scheme was proposed under an irregular WiNoC topology. The proposed routing scheme was based on an irregular mesh topology connected by the wireless links, as shown in Fig. 5. The topology described a heterogeneous SoC floorplan. Each processor tile in the MPSoCs contains a processor core and connection to the WiNoC. The RF nodes working as the WR are distributed across the chip. The transmission range of the RF node is T (one hop). Note that not all the RF nodes are connected to a processor tile in this topology. The RF nodes with wireless links construct the irregular mesh for wireless communication. This irregular mesh WiNoC topology can also be categorised into the pure wireless based topology.

3 Routing

After determining the WiNoC topology, the routing algorithm decides the path a message/packet will take from the source to its destination. The routing algorithm should distribute the traffic in an appropriate way based on the network topology such that the traffic in the network can be well balanced, and the latency and throughput of the network will be improved.

In the conventional wired NoC, the routing algorithms can be generally divided into three categories: deterministic, oblivious and adaptive. In a hybrid WiNoC design, the wired subnets can follow the convention routing algorithm. However, for the pure wireless based WiNoC, for example, McWiNoC, or in order to achieve an overall optimised routing for the hybrid WiNoC, specific routing algorithms for a given WiNoC topology are needed. Since the routing algorithms are strictly depending on the network topologies, we will introduce them based on the WiNoC topologies discussed in this survey.

3.1 Deterministic routing

3.1.1 Location-based routing: In a pure wireless link based topology, for example, McWiNoC, the topology is fixed before determining the routing algorithm. Therefore the traditional deterministic algorithm, such as the *X–Y* dimension-ordered routing (DOR), can be used. However, because of the special characteristics of the wireless communication, more efficient routing algorithms can be achieved in the WiNoCs. In [14], a deterministic routing algorithm, named location-based routing (LBR), was proposed for the McWiNoC. The LBR is based on the fact that before making the routing decision, the router node has already known all its neighbours within its transmission range (T). Then, if the router wants to send the packet to the destination node (D), it will send the packet to its neighbour (B) that has the shortest geographic distance to the destination D. In the 2D mesh-based McWiNoC, the geographic distance d_{BD} between nodes B and D can be calculated by the following equation

$$d_{BD} = \sqrt{(x_B - x_D)^2 + (y_B - y_D)^2} \quad (1)$$

To further optimise the computation overheads of the LBR algorithm, the authors proposed to divide the chip areas into four quadrants from a node's perspective. When the router node receives a packet, by checking the location of the destination D, only one quadrant containing node D will be selected. Then, only the neighbours in the selected quadrant will be calculated for the shortest path. To avoid the deadlock caused by the LBR, a restricted-turn oriented LBR was also proposed by prohibiting some types of turns to avoid the cycle. The authors also showed that LBR could be more efficient compared with the traditional *X–Y* routing.

3.1.2 Multiple tiers based routing: For the multiple tiers hybrid WiNoC, normally, routing algorithms should be chosen for each tier first. Then, other routing strategies are needed to determine whether and how the packets are transmitted among different layers. In [10], for the baseline CMesh shown in Fig. 3, the *X–Y* routing was used. For the WCube structures, the multidimensional DOR algorithm was adopted. In an *n*-level WCube, the packet is routed in a strictly decreasing dimensional order to ensure deadlock free. To avoid the deadlock caused by the hybrid routing cycle among the baseline CMesh and WCubes, the authors defined the UP and DOWN virtual channels to break the cycle for a 2-tier hybrid architecture. To determine whether a packet should be forwarded to the WCube or delivered in the CMesh, a minimum-latency routing was proposed by calculating and comparing the latency of different paths. Based on some assumptions on the network bandwidth, average packet size and router architecture, the authors proposed to use the WCube if the path of using the CMesh is more than three hop-count longer than that of using the WCube. Since the WCube topology is known after the network is constructed, the routing table can be statically configured in the base router. Therefore the minimum-latency routing is a deterministic routing algorithm.

3.1.3 Small-world-based routing: For the small-world-based topology, the routing within the subnet can be any of the traditional NoC routing algorithms according to the topology of each subnet. For example, for a 2D mesh subnet, the *X–Y* routing can be used. For the intersubnets transmission, the packet will be routed through the hubs, as shown in Fig. 4. Therefore whether using the wireless links or just taking the wired link along the ring is very important to the network performance. The authors proposed to compare the hop counts of different paths by only considering the paths with at most one wireless link to simplify the calculation. If two paths have the equal hop count, the path with the wireless link is chosen for less power dissipation. Note that similar to the minimum-latency routing in WCube, all the routing paths are predetermined and stored in the hub. Therefore the authors claimed that their small-world-based routing is deadlock free. To avoid the centralised evaluation of the shortest path at the source hub, the authors also proposed a distributed routing mechanism to check the shortest path at each node. The authors showed that the distributed routing has less hardware overhead, but worst network throughput compared with the centralised scheme.

3.2 Adaptive routing

As we mentioned above, in a hybrid WiNoC, the traditional NoC routing algorithm can be used for routing the packets within a wired subnet. However, for two nodes in two different subnets, the transmission may involve the wireless links, which makes the overall optimised routing difficult to achieve. For example, in the NePA-based hybrid topology, an adaptive minimal routing algorithm taking the buffer utilisation into account can be used within a wire-connected subnet. However, in their topology shown in Fig. 2, the subnets are still connected to each other by the wired links, whereas the wireless links established by the WRs are working as the highways (shortcuts). Therefore whether the packet will take or not take the highway is a difficult decision to make. Wang *et al.* [11] proposed an adaptive routing algorithm to route the packet with source and destination nodes from different subnets. The adaptive algorithm is based on the hop-count calculation/comparison and the buffer utilisation status. When routing the packet, both the travelling distance (hop count) for taking or not taking the wireless links are calculated and compared. To avoid congesting the wireless links, a δ is added to the wireless path in order to give it a lower priority for the network traffic balancing. The adaptive algorithm also takes the buffer utilisation into account. If the buffers at WRs are full, the buffer-full signals will be sent to the BR nodes to avoid the congestion. To avoid the deadlock caused by the hybrid path, the authors proposed to exempt the flits with source–destination pairs in the same horizontal/vertical coordinate as WR from taking the wireless path. By breaking the hybrid cycle, the routing algorithm is deadlock free.

3.3 Routing on irregular topology

In irregular WiNoC topologies, traditional routing algorithms may not work in certain scenarios. For example, in an irregular mesh shown in Fig. 5, traditional *X–Y* routing algorithm will cause dead ends for certain pairs of source and destination nodes (e.g. from node 12 to node 20). Therefore Wu *et al.* [26] proposed a distributed minimal table based routing scheme by utilising the geographic distance information. A turn class based buffer ordering scheme was further proposed to avoid the deadlock. The proposed routing can be also applied to the conventional irregular mesh topology connected by wirelines.

4 Flow control

4.1 Wormhole and virtual channels

Wormhole [27] and virtual channels [28] are two commonly used techniques in the flow control of conventional NoCs. The wormhole flow control breaks the packet into small units, called flits. Then, it cuts through the flits instead of the packets for better buffer utilisation. The virtual channels were first proposed to avoid deadlock [28]. They can also be used to mitigate the head-of-line blocking in the flow control, thus improving the network throughput.

In most WiNoC designs, the wormhole and virtual channels are still adopted for better performance and deadlock avoidance. In the NePA-based WiNoC, WCube WiNoC and the small-world-based WiNoC, wormhole flit flow control were used for performance improvement and low buffer requirement. In the small-world-based WiNoC, the conventional virtual channel scheme was used. In the NePA-based WiNoC, a routing-independent parallel buffer for virtual channel implementation [29] was adopted. For the deadlock avoidance, a scheme using two dedicated virtual channels for flits towards WR and destination nodes were also proposed. In the WCube WiNoC, as we discussed in the routing algorithms, two types of UP and DOWN virtual channels were defined to guarantee the deadlock free in the 2-tier hybrid architecture.

4.2 Multi-channels

Although some of the channel issues in the NoC design, such as the maximum channel load, are usually discussed in the category of the topology, the communication channels allocation, such as the idea of virtual channels, also affects the flow control mechanisms. Therefore we discuss the multi-channels design in WiNoCs in this flow control section.

Owing to the characteristics of the wireless communication, multi-channels techniques are commonly adopted in the WiNoC designs for performance improvement. Multiple nodes using multiple channels can communicate simultaneously, so the throughput and latency of the network will be improved. All the WiNoC designs we discussed above have adopted this technique. FDMA is a common method adopted for multi-channelling in WiNoCs. However, because of the technology and bandwidth limitation, usually we can have 16–24 channels for the on-chip wireless communication today [10–12]. In McWiNoC, the multi-channel arbitration can be benefited from the wired control lines between neighbour nodes. In [12], a carbon nanotube (CNT) antenna-based multi-channel techniques were discussed. By assigning multiple channels to a single wireless link, the link bandwidth will be increased. Based on the fact that the number of available channels per link is usually less than the flit width (e.g. 32 bits), Ganguly *et al.* [12] proposed a time division multiplexing scheme to send whole flit by using a small number of channels (e.g. four channels) simultaneously.

5 On-chip antennas

Suitable on-chip antennas are necessary to establish wireless links for WiNoCs. An on-chip antenna is always one of the most difficult components that can be integrated on-chip; the following research work has been done in contribution of different types of on-chip antennas.

5.1 Silicon integrated antennas

In [30], the on-chip wireless interconnections were first proposed for clock signal distribution. For an on-chip antenna at 15 GHz, the size is around 2 mm. To reduce the area overhead of the on-chip antennas, Lin *et al.* [7] proposed a zig-zag antenna with optimised power gain. The zig-zag monopole antennas of axial length 1–2 mm can achieve a communication range of about 10–15 mm [31], which is suitable for the on-chip long-distance communication.

As the CMOS technology is scaling down, the size and the cost of the antenna will decrease dramatically with the increasing operating frequency. In [10], a design of miniature on-chip antennas operating at the range of 100–500 GHz has been demonstrated. To minimise the substrate loss, they proposed that an on-chip antenna should be placed in a polyimide layer, which deposits on the top of the top silicon. Therefore most electro-magnetic energy can be confined within this low-loss dielectric layer. These kinds of antennas are capable of constructing the WiNoC with 1000 s nodes. The multiple tiers hybrid WCube is built based on these kinds of antennas.

5.2 UWB antennas

The design of UWB antennas was proposed for on-chip wireless communication in [8]. A CMOS UWB-based WiNoC design was recently proposed in [13]. The UWB can provide high bandwidth, low power and short-range communication for WiNoCs. Owing to the multi-channelling capability, the UWB can be shared among multiple nodes, which makes simultaneous communication possible and can improve the communication capacity as well as reduce the communication latency. However, the UWB-based antenna in [13] has a transmission range of 1 mm with a length of 2.98 mm. The short transmission range, compared with the current die area, requires multihop wireless communication across the chip, which makes it less efficient for long-distance communications. The McWiNoC- and MPSoCs-based irregular WiNoC are built based on the UWB antennas.

5.3 CNT antennas

All the antennas discussed above are operating in the millimetre wave range, so the sizes of these antennas should be on the order of a fewer millimetres. If the transmission frequency can be increased to Terahertz, the size of the antenna can be reduced correspondingly. CNT-based antennas have been proposed and studied to achieve Terahertz frequency range [32–34]. The CNT antennas operating in Terahertz frequency range can provide much higher communication data rates. Besides its small size, that is, the ultra-low area overhead, the virtually defect-free CNT antenna does not have the power loss problem caused by surface roughness and edge imperfection in conventional antennas. The CNT antenna can also have extremely high current density compared with copper, thus the high transmission power, which makes it suitable for long-distance communications. The radiation characteristics of the CNT antenna can be further improved by using an external laser source [32]. All these characteristics make the CNT antenna a good candidate to implement the wireless communication in WiNoCs. The small-world-based hybrid WiNoC is built based on the CNT antennas. However, the CNTs are still facing significant manufacturing challenges.

6 Reliability

The key reliability issue arising in the WiNoC design is to ensure the correctness of the wireless links transmission. For the hybrid WiNoC design, the reliability of the wired links can be guaranteed by adopting the protection schemes used in the conventional NoCs. However, because of the high manufacturing defect rates, process variability and operational uncertainties, the RF- or CNT-based wireless communication suffers from much less reliability compared with RC wire based conventional communication [35, 10]. Therefore error detection and correction schemes are urgently needed in the WiNoCs.

Since the WiNoC is still a recently emerging technique for the NoC implementation, few researchers have focused on the reliability enhancement of the WiNoCs. In [36], the reliability of the proposed hybrid WiNoC based on the small-world topology [12] was evaluated. The results showed that because of the inherent robustness of the small-world-based connectivity, their hybrid WiNoC has higher reliability compared the conventional implementation with wirelines.

Error control coding (ECC) is commonly used for improving the communication reliability in conventional NoCs [37, 38]. Research work [39, 40] proposed to employ the ECC to protect the intra- and inter-chip wireless communication. As the bit error rate (BER) is a key property to measure the reliability of the on-chip wireless channels and usually it is much higher than that of the wireline links, an energy-efficient ECC was proposed in [41] to improve the BER for intra-chip RF wireless communication. It concluded that the linear ECC like Hamming codes improves the BER at high signal-to-noise ratio (SNR). However, the performance will be degraded by employing the ECC at low SNR. Therefore the ECC is very efficient for the long range wireless communications.

In [42], a unified ECC scheme was proposed to improve the reliability of the small-world-based hybrid WiNoC. The evaluation of the SNR and BER was done on CNT antennas. It also compared the performance and energy efficiency of their proposed scheme with a conventional wired mesh NoC. The results demonstrated that the unified ECC scheme for the hybrid WiNoC can achieve low energy consumption, low network latency and almost similar reliability compared with the conventional NoCs.

7 3D-NoC and photonic NoC

Similar to WiNoC, 3D-NoC and photonic NoC are two emerging interconnect infrastructures in future NoC design and recent studies [24, 43, 44] have made some preliminary comparisons among these different infrastructures. Table 1 shows the comparisons among the WiNoC, 3D-NoC and photonic NoC, in terms of hardware requirement, communication bandwidth, power consumption, reliability and implementation challenge. Compared with 3D-NoC and photonic NoC designs, the WiNoC does not need physical interconnect layout. Therefore latency and energy dissipation in communication will be reduced in WiNoC. The WiNoC also can have a higher wireless bandwidth than the wired links. Although there are some reliability issues in 3D-NoC and photonic NoC designs, WiNoC has less reliability compared with the wired-based communication, which needs to be carefully addressed in future WiNoC design.

Table 1 Comparisons among the WiNoC, 3D-NoC and photonic NoC

	3D-NoC	Photonic NoC	WiNoC
hardware requirement	multiple-layer stacks	silicon photonic components	on-chip antennas
communication bandwidth	higher connectivity (vertical links)	high-speed optical links	high wireless channels
power consumption	shorter path (less hops)	low power in optical transmission	low power in wireless
reliability	TSV failure	crosstalk in photonic waveguides	less reliable in wireless
challenge	thermal issue because of high-power density	integration of on-chip photonic components	integration of on-chip antennas

8 Open research issues

Open research issues range from hardware implementation (on-chip antennas) to low power and reliable design. A few of these are listed below:

- *On-chip antennas:* Although some novel on-chip antennas implementations, for example, the CNT antennas, have been recently proposed and studies, there are still lots of challenges in manufacturing and integrating these on-chip antennas in terms of area, performance, energy overheads, especially as the number of on-chip cores is scaling up in the future. In [45], an antennaless WiNoC was proposed, which used inductive-coupling instead of antennas to do the wireless communication. Although the WiNoC proposed in [45] is more similar to a hybrid 3D-based architecture by using the inductive-coupling to replace the through silicon via (TSV) in traditional 3D-NoC, this new type of wireless communication may impact the future on-chip wireless design.
- *Topologies: pure wireless against hybrid.* Owing to the transmission range and bandwidth (number of channels available) limitation of today's on-chip wireless communication, the hybrid topologies seem dominating today's WiNoC design. Whether we will keep working on the hybrid design like today's wireless Internet architecture, or the pure wireless design will become easy to implement with the invention of new on-chip antennas and wireless communication technologies, are still open issues.
- *Routing and flow control:* As the size of the on-chip network is increasing, more complex topologies may be needed in future WiNoCs, especially for the hybrid topologies. Therefore the design of more efficient routing algorithms and flow control mechanisms will be another challenge.
- *Low power:* The wireless links are assumed consuming less power compared with the wired links, especially for these long-distance links. However, as the power is recognised as one of the most pressing constraints in today's chip design, low-power WiNoC designs are also essential.
- *Reliability*: The wireless links in WiNoC are less reliable compared with the wired links. Only few researches, such as protecting the wireless communication by ECC, have been conducted on improving the reliability of the WiNoCs. We definitely need more architectural and circuit level schemes to improve the reliability for future WiNoCs.

9 Conclusion

NoC architectures are becoming the dominant communication fabric in multi-/many-core chips. To solve the complex communication problems in NoCs as the number of on-chip cores is scaling up, the WiNOC architecture has been recently proposed and studied. In this paper, we have discussed the design of WiNoCs in five categories: topology, routing algorithm, flow control, on-chip antenna and reliability. We also described some open research issues in this field. Along with the current research work in WiNoCs, more insight into the design problems and more development in solutions to the open research issues are encouraged.

10 References

[1] Pande P.P., Grecu C., Jones M., Ivanov A., Saleh R.: 'Performance evaluation and design trade-offs for network-on-chip interconnect architectures', *IEEE Trans. Comput.*, 2005, **54**, (8), pp. 1025–1040

[2] Vangal S., Howard J., Ruhl G., ET AL.: 'An 80-tile 1.28tflops network-on-chip in 65 nm cmos'. Proc. IEEE Int. Solid-State Circuits Conf., February 2007, pp. 98–589

[3] Bell S., Edwards B., Amann J., ET AL.: 'Tile64 processor: a 64-core soc with mesh interconnect'. Proc. IEEE Int. Solid-State Circuits Conf., February 2008, pp. 88–598

[4] Bahn J.H., Lee S.E., Bagherzadeh N.: 'On design and analysis of a feasible network-on-chip (noc) architecture'. Proc. Fourth Int. Conf. Information Technology, April 2007, pp. 1033–1038

[5] Pavlidis V., Friedman E.: '3d topologies for networks-on-chip', *IEEE Trans. Very Large Scale Integr. (VLSI) Syst.*, 2007, **15**, (10), pp. 1081–1090

[6] Shacham A., Bergman K., Carloni L.P.: 'Photonic network-on-chip for future generations of chip multi-processors', *IEEE Trans. Comput.*, 2008, **57**, (9), pp. 1246–1260

[7] Lin J., Wu H.T., Su Y., ET AL.: 'Communication using antennas fabricated in silicon integrated circuits', *IEEE J. Solid-State Circuits*, 2007, **42**, (8), pp. 1678–1687

[8] Fukuda M., Saha P.K., Sasaki N., Kikkawa T.: 'A 0.18 μm cmos impulse radio based uwb transmitter for global wireless interconnections of 3d stacked-chip system'. Proc. Int. Conf. Solid State Devices and Materials, September 2006, pp. 72–73

[9] Nojeh A., Ivanov A.: 'Wireless interconnect and the potential for carbon nanotubes', *IEEE Des. Test Comput.*, 2010, **27**, (4), pp. 44–53

[10] Lee S.B., Tam S.W., Pefkianakis I., ET AL.: 'A scalable micro wireless interconnect structure for cmps'. Proc. 15th annual Int. Conf. Mobile Computing and Networking, September 2009, pp. 217–228

[11] Wang C., Hu W.H., Bagherzadeh N.: 'A wireless network-on-chip design for multicore platforms'. Proc. 19th Euromicro Int. Conf. Parallel, Distributed and Network-Based Processing (PDP), February 2011, pp. 409–416

[12] Ganguly A., Chang K., Deb S., ET AL.: 'Scalable hybrid wireless network-on-chip architectures for multicore systems', *IEEE Trans. Comput.*, 2011, **10**, (60), pp. 1485–1502

[13] Zhao D., Wang Y.: 'Sd-mac: design and synthesis of a hardware-efficient collision-free qos-aware mac protocol for wireless network-on-chip', *IEEE Trans. Comput.*, 2008, **57**, (9), pp. 1230–1245

[14] Zhao D., Wang Y., Li J., Kikkawa T.: 'Design of multi-channel wireless noc to improve on-chip communication capacity'. Proc. Fifth IEEE/ACM Int. Symp. Networks on Chip (NoCS), May 2011, pp. 177–184

[15] Benedetto M.-G.D., Giancola G.: 'Understanding ultra wide band radio fundamentals' (Prentice-Hall, 2004)

[16] Kim J., Balfour J., Dally W.J.: 'Flattened butterfly topology for on-chip networks'. Proc. Int. Symp. Microarchitecture, December 2007, pp. 172–182

[17] Ogras U.Y., Marculescu R.: 'Its a small world after all: noc performance optimization via long-range link insertion', *IEEE Trans. Very Large Scale Intergr. (VLSI) Syst.*, 2006, **14**, (7), pp. 693–706

[18] Watts D.J.: 'The dynamics of networks between order and randomness' (Princeton University Press, Princeton, NJ, USA, 1999)

[19] Buchanan M.: 'Nexus: small worlds and the groundbreaking theory of networks'. Technical Report, WW Norton and Company, 2003

[20] Teuscher C.: 'Nature-inspired interconnects for self-assembled large-scale network-on-chip designs', *Chaos*, 2007, **17**, (2), pp. 026106-1–026106-12

[21] Watts D.J., Strogatz S.H.: 'Collective dynamics of small-world networks', *Nature*, 1998, **393**, pp. 440–442

[22] Petermann T., Rios P.D.L.: 'Physical realizability of small-world networks', *Phys. Rev. E*, 2006, **73**, (2), pp. 026114-1–026114-4

[23] Deb S., Ganguly A., Chang K., ET AL.: 'Enhancing performance of network-on-chip architectures with millimeter-wave wireless interconnects'. Proc. 21st IEEE Int. Conf. Application-specific Systems Architectures and Processors (ASAP), July 2010, pp. 73–80

[24] Deb S., Chang K., Ganguly A., Pande P.: 'Comparative performance evaluation of wireless and optical noc architectures'. Proc. 2010 IEEE Int. SOC Conf. (SOCC), September 2010, pp. 487–492

[25] Kirkpatrick S., Gelatt C.D., Vecchi M.P.: 'Optimization by simulated annealing', *Science*, 1983, **220**, (4598), pp. 671–680

[26] Wu R., Wang Y., Zhao D.: 'Low-cost deadlock-free design of minimal-table rerouted xy-routing for irregular wireless nocs'. Proc. ACM/IEEE Int. Symp. Networks-on-Chip (NOCS), May 2010, pp. 199–206

[27] Dally W.J., Seitz C.L.: 'The torus routing chip', *J. Distrib. Comput.*, 1986, **1**, (3), pp. 187–196

[28] Dally W.J.: 'Virtual-channel flow control'. Proc. Int. Symp. Computer Architecture, 1990

[29] Bahn J.H., Bagherzadeh N.: 'Efficient parallel buffer structure and its management scheme for a robust network-on-chip(noc) architecture'. Proc. 13th Int. CSI Computer Conf., 2008

[30] Floyd B.A., Hung C.M., Kenneth K.O.: 'Intra-chip wireless interconnect for clock distribution implemented with integrated antennas, receivers, and transmitters', *IEEE J. Solid-State Circuits*, 2002, **37**, (5), pp. 543–552

[31] Zhang Y.P., Chen Z.M., Sun M.: 'Propagation mechanisms of radio waves over intra-chip channels with integrate antennas: frequency-domain measurements and time-domain analysis', *IEEE Trans. Antennas Propag.*, 2007, **55**, (10), pp. 2900–2906

[32] Kempa K., Rybczynski J., Huang Z., ET AL.: 'Carbon nanotubes as optical antennae', *Adv. Mater.*, 2007, **19**, (3), pp. 421–426

[33] Burke P.J., Li S., Yu Z.: 'Quantitative theory of nanowire and nanotube antenna performance', *IEEE Trans. Nanotechnol.*, 2006, **5**, (4), pp. 314–334

[34] Huang Y., Yin W.Y., Liu Q.H.: 'Performance prediction of carbon nanotube bundle dipole antennas', *IEEE Trans. Nanotechnol.*, 2008, **7**, (3), pp. 331–337

[35] Bahar R.I., Lau C., Hammerstrom D., ET AL.: 'Architectures for silicon nanoelectronics and beyond', *IEEE Comput.*, 2007, **40**, (1), pp. 25–33

[36] Ganguly A., Wettin P., Chang K., Pande P.: 'Complex network inspired fault-tolerant noc architectures with wireless links'. Proc. ACM/IEEE Int. Symp. Networks-on-Chip (NOCS), May 2011, pp. 169–176

[37] Bertozzi D., Benini L., DeMicheli G.: 'Low power error resilient encoding for on-chip databuses'. Proc. Conf. Design, Automation and Test in Europe, 2002, pp. 102–109

[38] Sridhara S., Shanbhag N.: 'Coding for system-on-chip networks: a unified framework', *IEEE Trans. Very Large Scale Integr. (VLSI) Syst.*, 2005, **13**, (6), pp. 655–667

[39] Rahaman M.S., Chowdhury M.H.: 'Improved bit error rate performance in intra-chip rf/wireless interconnect systems'. Proc. ACM/IEEE Great Lake Symp. VLSI, May 2008

[40] Rahaman M.S., Chowdhury M.H.: 'Bit-error-rate performance of inter-chip rf/wireless interconnect systems'. Proc. Int. Conf. Microelectronics, December 2008

[41] Rahaman M.S., Chowdhury M.H.: 'Energy efficiency of error control coding in intra-chip rf/wireless interconnect systems', *Microelectron. J.*, 2010, **41**, (1), pp. 33–40

[42] Ganguly A., Pande P., Belzer B., Nojeh A.: 'A unified error control coding scheme to enhance the reliability of a hybrid wireless network-on-chip'. Proc. IEEE Defect and Fault Tolerance Symp. (DFT), October 2011, pp. 277–285

[43] Carloni L.P., Pande P., Xie Y.: 'Networks-on-chip in emerging interconnect paradigms: advantages and challenges'. Proc. Int. Symp. Networks-on-Chip (NOCS), May 2009, pp. 93–102

[44] Ganguly A.: 'Towards a scalable and reliable wireless network-on-chip', *PhD dissertation*, Washington State University, 2010

[45] Matsutani H., ET AL.: 'A case for wireless 3d nocs for cmps'. Proc. 18th Asia and South Pacific the Design Automation Conf. (ASP-DAC), January 2013, pp. 23–28

Composite mechanisms for improving Bubble Rap in delay tolerant networks

Sweta Jain[1], Nikhitha Kishore[1], Meenu Chawla[1], Vasco N.G.J. Soares[2,3]

[1]*Department of Computer Science & Engineering, Maulana Azad National Institute of Technology, Bhopal, India*
[2]*Instituto de Telecomunicações, University of Beira Interior, Covilhã, Portugal*
[3]*Superior School of Technology, Polytechnic Institute of Castelo Branco, Castelo Branco, Portugal*
E-mail: nikhithakishore89@gmail.com

Abstract: Delay tolerant networks (DTNs) are a subset of mobile *ad hoc* networks where connections are sparse and intermittent. This often results in a network graph which is rarely connected which introduces a challenge in message forwarding because of a lack of end-to-end connectivity towards the destination. Recently, social-based forwarding algorithms are gaining popularity because of the social nature displayed by the node movements in a DTN, especially in application areas like the pocket switched networks. The social-based metrics like community, similarity, centrality etc. are used to determine the carrier to which a node has to forward its message. Composite methods are used to improve the performance of Bubble Rap social-based forwarding algorithm. In the proposed mechanism, a new social metric termed 'friendship' has been introduced along with a time-to-live (TTL)-based 'threshold' and acknowledgement (ACK) IDs. Real trace data and working day movement models are used for simulations in the opportunistic network environment simulator to demonstrate that the proposed algorithm gives better delivery ratio than the original Bubble Rap algorithm.

1 Introduction

Delay tolerant networks (DTNs) [1–3] can be described as sparse mobile *ad hoc* networks (MANETs) [4] where a lack of continuous end-to-end connectivity between the nodes makes the problem of routing messages a challenging task. However, DTNs find their own significance in an era dominated by the tremendous usage of wireless devices being used in a multitude of applications ranging from tactical networks, military application and sensor networks to disaster recovery. DTNs are characterised by their long transmission delays, frequent network partitions, intermittent connectivity and a lack of immediate end-to-end connectivity between source and destination, part of which can be attributed to factors such as power shutdown because of poor resource allocation, physical obstacles in the path of a DTN node's movement etc. [5]. Since network graph of such a network is hardly ever connected, routing using traditional MANET protocols like ad-hoc on demand distance vector routing (AODV), optimal secure routing protocol (OSR) etc. fails as all these protocols need a connected path between the source and the destination.

In this paper, focus is on a special application area of the DTNs known as the pocket switched networks (PSNs) [6] where mobile devices are carried by a mass of people who are involved in normal real-world activities. In such an environment, the routing can be improved a lot by better studying the human characteristics, that is, how often a person moves from one location to another, the people with whom a particular person meets frequently, his routines, the social groups in which the person has participation in etc. All the above-mentioned characteristics decide the movement pattern of the people carrying the mobile nodes and are known as the social characteristics [7]. One advantage of using social-based forwarding in PSNs is that social characteristics tend to be less unstable than the node mobility [7].

The 'Bubble Rap' social-based forwarding algorithm [8] and 'SimBet' [9] are the two most popular and earlier approach in forwarding based on social characteristics. Even though 'Bubble Rap' has been proved to be an efficient social-based forwarding technique in PSNs, it comes with a lot of limitations. First, the 'Bubble Rap' is basically based on the 'centrality' [10, 11] of the nodes which describes how much popular the nodes are in the network. However, centrality cannot estimate the strong relationship between a pair of nodes in the network which can be helpful in delivering messages to the destination quickly. Also in Bubble Rap, the long wait in the buffer until a popular node based on centrality is encountered can lead to either the expiry of the message if the message TTL is too short or buffer congestion if message TTL is too long. This is a serious problem in PSNs where buffer size is too limited. Another problem with 'Bubble Rap' is that it does not delete the delivered messages from the message buffer of other nodes. This leads to replication of messages that have been already delivered resulting in increased overhead and limited buffer space for new messages.

In this paper, composite mechanisms to improve Bubble Rap (CMs bubble rap) is proposed, which is an improvement to the already existing Bubble Rap algorithm. The main contribution of this paper is the inclusion of a new social metric termed as 'friendship' in the global 'bubble-up' phase of the 'Bubble Rap'. In addition, we include a message TTL-based threshold to limit the time for which the message waits for a popular node along with use of acknowledgments for messages that are known to be delivered for efficiently managing the buffer space. Finally, simulations are used to show that the proposed method can provide better delivery ratio, communication overhead and latency compared with the original Bubble Rap.

The paper is organised as follows. The related work is discussed in Section 2. In Section 3, the CM–Bubble Rap algorithm has been described in detail. Section 4 presents the evaluation of the new method and its comparison with the existing Bubble Rap algorithm using the Opportunistic Network Environment (ONE) simulator and Section 5 concludes this paper and presents the future work.

2 Related work

The DTN does not guarantee end-to-end connectivity, and because of this reason itself traditional MANET routing protocols fail here. Many routing approaches have been suggested in the literature for DTNs. The approaches can be categorised as naïve replication methods and utility-based forwarding methods.

The 'Epidemic [12] and Spray and Wait' [13] routing protocols are the two significant protocols in the category of naïve replication routing methods. In Epidemic routing, a node carrying messages forwards them to any other encountered node that does not possess the same message. This approach increases the delivery ratio only if the buffer size and bandwidth are not limited.

Moreover, the energy consumption is high because of larger number of relays. The Spray and Wait keeps a limit on the number of copies of a message to be forwarded as L. The source node sprays L copies to the first L encountered nodes. If destination is not met in the first L encounters, the L nodes forward it directly to the destination.

The 'ProPHET' [14] routing protocol is a good example of the utility-based forwarding. It measures how better a node can forward the message to the destination through a metric called the delivery predictability. This value is updated for each pair of nodes during each encounter. During an encounter, if the encountered node has better delivery predictability to destination than current node then the message is forwarded. Another family of routing algorithms that fall in this category is the social-based routing algorithms.

All approaches work best for a particular application area. For example, history-based encounters cannot be used in an area where node movements are random; it requires predictable movements [14]. Flooding will prove to be the worst approach in a large network with high node intensity. In the case of PSNs, where node movements are based on the human characteristics, routing can be done through a thorough study on the social metrics. A large network and limited number of buffer size characterise the PSNs. However, much work has not been done in social-based routing algorithms, but it is gaining interest in researchers with more and more people using mobile devices in our societies today.

Most social-based algorithms are based on the social metrics studied as a part of sociology [7]. A prior work in this field was done by Hui and Crowcroft [15] with a protocol termed as 'Label Routing'. The basic idea was that each message be tagged with labels that denote their community. The forwarding will be done only to the nodes in the same community. Since this requires very less information from the network, it is simple in its own functionality. However, this cannot assure good delivery ratio in all cases as nodes from the same community need not be well connected with another community member leading to delay in delivery.

The 'SimBet' [9] is another social-based routing protocol proposed by Daly and Haahr. The main idea was to find some 'bridge' nodes in the network which have high 'betweenness centrality' [16] (a centrality measure) and 'similarity' [7]. Although a node with high betweenness centrality can facilitate many connections between other nodes, similarity defines the common neighbours between two nodes. During an encounter, the current node calculates its betweenness and similarity utility to the encountered node. Then, it calculates a 'SimBet' utility which is a weighted combination of betweenness and similarity utility with the destination node. If the 'SimBet' utility of the encountered node with the destination is greater than this value, then forwarding is carried. However, because of the uncertainty of future encounters and underlying social graph, it is even possible that a node with high SimBet utility can fail to deliver messages.

The Bubble Rap social-based forwarding algorithm by Hui *et al.* [8] combines the notion of both community and centrality. The algorithm chooses popular nodes in the network to relay the messages to their destination communities and choose a popular node within the community to deliver the message to the destination. The popularity is measured using centrality metric. This approach assumes that nodes within the same community are well connected because of their common interests.

3 Proposed algorithm: CM–Bubble Rap

The 'Bubble Rap' social-based forwarding algorithm is one of such kind to take advantage of the community structure of the human society for the forwarding of messages in a PSN. It is based on two social metrics: 'community' and 'centrality'. However, using only centrality and community cannot exploit many other social ties between the nodes in a social network.

Bubble Rap finds the most popular node in the network to forward the messages to its own community member using centrality concept. However, there may be cases where a person may not be popular in the social network, but still has some ties with another person. For example, two people are best friends. In such cases, both need not belong to the same community or may not be popular in the network. However, their contacts are frequent. This type of social characteristics that exists between a pair of nodes is termed as 'friendship' [7, 17]. In DTNs, friendship can be defined between a pair of nodes. To be considered as friends of each other, two nodes need to have long-lasting and regular contacts/common interests. Capturing such a social characteristic can increase the delivery performance of the algorithm itself.

In this method called CM–Bubble Rap, friendship between two nodes is calculated using the average delivery predictability that was introduced in [18]. This is an improvement to the ProPHET routing algorithm [9] where the delivery predictability is used for forwarding the messages. Average predictability simply averages the past delivery predictabilities in each encounter to capture the overall network performance [18]. The main contribution of this paper is calculating the 'friendship' between two nodes using average delivery predictability.

Along with this, new metric CM–Bubble Rap uses two other mechanisms to improve the performance of Bubble Rap. To limit the time spent by a message waiting for a popular node, CM–Bubble Rap uses a message TTL-based threshold and to save the buffer space it exchanges ACK IDs. The CMs used by CM–Bubble Rap are described below:

(A) *Friendship between two nodes:* to calculate the amount of friendship between two nodes, the average delivery predictability is calculated between them. Each node i keeps delivery probability $P_{(i,\ j)} \in [0,\ 1]$ with every other node j it has encountered in the network. When node i encounters node j for the first time, the delivery predictability is calculated as in (1)

$$P_1(i,\ j) = P_{\text{init}} \text{ where } P_{\text{init}} \in [0,\ 1] \text{ is the initialisation constant} \tag{1}$$

Node i records this and time t_1 when the node i and j are encountered for the first time. When nodes i and j are encountered for the second time at time t_2, the delivery predictabilities are updated as in (2)

$$P_2(i,\ j) = P_1(i,\ j) + \left(1 - P_1(i,\ j)\right) \times P_{\text{init}} \tag{2}$$

Node i records this time t_2 and calculate the average delivery predictability between the past and the current encounters. This is calculated as in (3)

$$P_{\text{avg}(2)}(i,\ j) = \frac{(P_1(i,\ j) \times t_1 + P_2(i,\ j) \times t_2)}{t_1 + t_2} \tag{3}$$

Like this at the nth encounter, average delivery predictability can be calculated from node i to node j as in (4)

$$P_{\text{avg}(n)}(i,\ j) = \frac{(P_{\text{avg}(n-1)}(i,\ j) \times \sum_1^{n-1} t + P_n(i,\ j) \times t_n)}{\sum_1^n t} \tag{4}$$

If the message is not in its community and the node carrying the message encounters a node with a lower global rank, then the 'friendship' of the encountered node with the destination and that of current node with destination of the message are compared. If the encountered node has better 'friendship' with the destination of the message, then the message is forwarded. The idea is that messages when staying outside their community should not be held longer in the expense of obtaining a popular node. Instead wise decisions must be made and allowed to be forwarded for increasing the delivery ratio.

Message TTL threshold: when the message is not in its community, then when a node with lower global rank is encountered, the node calculates the remaining TTL of the message to be forwarded. In 'CM–Bubble Rap', a threshold parameter T_{ttl} is kept. The T_{ttl} is n % of the initial TTL of a message. This parameter n can be called as the flood control parameter because the value of n controls the level of flooding in the network. The value n is a configurable parameter. For limiting the flooding, we keep this parameter as 25 always. During message forwarding if the message's remaining TTL is less than the parameter T_{ttl}, then the message will be forwarded irrespective of the ranking of the encountered node. The idea behind this is that if the message has only less time to expire, then there is no need to naively hold the message until a popular node is encountered.

(B) *Exchange of ACK ID:* the 'distributed Bubble Rap' deletes the messages from the original carrier once they have entered the local community. This method uses the concept of ACKs to delete those messages from the buffer which are known to be already delivered. Since in PSNs the buffer size is too low, efficient buffer management is very important for a better delivery ratio. Thus, if the buffer is full, a message which may be important will have to be dropped when another message arrives. This will have a serious impact on the delivery ratio. Thus, deleting those messages which are already delivered can save the important messages from being dropped.

We have used 'distributed *K*-Clique community detection algorithm' [19] which was used in Bubble Rap [8] for detecting communities. The node centralities are calculated by using 'C-Window centrality' [8] which approximates the cumulative degree of the past windows (e.g. 6 h).

3.1 CM–Bubble Rap forwarding algorithm

The following assumption is added with those made in [8]. Each node i stores a table to calculate the average delivery probability $P_{avg(i,\ j)}$ with every other node j in the network. This calculates the 'friendship' between a pair of nodes in the network.

Let us consider that a node A has a message destined for node D whose community is C_D. When node A encounters node B, ACKs are exchanged. Each node updates their local community, average delivery probability and centrality measures. Those messages whose ACKs have arrived are deleted from each node's buffer. After these steps, nodes A and B look forward to message exchange. If node A is of the same community as node D, that is, node A belongs to community C_D, then node A checks whether node B is from the same community C_D. If yes, then the local centrality value of node B is checked. If local centrality ranks of node B are higher in C_D than that of node A, then send the message to node B. If node B is not in the same community, then the message will be held within A itself.

Now, let us consider a situation where node A is not in the community C_D. Here also, the messages that are known to be delivered are deleted from the buffer. In this case, it checks whether node B is of the same community or if it is not in C_D, then it checks whether its global rank is better than that of node A. Then, it forwards the message to node B. Now, if node B's global centrality rank is less than node A, then in such a case instead of simply rejecting the message to node B as is done in Bubble Rap, CM–Bubble Rap implements a new strategy. As already mentioned, we keep a threshold parameter called T_{ttl}. If node B has a lower global centrality value, then the remaining TTL of the message in node A, M_{rem} is calculated. If the M_{rem} is less than T_{ttl}, CM–Bubble Rap assumes that since there is not so much time for the message to expire or it has spent a lot of time in the buffer. Thus, this message is forwarded to node B increasing the possibility of the message delivery. In addition, the average delivery probability of A with D, $P_{avg(A,\ D)}$ is compared with that of B with D, $P_{avg(B,\ D)}$. If $P_{avg(B,\ D)} > P_{avg(A,\ D)}$, then irrespective of the global ranking of the encountered node B, we forward the message to node B.

Table 1 Heterogeneous scenario settings

Scenario parameters	Values
simulation time	700 k s
number of nodes	150
number of node groups	17 (8 groups of people and 9 groups of vehicle)
transmission speed	250 kbps
transmission range	10 m
buffer size	2 M
message size	1–100 kb
vehicle_group1	movement model = shortest path based movement model number of nodes = 6 wait time = 100–300 s speed = 7–10 m/s
vehicle_group2–vehicle_group9	movement model = bus movement model number of nodes = 2/group wait time = 10–30 s speed = 7–10 m/s
people_group1–people_group8	movement model = working day movement model number of nodes = 16/group wait time = 0 s speed = 0.8–14 m/s

The pseudo code for the algorithm CM–Bubble Rap is summarised in Table 1 (Fig. 1). The method 'encounter' *(A, B, n)* is called whenever there is an encounter between the two nodes A and B. The value of n is also passed to the method 'encounter'. The 'compute_AverageDeliveryPredictability' [$P(A,\ B)$] is the function to update the average delivery predictabilities for every encounter and is explained in Table 2 (Fig. 2). The lines 6, 7, 8, 9, 26, 27 and 28 in Algorithm 1 (Fig. 1) are the additions of CM–Bubble Rap to the original Bubble Rap protocol.

4 Evaluation

In this section, the proposed algorithm is evaluated under various scenarios, the results are then presented and analysed. The 'CM–Bubble Rap' is compared with the traditional Bubble Rap in terms of delivery probability, overhead ratio and latency median. Two scenarios are considered: (i) based on the working day movement model and (ii) based on real human traces to evaluate the results.

4.1 Evaluation metrics

The metrics that have been used across the simulations are: delivery ratio (DR), average latency (ALat) and overhead ratio (OR). Here, DR and ALat are meant to evaluate the effectiveness of the routing protocol in terms of delivery and OR gives a measure of the resource consumption/friendliness of the protocol. Higher the OR less resource friendly the protocol; higher the DR and a small ALat indicates high effectiveness of the protocol.

4.2 Experimental setup

The experiments have been performed in the ONE simulator [20, 21]. The heterogeneous scenario consists of 150 nodes simulating city life. The Helsinki city map is used for the simulations. The

```
Algorithm 1: Procedure CM-Bubble rap: encounter(A,B,n)
for each message m ∈ Buffer_A do
   if m.ID ∈ deliveredMessagesID OR TTL(m) expires then
      A.removeFromBuffer(m)
   end if
end for
compute_AverageDeliveryPredictability(A,B)
for each message m ∈ Buffer_A do
   T_ttl ← InitTTL(m) × n/100
   M_rem ← RemTTL(m)
   D←m.Destination
   if B = D then
      B.addInBuffer(m)
   else
      if D ∈ A.localCommunity AND D ∈ B.localCommunity then
         if B.localCentrality > A.localCentrality then
            B.addInBuffer(m)
         end if
      else
         if D ∉ A.localCommunity
            if D ∈ B.localCommunity then
               B.addInBuffer(m)
            else
               if B.globalCentrality > A.globalCentrality then
                  B.addInBuffer(m)
               else
                  if B.globalCentrality < A.globalCentrality then
                     if M_rem < T_ttl OR P_avg(A,D) < P_avg(B,D) then
                        B.addInBuffer(m)
                     end if
                  end if
               end if
            end if
         end if
      end if
   end if
end for
```

Fig. 1 *Pseudo code for the algorithm CM–Bubble Rap*

nodes are grouped into eight different groups of people and nine different groups of vehicles. All vehicle groups use the 'bus movement model', except one group which uses the 'shortest path-based movement model'. The people follow the 'working day movement model' [20]. The settings of this scenario are given in Table 1. In the case of 'working day movement model', each group was given different meeting locations, offices and home locations. The parameters used in common for all the groups are mentioned in Table 2. In total, an external load file was used to generate a total of 1225 messages for the simulation.

Along with using movement models, the real life traces are used to compare the new protocol with Bubble Rap. For this purpose, Cambridge traces have been used which consist of encounter traces of 36 nodes [22]. The data in the traces were accumulated in various locations for two months for the Cambridge students as a part of the Haggle project. For the Cambridge dataset, an external load file was also used to generate 1225 messages. The settings are given in Table 3.

```
Algorithm 2: Function compute_AverageDeliveryPredicatability(A,B)
        StoreToList(N,P_N(A,B),P_avg(N)(A,B), t_N) is a list for each node A
for a node B that it encounters,
        where
            N is the number of encounter, initially N = 0,
            P_N(A,B) is the N^th encounter delivery predictability,
            P_avg(N)(A,B) is the N^th encounter average delivery predictability,
            t_N is the time of encounter between A and B
        P_init is the initial delivery probability
if N = 0
   t_N = getCurrentTime()
   N = N + 1
   P_N(A,B) = P_init
   P_avg(A,B) = 0
   A.StoreToList(N, P_N(A,B), P_avg(N)(A,B), t_N)
else
   N = N + 1
   P_N(A,B) = P(N-1)(A,B) + (1 – P(N-1)(A,B)) × P_init
   P_avg(N)(A,B) = (P_avg(N-1)(A,B) × σ ^-1 t + P_N(A,B) × t_N) / Σ_1^N t
   A.StoreToList(N, P_N(A,B), P_avg(N)(A,B), t_N)
end if
```

Fig. 2 *Pseudo code for the function compute_AverageDeliveryPredictability (A, B)*

Table 2 Working day movement model settings

Scenario parameters	Values
number of offices	50
working day length	28 800 s (= 8 h)
probability to do evening activity	50%
number of meeting spots	10
the coefficient for the Pareto distribution controlling pause time inside office	0.5
minimum pause time inside office	10 s
maximum pause time inside office	100 000 s
size of the office	100 m
standard deviation for the normal distribution controlling differences in schedule nodes have	7200
minimum groups size for evening activity	1
maximum groups size for evening activity	3
minimum pause time after evening activity	3600 s
maximum pause time after evening activity	7200 s
probability that the node owns a car	50%

Table 3 Cambridge traces details

Scenario parameters	Values
simulation time	1 036 800 s
number of nodes	36
buffer size	2 M
message size	1–100 kb

Since the comparisons are made against Bubble Rap, those message TTLs are chosen in which Bubble Rap proves to give the best results. Hence, the message TTL is varied with values of 24 h (1 day), 48 h (2 days), 96 h (4 days), 168 h (1 week) and 504 h (3 weeks). For each value of message, TTL simulations were run five times each with different random number generator seeds. The metrics evaluated are the average of the five runs. As already mentioned, '*K*-Clique community detection' algorithm is used to detect community structures and 'C-Window' centrality to calculate node centralities. The value of *k* in the 'K-Clique community detection' algorithm is chosen to be 5 on the basis of the simulations in the Bubble Rap paper for which the best results appeared in [8].

4.3 Results

The delivery ratio results for the heterogeneous scenario and the Cambridge trace results are discussed in this section. Figs. 3*a* and *b* show the comparisons for the delivery ratio of CM–Bubble Rap and Bubble Rap for various message TTLs in heterogeneous and Cambridge trace scenarios, respectively.

From the graphs, it is clear that for the heterogeneous scenario, CM–Bubble Rap provides an average increase of 40% in the delivery ratio with the original Bubble Rap. It may be observed that as the message TTL increases Bubble Rap has a considerable degradation in its delivery ratio. The reason behind this is that as TTL increases, the message stays more time in the buffer. In addition, since most nodes have heterogeneous centralities in the network, the probability of message forwarding is more when the node encounters occur. Thus, with higher TTL values, the message

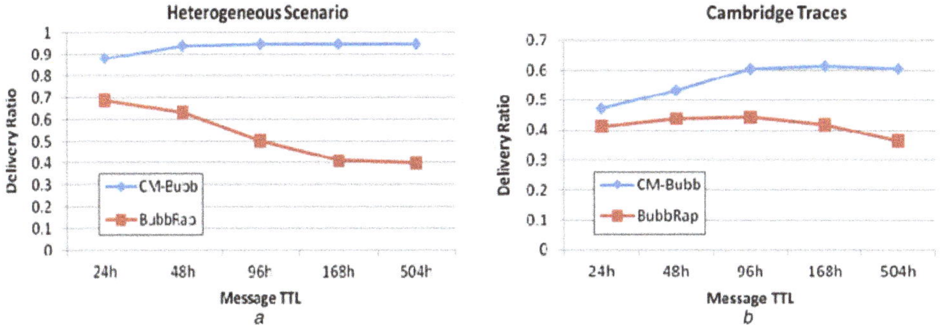

Fig. 3 *Comparisons for the delivery ratio of CM–Bubble Rap and Bubble Rap*
a DR for heterogeneous scenario
b DR for Cambridge traces

stays for more time in the buffer, and here since the node centralities differ highly, the message replications also increase. Since the buffer size is limited in PSNs, such a situation leads to a higher drop rate of messages. The higher drop rates lead to a lower delivery ratio. However, it can be seen that these problems are solved in CM–Bubble Rap with the results presented. CM–Bubble Rap does not solely rely on centrality, but also on friendship metric. This enables CM–Bubble Rap to make wiser decisions than the Bubble Rap which leads to its increase in the delivery ratio. Also it can be seen that as the message TTL increases, the delivery ratio does not decrease considerably, rather it remains a constant. The main reason is that, in CM–Bubble Rap, buffer is properly managed by means of exchanging ACK IDs. This means that it deletes those messages that are already delivered to prevent the drop of other messages that are not yet delivered. In addition, after a particular threshold, the messages are no longer kept in the buffer, rather they are forwarded. Thus, even if their message TTL increases, they do not congest the buffer space by staying for a long time because of the threshold T_{ttl}.

Now looking into the Cambridge traces, it can be inferred that the CM–Bubble Rap is still able to provide a better delivery ratio, with an average difference of 15% with that of Bubble Rap. The original Bubble Rap does not decrease quickly with an increase in message TTL here. This is because the probability of the nodes in these traces to encounter another node with a higher centrality value is lesser. The devices are carried by students and they do not vary much in their centrality. Thus, replication is very low, and therefore there is no buffer congestion which means a low rate in the dropping of messages. A message with a TTL of 96 h is enough to reach the destination without much replication. This can be explained because the number of communities detected by '*K*-Clique' can cover almost all the nodes in the traces because the number of nodes is just 36. However, increasing the TTL value does not help as this leads to limitation in buffer space, again leading to a low delivery ratio. Owing to the wiser forwarding decisions taken by CM–Bubble Rap, the delivery ratio is more than that of Bubble Rap. This shows the advantage of using friendship along with centrality for forwarding decisions. In addition, since a message TTL-based threshold plus exchange of ACK IDs are performed, the delivery ratio does not decrease for high TTL values.

The results of the paired *T*-tests between the delivery probabilities of CM–Bubble Rap and Bubble Rap in both heterogeneous scenario and Cambridge traces are given in Table 4. The difference mean between CM–Bubble Rap and Bubble Rap in the Cambridge traces is ~0.15 and that in the heterogeneous scenario is ~0.40. The positive value in difference clearly indicates that CM–Bubble Rap performs far better than Bubble Rap. Also the *P*-values which are far below 0.01 indicate that the results are statistically significant.

Figs. 4*a* and *b* show the overhead ratios of CM–Bubble Rap and Bubble Rap in both heterogeneous scenario and Cambridge traces, respectively, for varying message TTL values.

When analysing the overhead ratios between our proposed method and the original Bubble Rap in the heterogeneous scenario, it can be understood that the CM–Bubble Rap has a very small overhead which is almost constant throughout the varying message TTL. In Bubble Rap, the communication overhead increases with an increase in the message TTL. The main reason is that the Bubble Rap does not delete the replicas of those messages which are already delivered from the other nodes' buffer. Thus, a longer TTL means, the replicas of those messages already delivered continue to replicate for a longer time staying in the nodes' buffer. This significantly increases the overhead of the algorithm. The wiser decision taken by CM–Bubble Rap requires very less communication cost for the messages to reach the destination. By inclusion of the friendship metric better social ties were discovered. The CM–Bubble Rap could discover nodes that can easily take the message to the destination rather than relaying to a number of nodes. In addition, since the ACK IDs of those messages known to be delivered are deleted from the nodes' buffer, replication of delivered messages does not occur which can reduce the overhead. In addition, since after a particular threshold, the messages are forwarded to the other nodes staying in a particular node for a long time and replicating reduces decreasing the overhead further. The same reasoning can be provided for the Cambridge traces also.

Table 4 Results of the paired *T*-test between delivery ratios of CM–Bubble Rap and Bubble Rap

Protocols	*t*-values	Differences in mean	Minimum differences with 95% CI	Maximum differences with 95% CI	*P*-values	Statistical significances
			Cambridge traces			
CM against Bubble Rap	4.594851186	0.151204	0.06072	0.24146	0.005	significant
			Heterogeneous scenario			
CM against Bubble Rap	5.941736422	0.406995	0.193195	0.5489	0.002	significant

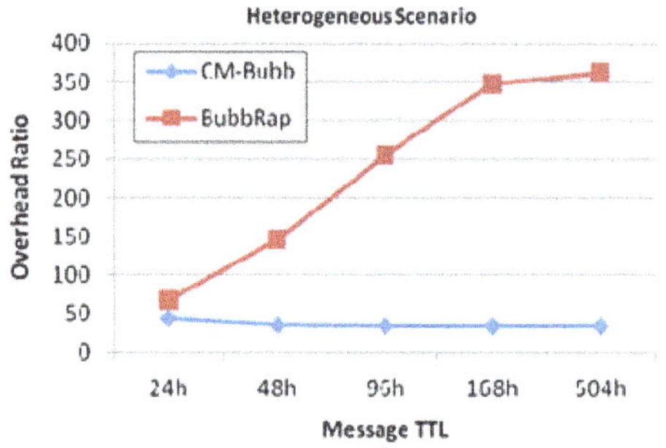

a

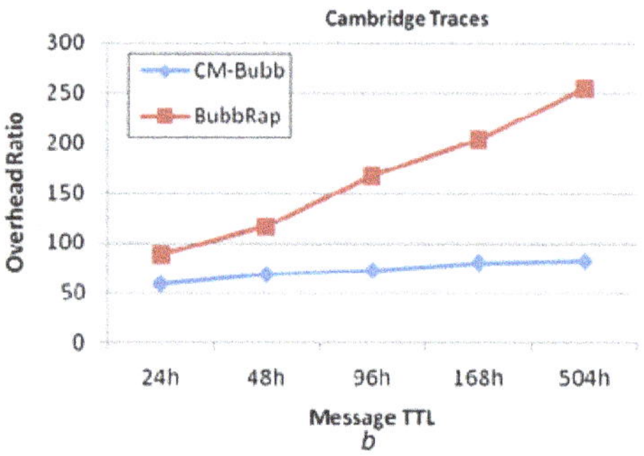

b

Fig. 4 *Overhead ratios of CM–Bubble Rap and Bubble Rap*
a OR for heterogeneous scenario
b OR for Cambridge traces

The results for the paired *T*-test of the overhead ratios between the two algorithms, CM–Bubble Rap and Bubble Rap are shown in Table 5. The difference in mean between CM–Bubble Rap and Bubble Rap is −93.61 for Cambridge traces and is −199.43 in case of heterogeneous scenario. This vividly underlines that the new forwarding decisions has significantly decreased the overhead cost of the algorithm. The *P*-values that are near to 0.01 also shows that the results are statistically significant.

The ALat comparison between CM–Bubble Rap and Bubble Rap in heterogeneous scenario and Cambridge traces are shown in Figs. 5*a* and *b*, respectively, which provides an estimate of the time taken by the protocols to deliver the messages to the destination.

With heterogeneous scenario, the CM–Bubble Rap has a significantly low latency than that of the Bubble Rap. This again proves that using friendship along with centrality can improve the delivery in the Bubble Rap. With new forwarding decisions, the CM–Bubble Rap can now find more better relays to deliver the messages more quickly. The Bubble Rap has an increase in the latency with message TTL because it cannot discover strong social ties as CM–Bubble Rap does and requires more replicas and time to deliver the message to the destination. However, it can be seen that in the real human traces difference between the new method and Bubble Rap in terms of ALat is not that much wide. This is because the users are students and the friendship ties can be almost homogeneous not rendering much difference in friendship between two nodes. Thus, the centrality will be showing a certain overhand over friendship. In addition, in such cases, the probability to find nodes that have greater centrality is also less which renders less replication. Therefore much difference is not reflected here although still CM–Bubble Rap has a less latency than Bubble Rap because of the message TTL-based threshold and exchange of ACK IDs.

4.4 *Confidence interval*

As already mentioned above, the simulation results are averaged over five runs per message TTL value. To prove that the results are more robust, confidence intervals with 95% of significance level of the delivery ratios attained by the CM–Bubble Rap in both heterogeneous scenario and real life traces are also presented. Tables 6 and 7 show the confidence intervals for the five runs for

Table 5 Results of the paired *T*-test between overhead ratios of CM–Bubble Rap and Bubble Rap

Protocols	*t*-values	Differences in mean	Minimum differences with 95% CI	Maximum differences with 95% CI	*P*-values	Statistical significances
			Cambridge traces			
CM against Bubble Rap	3.61315132	−93.6124	−172.8725	−28.81116	0.011	significant
			Heterogeneous scenario			
CM against Bubble Rap	3.399329996	−199.431476	−328.13624	−23.3983	0.0136	significant

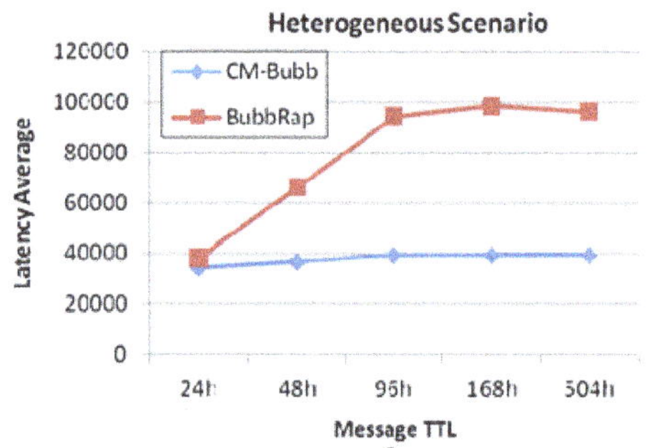

a

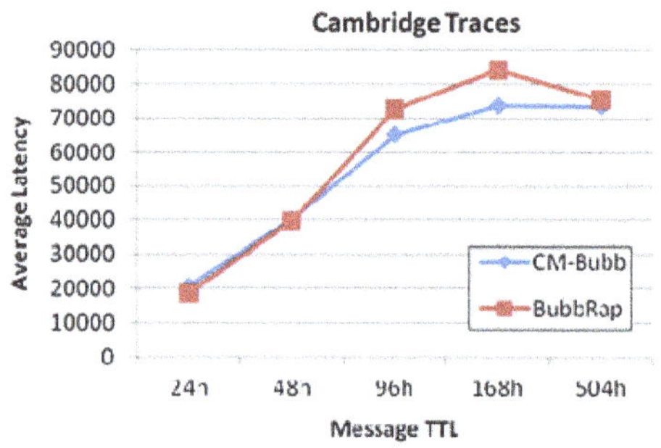

b

Fig. 5 *ALat comparison between CM–Bubble Rap and Bubble Rap*
a ALat in heterogeneous scenario
b ALat in Cambridge traces

Table 6 Confidence interval of delivery ratios for CM–Bubble Rap: Cambridge traces

Message TTL, h	Average	Variance	Confidence interval	DR minimum	DR maximum
24	0.87952	0.000152617	0.0153393	0.8669	0.8955
48	0.93894	0.000109673	0.013003311	0.9224	0.9502
96	0.94626	5.2388×10^{-5}	0.008987109	0.9363	0.9567
168	0.94646	7.1673×10^{-5}	0.010511916	0.9339	0.9576
504	0.94646	7.1673×10^{-5}	0.010511916	0.9339	0.9576

Table 7 Confidence interval of delivery ratios for CM–Bubble Rap: heterogeneous scenario

Message TTL	Average	Variances	Confidence intervals	DR minimum	DR maximum
24	0.47216	4.5703×10^{-5}	0.008394145	0.462	0.4808
48	0.53208	9.5627×10^{-5}	0.012142115	0.5167	0.5412
96	0.60378	0.000272387	0.02049261	0.5747	0.6139
168	0.61304	0.000110768	0.013068064	0.6057	0.631
504	0.60422	0.000360737	0.023583021	0.5722	0.6204

varying message TTL in both experiments. The results show that the proposed method is more robust.

5 Conclusion and future work

This paper tries to overcome some limitations of Bubble Rap social-based forwarding in PSNs by using different mechanisms. The main contribution of this paper is the calculation of 'friendship' between a pair of nodes which was not used in the original Bubble Rap algorithm. Average delivery predictabilities were used to calculate the level of friendship between two nodes. In addition, the naïve holding of messages in the buffer until a popular node is not encountered was removed by using a message TTL-based threshold which forwards the messages from the buffer after a particular threshold. The exchange of ACK IDs to delete those messages which are deleted provides efficient buffer management in an area like PSNs where the buffer size is limited. Through various experiments it was proved that this new mechanism provided far better delivery ratio and overhead when compared with the Bubble Rap which solely relies on the centrality for forwarding the messages.

6 References

[1] Fall K.: 'A delay-tolerant network architecture for challenged internets'. ACM SIGCOMM, August 2003

[2] IRTF: RFC 4838 – DTN Architecture, 2007. Available at http://www.ietf.org/rfc/rfc4838.txt

[3] IRTF: Delay Tolerant Networking Research Group, 2009. Available at http://www.dtnrg.org

[4] Perkins C.: 'Ad hoc networking' (Addison-Wesley, 2001)

[5] Daly E.M., Haahr M.: 'The challenges of disconnected delay-tolerant MANETs', *Ad Hoc Netw., Elsevier*, 2010, **8**, pp. 241–250

[6] Pietilainen A.-K., Diot C.: 'Social pocket switched networks'. IEEE INFOCOM Workshops 2009, Rio de Janero, 19–25 April 2009

[7] Zhu Y., Xu B., Shi X., Wang Y.: 'A survey of social-based routing in delay tolerant networks: positive and negative social effects', *IEEE Commun. Surv. Tutor.*, **15**, (1), pp. 387–401

[8] Hui P., Crowcroft J., Yoneki E.: 'Bubble rap: social-based forwarding in delay tolerant networks', *IEEE Trans. Mob. Comput.* , 2011, **10**, pp. 1576–1589

[9] Daly E.M., Haahr M.: 'Social network analysis for routing in disconnected delay-tolerant manets'. MobiHoc '07 Proc. Eighth ACM Int. Symp. Mobile Ad-hoc Networking and Computing, 2007

[10] Freeman L.C.: 'A set of measures of centrality based on betweenness', *Sociometry*, 1977, **40**, (1), pp. 35–41

[11] Freeman L.C.: 'Centrality in social networks: conceptual clarification', *Soc. Netw.*, 1979, **1**, (3), pp. 215–239

[12] Vahdat A., Becker D.: 'Epidemic routing for partially connected ad hoc networks'. Technical Report, CS-200006, Duke University, April 2000

[13] Spyropoulos T., Psounis K., Raghavendra C.S.: 'Spray and wait: an efficient routing scheme for intermittently connected mobile networks'. ACM Workshop on Delay-tolerant Networking, 2005

[14] Lindgren A., Doria A., Schelen O.: 'Probabilistic routing in intermittently connected networks'. Proc. IEEE Infocom, April 2006

[15] Hui P., Crowcroft J.: 'How small labels create big improvements'. Int. Workshop on Intermittently Connected Mobile Ad-hoc Networks in Conjunction with IEEE PerCom 2007, 19–23 March 2007

[16] Newman M.J.: 'A measure of betweenness centrality based on random walks', *Soc. Netw.*, 2005, **27**, (1), pp. 39–54

[17] McPherson M., Smith-Lovin L., Cook J.M.: 'Birds of a feather: homophily in social networks', *Annu. Rev. Sociol.*, 2001, **27**, pp. 415–444

[18] Xue J.-F., Li J.: 'Advanced PROPHET routing in delay tolerant network'. IEEE Int. Conf. Communication Software and Networks, 2009

[19] Hui P., Yoneki E., Chan S.-Y., Crowcroft J.: 'Distributed community detection in delay tolerant networks'. MobiArch '07 Proc. Eighth ACM Int. Symp., 2007

[20] Keränen A., Ott J., Kärkkäinen T.: 'The ONE simulator for DTN protocol evaluation'. SIMUTools 2009, Rome, Italy, 2009

[21] Homepage of Opportunistic Network Environment (ONE). Available at http://www.netlab.tkk.fi/tutkimus/dtn/theone

[22] Homepage of CRAWDAD archive. Available at http://www.crawdad.org/

21

Analysis of the spatial layer discrete cosine transform coefficient distribution and its application to rate model for H.264/SVC encoder

Szu-Wei Lee

Graphic Software Development and Validation Team, Visual and Parallel Computing Group, Intel, Santa Clara, CA, USA
E-mail: swli0311@yahoo.com.tw

Abstract: Knowledge of the discrete cosine transform coefficient distribution (DCT-DIST) is important for the encoder design. For example, rate control relies on this knowledge to estimate a possible bit rate and then decide proper coding parameters before the actual encoding task is performed. Therefore the rate control performance is fairly dependent on how accurately the DCT-DIST is modelled. The spatial enhancement layer (SL) DCT-DIST for H.264 scalable video coding (SVC) is studied in this Letter. SL DCT-DIST knowledge is furthermore used to derive a novel rate model. Our results can help design a proper rate control module for the H.264/SVC encoder.

1 Introduction

Recently, the scaling video coding (SVC) extension of H.264/AVC [1] has been standardised, which allows a video sequence to be coded into one base layer (BL) and several enhancement layers so as to support temporal, spatial and quality scalability. In this Letter, the H.264/SVC spatial scalable encoder is taken into consideration, and the distribution of the spatial enhancement layer (SL) discrete cosine transform (DCT) coefficients will be studied. Moreover, knowledge of the SL DCT coefficient distribution (DCT-DIST) will be applied to develop a novel rate model to estimate the SL bit rate. Our study of the SL DCT-DIST and the proposed rate model could help design a proper rate control algorithm for the H.264/SVC spatial scalable encoder.

The DCT-DIST has been studied over the past few decades. Gaussian or Laplacian distributions were popular to estimate the DCT-DIST in practice. However, it is observed that the DCT-DIST in image and video applications could differ from Gaussian or Laplacian distributions in most cases. Kamaci *et al.* [2], therefore, proposed that Cauchy distribution is more accurate to model the DCT-DIST for H.264/AVC. Although the DCT-DIST can be well estimated by Cauchy distribution, the result may not be able to apply to H.264/SVC when coding the SL DCT coefficients since the video signal could be predicted by the reconstructed signal of the reference layer, which can be the previous coded SL or the BL. To be more specific, the SL DCT coefficients contain the quantisation noise because of the quantisation effect in the reference layer, and the distribution of the quantisation noise in the reference layer could affect the SL DCT-DIST. Recently, Liu *et al.* [3] proposed a rate control algorithm for H.264/SVC. The SL DCT-DIST is assumed to be Cauchy distributed. Although their algorithm could provide good rate control results, the key assumption that the SL DCT-DIST is Cauchy distributed is not discussed in detail while it was verified experimentally only. In other words, there is no clue as to why the SL DCT-DIST is Cauchy distributed in the previous work.

The SL DCT-DIST for H.264/SVC will be investigated in this Letter. Furthermore, one application of SL DCT-DIST knowledge to bit rate modelling will be presented. There are three main contributions in our work. First, we provide the mathematical derivation to show why the SL DCT-DIST is Cauchy distributed in detail. Second, a novel SL rate model is proposed. The proposed SL rate model is more accurate and simpler as compared with the previous rate model [3]. Finally, several experiments will be conducted to verify the accuracy of the proposed SL rate model through various sizes of sequences while high-definition sequences were not considered in [3]. We conclude that the proposed SL rate model can provide fairly accurate estimation results for various sizes of bit streams.

2 Overview of H.264 spatial scalable encoder

Without loss of generality, we consider the case that two layers of bit streams (i.e. one BL and one SL) are generated by the H.264/SVC encoder. Under such a scenario, the BL must be used as the reference layer when coding the SL bit stream.

To improve the coding efficiency, the H.264/SVC encoder allows either the H.264/AVC conventional coding scheme or the so-called inter-layer prediction coding scheme to code macro blocks (MBs) in the SL. When the H.264/AVC conventional coding scheme is selected, all H.264/AVC existing coding methods can be used to generate the prediction signal, and the current coding MB can be coded as either the intra- or the inter-mode. On the other hand, the reconstructed co-located MB in the BL is used as the prediction signal when the inter-layer prediction coding scheme is selected. MB coding information, such as MB partition modes, motion vectors, reference frame indices and so on, of the current coding MB is derived from the co-located MB in the BL, and the current coding MB is marked as a base mode MB. Notably, when decoding a base mode MB, the motion compensation process only needs to be performed in the target decoding layer because MB coding information can be directly derived from the corresponding reference layer. This is so-called single-loop decoding. More information can be found in [1].

Dataflow of the H.264/SVC spatial scalable encoder using the inter-layer prediction coding scheme is shown in Fig. 1. Table 1 also summarises the symbols employed in this Letter. Let X be the source video signal. Its down-scaled video signal (X_{BL}), which contains the low-frequency component of X, is first fed

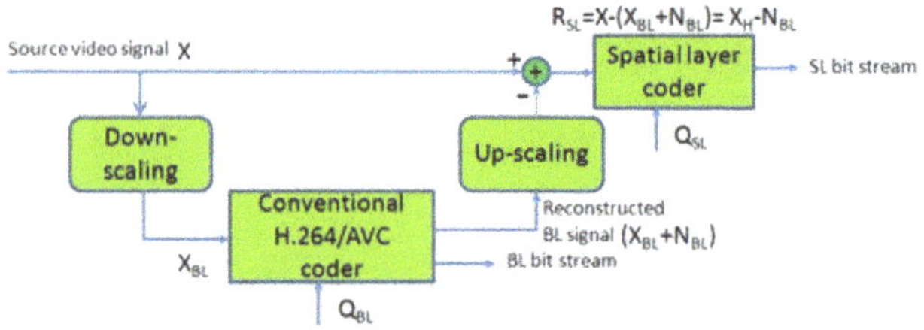

Fig. 1 *Dataflow of H.264/SVC spatial scalable encoder using the inter-layer prediction coding scheme*

Table 1 Summary of symbols

Symbols	Definitions
DCT	discrete cosine transform
DCT-DIST	DCT coefficient distribution
BL	base layer
SL	spatial enhancement layer
MB	macro block
X	source video signal
X_{BL}	low-frequency component of X
X_{H}	high-frequency component of X
Q_{BL}	BL quantisation step size
Q_{SL}	SL quantisation step size
N_{BL}	BL quantisation noise
R_{SL}	SL residual signal
PDF	probability density function
$f_{N_{\mathrm{BL}}}(.)$	PDF for the distribution of N_{BL}
$f_{\mathrm{SL}}(.)$	PDF for the SL DCT-DIST
μ_N	Cauchy scale parameter for the distribution of N_{BL}
$\mu_{X_{\mathrm{H}}}$	Cauchy scale parameter for the DCT-DIST of X_{H}
μ_{SL}	Cauchy scale parameter for the SL DCT-DIST
B	bit number
ρ	percentage of zero quantised DCT coefficients
θ	ρ-domain rate model parameter
B_{SL}	SL bit number
ρ_{SL}	percentage of zero quantised SL DCT coefficients
θ_{SL}	SL ρ-domain rate model parameter
η_{SL}, γ, δ	proposed SL rate model parameters

into the conventional H.264/AVC coder with the BL quantisation step size (Q_{BL}) to generate the BL bit stream. Then, the reconstructed BL signal, which includes not only X_{BL} but also the quantisation noise (N_{BL}) because of the quantisation effect by Q_{BL}, is up-scaled and used as the prediction signal of the source video signal, X. Finally, the SL residual signal (R_{SL}), which is the difference between X and the reconstructed BL signal, is coded by the SL coder with the SL quantisation step size (Q_{SL}) to produce the SL bit stream. Notably, R_{SL} contains not only the high-frequency component of X (denoted by X_{H}) but also N_{BL}.

3 SL DCT coefficient distribution

The H.264/SVC spatial scalable encoder using the inter-layer prediction coding scheme and the DCT-DIST of R_{SL} (i.e. SL DCT-DIST) are taken into consideration in this section. Since R_{SL} contains both X_{H} and N_{BL}, their distributions will be considered separately. We first borrow the DCT-DIST result in [2] to derive the distribution of N_{BL}. Then, the SL DCT-DIST is derived accordingly.

Lemma 1: Distribution of the BL quantisation noise is Cauchy distributed.

Proof: The probability density function (PDF) of the quantisation noise can be expressed as a function of the quantisation step size and the characteristic function of the source coding signal [4]. Consider the conventional H.264/AVC coder with the BL quantisation step size, Q_{BL}. Since the quantisation is performed in the DCT domain, the PDF for the distribution of N_{BL} (i.e. $f_{N_{\mathrm{BL}}}(q)$) can be written as follows

$$f_{N_{\mathrm{BL}}}(q) = \frac{1}{Q_{\mathrm{BL}}} + \frac{1}{Q_{\mathrm{BL}}}\sum_{k\neq 0}\Phi\left(\frac{2\pi k}{Q_{\mathrm{BL}}}\right)\mathrm{e}^{(-\mathrm{j}2\pi kq/Q_{\mathrm{BL}})} \quad (1)$$

where $\Phi(.)$ is the characteristic function of the BL DCT coefficients. Since the BL DCT-DIST is approximately Cauchy distributed [2], its characteristic function can be expressed by

$$\Phi(\omega) = \mathrm{e}^{-\mu|\omega|} \quad (2)$$

where μ is the scale parameter of the Cauchy distribution function. Substituting (2) into (1) yields

$$f_{N_{\mathrm{BL}}}(q) = \frac{1}{Q_{\mathrm{BL}}}\frac{\alpha^2 - 1}{\alpha^2 + 1 - 2\alpha\cos((2\pi/Q_{\mathrm{BL}})q)} \quad (3)$$

where $\alpha = \mathrm{e}^{(2\pi/Q_{\mathrm{BL}})\mu}$.

Considering the Taylor approximation of the cosine function (i.e. $\cos(x) \simeq 1 - (1/2)x^2$) and substituting it into (3) yields

$$f_{N_{\mathrm{BL}}}(q) \simeq \tau\frac{\beta}{\beta^2 + 4\pi^2 q^2} \quad (4)$$

where $\beta = Q_{\mathrm{BL}}\cdot(\alpha - 1)\cdot\alpha^{-0.5}$ and $\tau = (\alpha + 1)\cdot\alpha^{-0.5}$.

Equation (4) suggests that the distribution of the BL quantisation noise, N_{BL}, is approximately Cauchy distributed. Thus, its PDF can be written as

$$f_{N_{\mathrm{BL}}}(q) = \frac{1}{\pi}\frac{\mu_N}{\mu_N^2 + q^2} \quad (5)$$

where μ_N is the Cauchy scale parameter and it can be expressed as a function of Q_{BL}. □

Lemma 2: SL DCT-DIST is Cauchy distributed.

Proof: Since X_{H} and N_{BL} result from the high- and low-frequency components of X, it is reasonable to assume that X_{H} and N_{BL} are independent. Furthermore, the DCT-DIST of X_{H} is approximately Cauchy distributed [2], and the distribution of N_{BL} is Cauchy distributed, too, as shown in Lemma 1. We conclude that the SL DCT-DIST, which is the DCT-DIST of R_{SL}, is Cauchy distributed because $R_{\mathrm{SL}} = X_{\mathrm{H}} - N_{\mathrm{BL}}$ and the sum of two independent Cauchy random variables is Cauchy distributed [5]. Notably, the scale parameter for the sum of two independent Cauchy random variables is the sum of individual scale parameters for these two Cauchy random variables. Therefore the PDF of the SL DCT-DIST can be expressed as

$$f_{\mathrm{SL}}(x) = \frac{1}{\pi}\frac{\mu_{\mathrm{SL}}}{\mu_{\mathrm{SL}}^2 + x^2} \quad (6)$$

where $\mu_{\mathrm{SL}} = \mu_{X_{\mathrm{H}}} + \mu_N$, and $\mu_{X_{\mathrm{H}}}$ is the Cauchy scale parameter for the DCT-DIST of X_{H}. Since μ_N can be written as a function of Q_{BL} (i.e. $\mu_N = f(Q_{\mathrm{BL}})$), μ_{SL} can be further rewritten as a function of Q_{BL} too. That is

$$\mu_{\mathrm{SL}} = \mu_{X_{\mathrm{H}}} + f(Q_{\mathrm{BL}}) \quad (7)$$

Notably, our result of the Cauchy scale parameter for the SL DCT-DIST, μ_{SL}, is consistent with the experimental result as shown in [3], where μ_{SL} is simply treated as a linear function of Q_{BL}. □

4 SL bit rate model

Consider the ρ-domain rate model [6] that the coded bit number (B) is expressed as

$$B = \theta(1 - \rho) \quad (8)$$

where θ is the rate model parameter, and ρ is the percentage of zero quantised DCT coefficients.

Equation (8) suggests that the coded bit number is considered being proportional to the percentage of non-zero quantised DCT coefficients. Similarly, the SL bit number (B_{SL}) can be modelled as a function of the percentage of non-zero quantised SL DCT coefficients. That is

$$B_{SL} = \theta_{SL}(1 - \rho_{SL}) \tag{9}$$

where θ_{SL} is the SL rate model parameter, and ρ_{SL} is the percentage of zero quantised SL DCT coefficients.

The percentage of zero quantised SL DCT coefficients, ρ_{SL}, can be calculated by

$$\rho_{SL} = \int_{-0.5 \cdot Q_{SL}}^{0.5 \cdot Q_{SL}} f_{SL}(x)\,dx \tag{10}$$

where $f_{SL}(\cdot)$ is the PDF of the SL DCT-DIST.

Since the SL DCT-DIST is Cauchy distributed as shown in Lemma 2, (9) can be rewritten as

$$B_{SL} = \theta_{SL}\left[1 - \frac{2}{\pi}\tan^{-1}\left(\frac{Q_{SL}}{2\,\mu_{SL}}\right)\right] \tag{11}$$

Considering the Taylor expansion of $\tan^{-1}(x)$ (i.e. $\tan^{-1}(x) = \sum_{n=0}^{\infty}((-1)^n/(2n+1))x^{2n+1}$) and substituting it into (11) yields

$$B_{SL} = \theta_{SL}\left[1 - \frac{2}{\pi}\sum_{n=0}^{\infty}\frac{(-1)^n}{(2n+1)}\left(\frac{Q_{SL}}{2\,\mu_{SL}}\right)^{2n+1}\right] \tag{12}$$

Furthermore, since μ_{SL} can be expressed as a function of Q_{BL} as shown in (7), μ_{SL} in (12) can be replaced by its Taylor approximation. That is

$$\mu_{SL} \simeq \mu_{X_H} + c_1\,Q_{BL} + c_2\,Q_{BL}^2 \tag{13}$$

As a result, (12) can be expressed as

$$B_{SL} \simeq \theta_{SL}$$
$$\left[1 - \frac{2}{\pi}\sum_{n=0}^{\infty}\frac{(-1)^n}{(2n+1)}\left(\frac{0.5\,Q_{SL}}{\mu_{X_H} + c_1\,Q_{BL} + c_2\,Q_{BL}^2}\right)^{2n+1}\right] \tag{14}$$

Finally, we simplified the above equation and used the following approximate equation as the proposed SL rate mode

$$B_{SL} \simeq \eta_{SL}\,Q_{SL}^{\gamma}\,Q_{BL}^{\delta} \tag{15}$$

where η_{SL}, γ and δ are the proposed SL rate model parameters.

As compared with the previous rate model proposed in [3], the SL bit number was modelled by two different equations, and the selection of a proper equation is dependent on Q_{BL} and Q_{SL}. The proposed rate model, however, is much simpler for a single equation is used to estimate the bit rate. The accuracy of the proposed SL bit rate model will be examined in the next section.

5 Experimental result

The proposed SL rate model was implemented with JSVM 9.19.15. In our experiment, all test bit streams contain two layers, where BL and SL are of QCIF-CIF, CIF-4CIF and 540P-1080P resolutions. As shown in (15), the proposed model contains three parameters (i.e. η_{SL}, γ and δ). These three parameters have to be decided first so that the proposed model can be used to estimate a possible SL bit rate. η_{SL} is updated by the least mean square approach per rate control unit. The rate control unit is one of the encoding parameters and it may have one or several MBs. On the other hand, γ and δ are fixed in our experiment and were determined by the following training process: (i) Foreman and Mobile sequences were selected to generate several training bit streams of QCIF-CIF resolution whose BL and SL were coded by different QPs (QP = 15, 25, 35 and 45). (ii) Coded SL bit numbers and QPs for these training bit streams were used to determine γ and δ by the least square approach. These trained model parameters are

$$\gamma = -0.6846, \quad \delta = 0.7769 \tag{16}$$

For model verification, each test bit stream contains 15 frames. Rate control was enabled to code the BL under different bit rate constraints. The rate control unit has exactly 11, 22 and 120 MBs for QCIF-CIF, CIF-4CIF and 540P-1080P bit streams, respectively. All SL bit streams were coded with *default_base_mode_flag* equal to one. This implies that the inter-layer prediction coding scheme is always selected in the encoder. The percentage error between the actual coded SL bit number and the estimated SL bit number calculated by the proposed SL rate model is listed in Table 2.

Table 2 Accuracy of the proposed SL rate model

QCIF-CIF bit stream	Bit rate, bits/s	Error, %
foreman	32 K	5.09
bus	48 K	3.30
mobile	56 K	5.69
coast	72 K	5.74
tempete	88 K	6.10
CIF-4CIF bit stream	bit rate, bits/s	error, %
harbour	384 K	2.55
city	512 K	2.43
540P-1080P bit stream	bit rate, bits/s	error, %
blue sky	2048 K	2.06
sunflower	6400 K	1.69
rush hour	8120 K	1.61

It can be seen that the proposed SL rate model can provide good estimation results of the SL bit rate. As compared with the experimental results in [3], the proposed model can provide even more accurate data for various bit streams.

6 Conclusion

The SL DCT-DIST and the application of SL DCT-DIST knowledge to bit rate modelling were studied in this Letter. We first showed mathematically how the SL DCT-DIST is Cauchy distributed. After that a novel SL rate model was proposed. Experimental results show that the proposed SL rate model could result in a fairly good estimation to the SL bit rate. As compared with the previous work, the proposed rate model is more accurate and simpler. Our results can help design a good bit rate control algorithm for the H.264/SVC spatial scalable encoder.

7 Acknowledgment

The author would like to thank the Intel for supporting this work.

8 References

[1] Schwarz H., Marpe D., Wiegand T.: 'Overview of the scalable video coding extension of the H.264/AVC standard', *IEEE Trans. Circuits Syst. Video Technol.*, 2007, **17**, (8), pp. 1103–1120

[2] Kamaci N., Altunbasak Y., Mersereau R.M.: 'Frame bit allocation for the H.264/AVC video coder via Cauchy-density-based rate and distortion models', *IEEE Trans. Circuits Syst. Video Technol.*, 2005, **15**, (8), pp. 994–1006

[3] Liu J., Cho Y., Guo Z., Kuo C.C.: 'Bit allocation for spatial scalability coding H.264/SVC with dependent rate-distortion analysis', *IEEE Trans. Circuits Syst. Video Technol.*, 2010, **20**, (7), pp. 967–981

[4] Sripad A.B., Snyder D.L.: 'A necessary and sufficient condition for quantization errors to be uniform and white', *IEEE Trans. Acoust. Speech Signal Process.*, 1977, **25**, (5), pp. 442–448

[5] Wikipedia the free encyclopedia: 'Cauchy distribution'. Available at http://www.en.wikipedia.org/wiki/Cauchy_distribution

[6] He Z., Kim Y.K., Mitra S.K.: 'Low delay rate control for DCT video coding via ρ-domain source modeling', *IEEE Trans. Circuits Syst. Video Technol.*, 2001, **11**, (8), pp. 928–940

22

Space-time QAM wireless MISO systems employing differentially coded in-/out-FECC SCQICs over slow-fading Jakes scattering mobile radio links

Ardavan Rahimian[1], Farhad Mehran[1], Robert G. Maunder[2]

[1]School of Electronic, Electrical and Computer Engineering, University of Birmingham, Birmingham B15 2TT, UK
[2]School of Electronics and Computer Science, University of Southampton, Southampton SO17 1BJ, UK
E-mail: rahimian@ieee.org

Abstract: This study presents research that supplements and extends the previous works on design of space-time fully systematic unpunctured (FSU) serial concatenation of quadratic interleaved codes (SCQICs). The requirements for efficient design of the forward error correction (FEC) codecs motivated potential information-theoretic studies for enjoying the development of low-complex system components within the FEC encoder/decoder for securing the transmission reliability. Inspired by this motivation, this study not only provides design guidelines to achieve better bit error rate performance in terms of the major design factors of FSU-SCQICs, that is, component code constraint length and trellis structure, and FEC rate, but also estimates the gain gaps of different quadratic permutation (QP) structures in two crucial untouched aspects: (i) signal-to-noise ratio-region comparison on the optimality and (ii) investigation on the structural parameters of QPs, that is, cyclic shift and primitive factor.

1 Introduction

One of the most important design issues concerning research studies on the space-time turbo codes (STTCs) [1] aims to improve the key performance factors that have direct determinant effect on the overall bit error rate (BER) performance. Hence, it is vital to design the effective permutation arrays in regard to a waterfall-region of the BER curve and the region that exhibits a much shallower slope, that is, flare region. In particular, for the STTCs that enjoy cascaded systematic recursive convolutional codes (SRCCs) for yielding relatively high minimum distance, interleaver design issues are not reaching the state of maturity [2]. From theoretical design aspects, algebraic constructions are attracting particular interest for performing the scrambling/unscrambling functions since they yield provisioning coding gain in critical regions of the BER performance curve and enable possibility of analysis and compact representation. An additional enhancement which opens the way for their practical utilisation is their considerable lower implementation complexities.

2 Research contribution

It has been previously shown that serial turbo-like codes employing the quadratic permutations (QPs), that is, the serial concatenation of quadratic interleaved codes (SCQICs), result in outstanding coding gains [3–6]. This paper aims to supplement and extend the pivotal system design guidelines of our previous works in order to achieve a flexible SCQIC framework in terms of other crucial contradictory performance factors, so as to address the proposed SCQICs in co-operation with the highly matured technologies (e.g. bit-level space-time codes) employing the high-throughput modulation along with other specified features; which have been briefly outlined as follows:

(1) *Component code structure*: Inspired by the motivation of finding promising generator polynomials which result in maximisation of component code minimum free distance [7]; this paper investigates the gain gap limits between provisioning choices of the convolutional component code (CCC) trellis structures for different feasible constraint lengths, with respect to the tolerable incurred decoding complexity, that is, the widely used range $L \leq 5$.
(2) *FEC rate*: In terms of providing both coding gains and spectral efficiencies, flexible FEC code rate enables adaptation of provision signal-to-noise ratio (SNR) operating point (which yields minimal BER) with respect to the bandwidth limitations of wireless channels. Hence, the investigation has been further broadened for lower code rates than typical 0.33, which is widely used for wireless transmissions.
(3) *Differential coding (DC) inside-/outside-forward error correction codec (FECC)*: There exist a wide range of advantages and disadvantages as a result of employing DC as the phase-ambiguity resolution technique [8] for high-throughput widely used 16-quadrature amplitude modulation (QAM) transmissions. One important issue of great theoretical and practical interest is to gain the benefits of DC technique while avoiding unnecessary BER degradation (because of the double-error phenomenon) by applying the DC external to the FECC, rather than modulating differentially coded version of FECC, that is, DC inside-FECC [9]. The studies (1) and (2) of this outlined list are supplemented by comparing the sacrificed gain gaps with respect to employing DC internal or external to FECC or the same system model based on the coherent detection.
(4) *Optimality on SNR region(s)*: Studies on comparison of different permutation algorithms to be employed in FEC-coded systems reveal that the superior BER performance of the selected one may yield inferior performance as interleaver size changes significantly. For example, for the parallel concatenated convolutional coding (PCCCing), although pseudo-random interleavers have been shown to have superior performance to the block interleavers for the large frame lengths, in many cases their inferior performance in the low frame sizes has been confirmed [10]. Hence, the inevitable investigation for the gain gaps between SCQICs with respect to reference of comparison, that is, the same system using randomly generated scramblers/unscramblers, are carried out in order to ensure that the gap is not significant; i.e. not yielding unexpected high probability of bit error and hence changing the unexpected level of optimality in particular SNR region.
(5) *Structural parameters of QPs*: Monte Carlo computer simulations have been conducted in order to potentially predict the performance of SCQICing when the QP-vectors utilise different multiplicative factors and cycle shifts.

The system investigations have been thoroughly conducted in a rich isotropic scattering mobile radio channel based on the Jakes

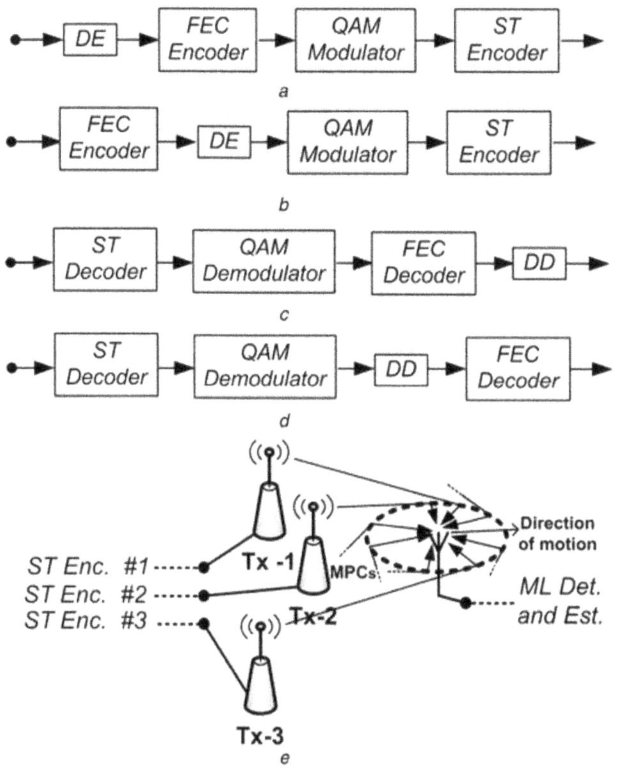

Fig. 1 *STTC encoder/decoder*
a External-FECC-DC SCQIC encoder
b Internal-FECC-DC SCQIC encoder
c DD-out-FECC iterative SCQIC decoder
d DD-in-FECC iterative SCQIC decoder
e Conceptual representation of Jakes scattering wireless MISO mobile radio link

Doppler power spectrum [11]; that is, where a scatter ring is placed around the receiver in order to model and analyse the multipath components, and the isotropic continuum of arriving components is approximated by the plane radiowaves arriving at the uniformly spaced azimuthal angles.

3 Wireless system model

3.1 *STTC encoder/decoder*

Fig. 1 outlines the wireless system architecture operating in portrayed conceptual representation of uniformly distributed scattering. For the DC outside-FECC scheme, that is, Fig. 1*a*, the differential encoder (DE) is fed with a *(N/2)*-bit information frame $u = (u_0, \ldots, u_k)$ based on the Bernouli distribution, with the pmf $f(k; p) = p$ if $k = 1$, $(1 - p)$ if $k = 0$, and 0 otherwise. The outer SRCC encoder is fed by either of the DC-coded bit stream (for the case of external DC) or information frame for coherent and internal DC-based systems. The selected half-rate outer SRCC trellis structures for $r \leq 0.33$ and $L \leq 5$ are tabulated in Table 1. The interleaver applies the QP on the bit sequences obtained from the previous step, which is originally proposed in [12, 13] and further developed for serial turbo-like codes in [3–6]. Let $x_{oc.} = (x_0, \ldots, x_{N-1})$ be a sequence in $\{0, 1\}^N$ which is the N-bit outer system coded sequence. The QP-based interleaver maps the $x_{oc.}$ to a sequence $\hat{x}_{oc.} = (x_0, \ldots, x_{N-1})$ according to Theorems I and II of [12] (examples are given in the next section). The inner SRCC encoder is fed with $\hat{x}_{oc.}$, that is, an N-bit block containing the QP-permutation sequence of outer coded bit stream. The selected inner SRCC trellis structures with $r^{i.} \in \{0.66, 0.5, 0.4\}$ for $r \leq 0.33$ and $L \leq 5$ are tabulated in Table 1. For the DC inside-FECC scheme, that is, Fig. 1*b*, the DE is fed with a $(N/r^{o.})$-bit SCQIC-coded sequence. Compared with this scheme, the major advantage of using the DC external to the FECC is that the double-error phenomenon because of the consecutive errors has been eliminated. However, since the system resolution performance depends on the synchroniser circuit of the channel decoder, applying the DC outside FECC is not appropriate for the application to burst mode [9]. The resultant SCQIC can readily be applied to 16-QAM modulation in order to provide a good solution in terms of both power and bandwidth efficiencies. Meanwhile, the low-operating SNR of the resultant STTC increases the transmission robustness to interference and distortion. The QAM modulator maps the DC-SCQICs of $(N/r^{o.})$-bit blocks to the symbol $\aleph_k$ from the 2^Q-ary constellation set $S = \{\alpha_1, \alpha_2, \ldots, \alpha_{2Q}\}$, where α_i corresponds to the bit pattern $s_i = [s_{i,1}, s_{i,2}, \ldots, s_{i,Q}]$ with $s_{i,j} \in \{0, 1\}$. Space-time encoding process follows the wireless transmission sequences for the triplet-antenna systems derived in (37) of [14].

Table 1 CCC structures

Constraint length, L	Overall rate, r	Outer trellis structure and rate, $r^{o.} = 1/2$	Inner trellis structure and rate, $r^{i.}$
3	1/3	$G_3^{o.}: \{111, 101\}_2$	$r^{i.} = 2/3 \quad G_3^{i.}: \begin{Bmatrix} 111, 101, 000; \\ 000, 111, 101 \end{Bmatrix}_2$
	1/4	same as $G_3^{o.}$	$r^{i.} = 2/4 \quad G_3^{i.}: \begin{Bmatrix} 111, 101, 011, 000; \\ 000, 111, 101, 011 \end{Bmatrix}_2$
	1/5	same as $G_3^{o.}$	$r^{i.} = 2/5 \quad G_3^{i.}: \begin{Bmatrix} 111, 101, 011, 010, 000; \\ 000, 111, 101, 011, 010 \end{Bmatrix}_2$
4	1/3	$G_4^{o.}: \{1111, 1101\}_2$	$r^{i.} = 2/3 \quad G_4^{i.}: \begin{Bmatrix} 1111, 1101, 000; \\ 000, 1111, 1101 \end{Bmatrix}_2$
	1/4	same as $G_4^{o.}$	$r^{i.} = 2/4 \quad G_4^{i.}: \begin{Bmatrix} 1111, 1101, 1011, 000; \\ 000, 1111, 1101, 1011 \end{Bmatrix}_2$
	1/5	same as $G_4^{o.}$	$r^{i.} = 2/5 \quad G_4^{i.}: \begin{Bmatrix} 1111, 1101, 1011, 1001, 000; \\ 000, 1111, 1101, 1011, 1001 \end{Bmatrix}_2$
5	1/3	$G_5^{o.}: \{11111, 10001\}_2$	$r^{i.} = 2/3 \quad G_5^{i.}: \begin{Bmatrix} 11111, 10001, 00000; \\ 00000, 11111, 10001 \end{Bmatrix}_2$
	1/4	same as $G_5^{o.}$	$r^{i.} = 2/4 \quad G_5^{i.}: \begin{Bmatrix} 11111, 10001, 10011, 00000; \\ 00000, 11111, 10001, 10011 \end{Bmatrix}_2$
	1/5	same as $G_5^{o.}$	$r^{i.} = 2/5 \quad G_5^{i.}: \begin{Bmatrix} 11111, 10001, 10011, 10111, 00000; \\ 00000, 11111, 10001, 10011, 10111 \end{Bmatrix}_2$

At the receiver, for detecting symbols of the system, the maximum-likelihood (ML) detector amounts to minimise the decision metrics given in [15], and the resultant sequences are inserted in demodulator block. At the demodulator, the QAM constellations are demapped in order to form the system blocks of SCQIC-coded bit stream, that is, demaps the symbol $\hat{\aleph}_k$ from the 2^Q-ary constellation set S to DC-SCQICs of $(N/r^{o\cdot})$-bit blocks. For the case of the DC internal to FECC, the detected symbols are inserted into the differential decoder (DD) for applying the inverse function of DE, whereas for DC external to FECC the input to the DD is the iteratively decoded SCQIC bit sequences. The demodulator's output has also been computed based on the hard-decision. The system channel values (i.e. soft-decision log-likelihood ratios (LLRs)) have been computed by a separate function at the FEC decoder at the initialisation step of the system iterative decoding.

The DD bit stream inserted into the system SCQIC decoder for recovering the transmitted bits follows near-ML iterative decoding procedure developed for the SCQICs in [6]. Herein, constituent decoders corresponding to SRCC trellis structures iteratively exchange extrinsic information between themselves at the core of the SCQIC iterative decoder architecture. The targeted investigations have been systematically carried out based on the Log-maximum a posteriori (MAP) decoders [8], but the effects of using other decoders for SCQICs are addressed in [6]. For necessary deinterleaving at the decoder architecture, the re-arranged QP feature will be more exploited in Section 4; wherein the samples of the bit streams are scrambled and unscrambled.

3.2 Wireless system specifications

We assume rich scattering environments where the receiver antenna responds to each transmitter antenna system through a statistically independent fading coefficient [16], given the wireless transmission is performed using the Gray-coded (GC)-16-QAM. The selected set of the trellis structures for $L \leq 5$ are tabulated in Table 1, supplemented by all extended variants for the FEC rates 1/3, 1/4 and 1/5, that is, every source data bit (or DE data bit in out-DC-FECC) is mapped into three, four and five symbols, respectively. We assume that both the outer and inner SRCCs are terminated by $L-1$ tail bits in order to drive the encoders to all-zero state. For the half-rate space-time coded system, the measured Doppler spectrum of targeted isotropic scattering mobile radio channel is compared with theoretical Jakes spectrum as in Fig. 2. The signal amplitude received by Rx is affected by the presence of slow-fading Rayleigh flat-fading mechanism and corrupted by additive noise, and follows the channel matrix descriptions as in [2]. The channel modelling analysis has employed the stochastic system models, and considers the generic probability density function (PDF) of the field strength in the intended wireless link. Therefore, the bandlimited impulse response for the three slow-fading links is depicted in Fig. 3, in order to clearly demonstrate the link's impulse response in the proposed analytical case of 3 × 1 links along with the given specifications without affecting the overall system BER performance. It is also assumed that the stationary scatters are on average distributed uniformly on a circle around the receiver with omnidirectional antenna, with the same average power from all directions. Invoking the central limit theorem, we can assume that power from one direction is complex Gaussian distributed. In addition, the technical issues related to the intended stochastic multiple antenna channel modelling tackles the same procedure as given in [2, 6].

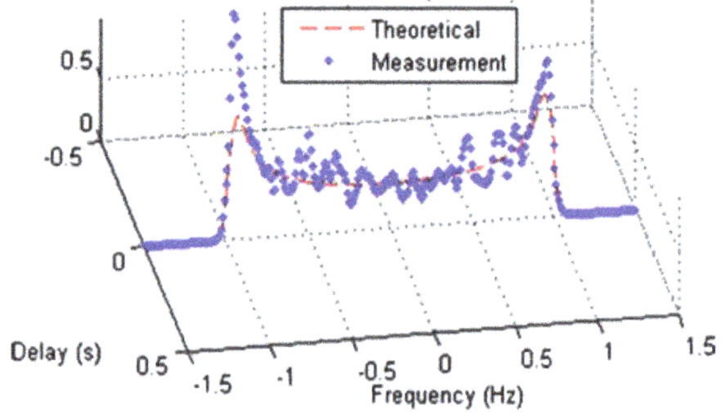

Fig. 2 *Computed theoretical against measured Jakes Doppler spectra of a path against path delay(s) in one branch of wireless MISO mobile radio link*

If we assume that the receiver moves with a certain speed, the number of components coming from each direction will take the tub form of Jakes spectrums, given analytically by $S_D(v) = (1/\pi.v_{max}).(1-(v/v_{max})^2)^{-1/2}$ for $-v_{max} \leq v \leq v_{max}$ [17], where f_{max} is the maximum Doppler-shift. The scatters are equally distributed in all directions and the receiver is also moving. The radio signal from all directions has on average the same power and delay. The spectrum can be interpreted such that infinite contributions, which are very small, come from 0 and 180°, which creates two singularities in v_{max} and $-v_{max}$.

4 Numerical results and discussion

In this section, representative wireless system performance evaluation results have been used in order to characterise the attainable

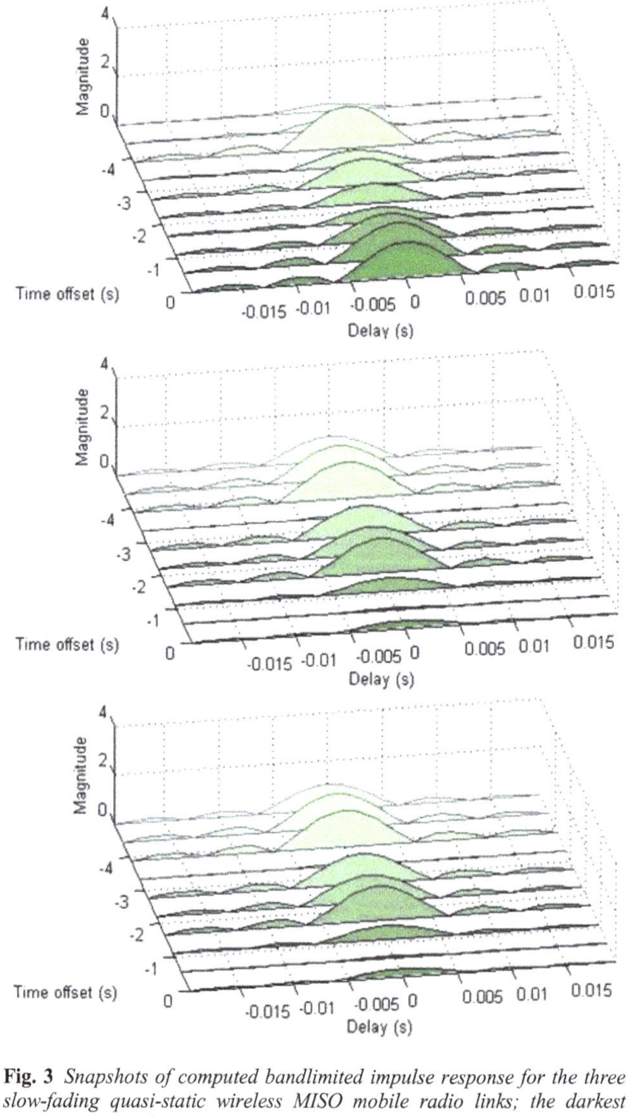

Fig. 3 *Snapshots of computed bandlimited impulse response for the three slow-fading quasi-static wireless MISO mobile radio links; the darkest curve presenting the current response at BER = 1.778e − 4 and E_b/N_0 = 4 dB*

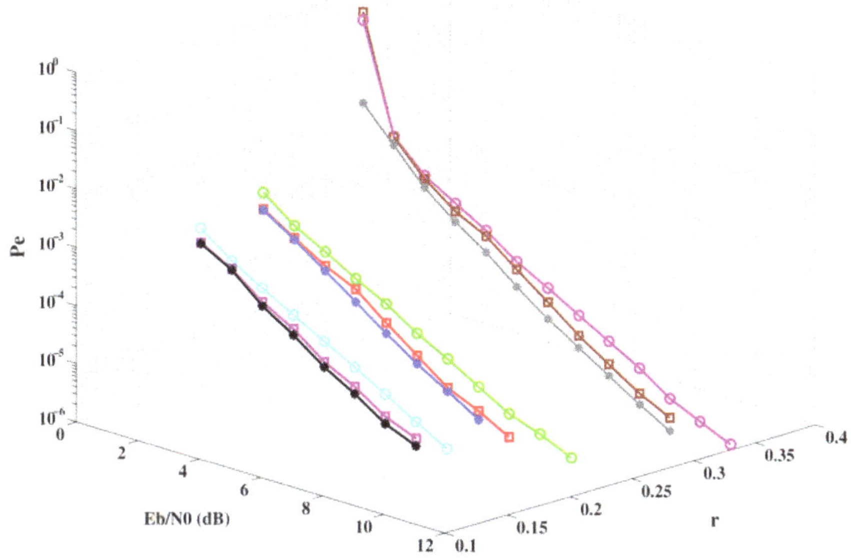

Fig. 4 *Comparison of BER against E_b/N_0 against r for the coherent wireless systems*

performance of the coded multiple-input–single-output (MISO) systems from the aspects outlined in Section 1. In Figs. 4–6, the BER is shown against the energy per bit/noise power spectral density (E_b/N_0) and the FEC rate (r) for coherent, out-FECC DC and in-FECC DC systems. Herein, the lines with circle, square and star markers present the performance curves associated with employing $L=3$, 4, and 5. The particular generator polynomials are chosen so that the designed SCQIC does not have catastrophic error propagation and has high free distance for the selected FEC rates and the constraint lengths. Meanwhile, FEC rates and component code constraint lengths are chosen from among the most widely used range for provisioning overall design tradeoff. As it can be seen, the huge difference in performance that can result from extra parity bit as a result of lowering FEC rate is on average of 2 dB when the comparison is made for fixed-length CCCs with the same trellis structure.

For $r=1/3 \rightarrow 1/5$ coding gains of up to $\simeq 4$ dB are achieved at the expense of spectral efficiency degradations because of two extra parity bits.

The performance improvement due to employing the CCCs with the higher L becomes not outstandingly significant, and suggests to limit the $L<6$ in order to avoid the considerable increase in system decoding complexities which does not result in worthwhile BER reduction, neither in waterfall-region, nor in flare region; that is, based on [8], increasing the L by a factor one (e.g. $L: 3 \rightarrow 4$ or $L: 4 \rightarrow 5$) requires around two times more additions, multiplications and maximisation operations when Max-Log-MAP algorithm is employed at the iterative decoder, while yield performance gains around 1 dB according to Figs. 4–6 for attaining BER $\simeq 10^{-6}$. As far as the Log-MAP algorithm is a more interesting candidate to be used for yielding performance close to original MAP decoders by using correction function, the number of lookups required for

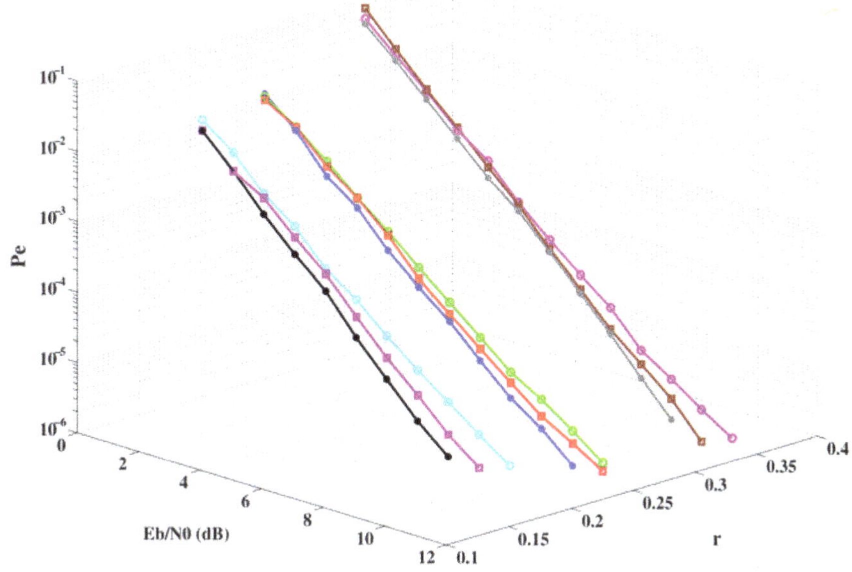

Fig. 5 *Comparison of BER against E_b/N_0 against r for the wireless system with out-DC-FECC*

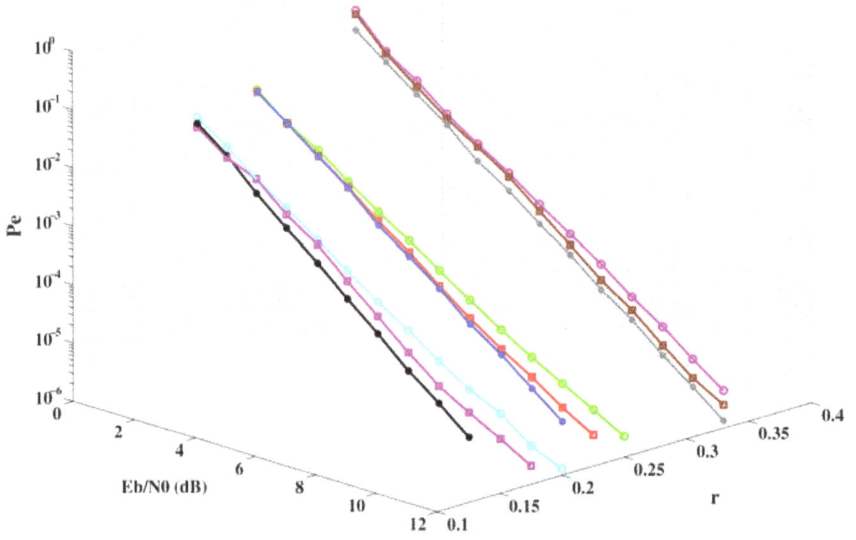

Fig. 6 *Comparison of BER against E_b/N_0 against r for the wireless system with in-DC-FECC*

computing correction function brings further operations for the practical use of high constraint lengths.

As be seen from the performance curves, since for DC-in-FECC one erroneously detected phase will cause two successive false symbols even if the next phase is received correctly, that is, double-error phenomenon, the error burst causes serious BER degradation of above 2 dB compared with the coherent systems. For reducing the highly correlated errors, in DC-out-FECC scheme, the SCQIC decoder does not encounter the double-error phenomenon and only the BER of SCQIC decoder output is doubled. Hence, doubling the SCQIC decoder BER yields a smaller coding gain loss at the expense of using synchroniser circuit for SCQIC decoder, which is not suitable for applications where long delay for resolving the phase ambiguity cannot be tolerated [9].

The effects of changing the structure of the QP-vectors and their lengths have been examined in this part. For instance, according to design Theorems I and II of [12], the QP-vectors for $N = 16$ and 32 are given as follows, where $\pi_N^{(C,F)}$ denotes the vector associated with QP-length N, with cyclic shift C and multiplicative factor F (see equation at the bottom of the next page).

Hence, for instances of 16-/32-bit permutation lengths, the QP-based interleaver maps generated sequences from half-rate outer SRCC trellis structures for $r \le 0.33$ and $L \le 5$, for example

$$\rightarrow \bar{v} = \langle 1110001010011000 \rangle_{16}$$
$$\rightarrow \bar{v} = \langle 11100010100110001110001010011000 \rangle_{32}$$

Into the reordered sequences of w_C^F given as follows:

For π_{16} (see equation at the bottom of next page)

For π_{32}

(see equation at the bottom of next page)

For the case of message passing between the component decoders corresponding to outer/inner SRCC trellis structures, the input of outer decoder from the output of inner decoder utilises the inverse of this function at the deinterleaver. Instances of the QP-based deinterleaving are given as follows, where the $\tilde{w}_C^F$ is an

$$\pi_N^{(C,F)}:\begin{cases} \pi_{16}:\begin{cases} \pi^{1,F\in\{1:15\}_{\text{odd}}}:\begin{cases} \pi^{1,1} = [4,\ 15,\ 7,\ 14,\ 13,\ 11,\ 3,\ 1,\ 9,\ 16,\ 10,\ 5,\ 8,\ 12,\ 6,\ 2] \\ \cdots \\ \pi^{1,15} = [12,\ 6,\ 10,\ 13,\ 8,\ 2,\ 9,\ 1,\ 15,\ 7,\ 5,\ 4,\ 11,\ 3,\ 14,\ 16] \end{cases} \\ \cdots \\ \pi^{15,F\in\{1:15\}_{\text{odd}}}:\begin{cases} \pi^{15,1} = [6,\ 2,\ 4,\ 15,\ 7,\ 14,\ 13,\ 11,\ 3,\ 1,\ 9,\ 16,\ 10,\ 5,\ 8,\ 12] \\ \cdots \\ \pi^{15,15} = [14,\ 16,\ 12,\ 6,\ 10,\ 13,\ 8,\ 2,\ 9,\ 1,\ 15,\ 7,\ 5,\ 4,\ 11,\ 3] \end{cases} \end{cases} \\ \pi_{32}:\begin{cases} \pi^{1,F\in\{1:31\}_{\text{odd}}}:\begin{cases} \pi^{1,1} = [4,\ 15,\ 7,\ 14,\ 32,\ 11,\ 30,\ 26,\ 25,\ 16,\ 31,\ 6,\ 24,\ 28,\ 22,\ 1,\ 17,\ 8,\ 18,\ 13,\ 29,\ 20, \\ \quad 3,\ 9,\ 12,\ 23,\ 10,\ 5,\ 21,\ 19,\ 27,\ 2] \\ \cdots \\ \pi^{1,31} = [7,\ 15,\ 13,\ 29,\ 24,\ 11,\ 22,\ 25,\ 31,\ 14,\ 5,\ 21,\ 16,\ 26,\ 17,\ 1,\ 12,\ 6,\ 10,\ 28,\ 3,\ 18, \\ \quad 9,\ 8,\ 4,\ 23,\ 2,\ 20,\ 27,\ 19,\ 30,\ 32] \end{cases} \\ \cdots \\ \pi^{31,F\in\{1:31\}_{\text{odd}}}:\begin{cases} \pi^{31,1} = [27,\ 2,\ 4,\ 15,\ 7,\ 14,\ 32,\ 11,\ 30,\ 26,\ 25,\ 16,\ 31,\ 6,\ 24,\ 28,\ 22,\ 1,\ 17,\ 8,\ 18,\ 13,\ 29, \\ \quad 20,\ 3,\ 9,\ 12,\ 23,\ 10,\ 5,\ 21,\ 19] \\ \cdots \\ \pi^{31,31} = [30,\ 32,\ 7,\ 15,\ 13,\ 29,\ 24,\ 11,\ 22,\ 25,\ 31,\ 14,\ 5,\ 21,\ 16,\ 26,\ 17,\ 1,\ 12,\ 6,\ 10,\ 28, \\ \quad 3,\ 18,\ 9,\ 8,\ 4,\ 23,\ 2,\ 20,\ 27,\ 19] \end{cases} \end{cases} \end{cases}$$

arbitrary sequence

$$\tilde{w}_5^5 = \langle 1110001010011000\rangle_{16} \xrightarrow{\pi^{-1}} \langle 0001010111000101\rangle_{16}$$
$$\tilde{w}_{21}^{21} = \langle 11100010100110001010001010010101\rangle_{32} \xrightarrow{\pi^{-1}} \langle 10101001000000001100011011110101\rangle_{32}$$

As it has been shown, for the case of permutations with different *Fs* and *Cs*, different randomisation mappings can be constructed. In Fig. 7, these variations have been investigated for the high gain $N = 1024$ system, where the F and C are chosen in mid-range of different size categories. So, when either of F or C is varying, the other is fixed to its mid-range value. As it can be seen, as far as the gain gaps are negligible, the selection of structural parameters for construction of QPs makes no significant BER superiority/inferiority.

In the classic turbo coding and its subsequent contributions, exceptional coding gains have been reported to be attained in the case of employing long permutations, that is, N exceeds 10 000 bits such as, for example, Berrou's classic half-rate PCCCing with $N = 65\,536$ bits. However, since the extra complexity and delay are not jus-

$$w_{1:15}^{1}:\left\{\begin{array}{l}
\langle 001010111000010 1\rangle\\
\langle 0101011100001010\rangle\\
\langle 1010111000010100\rangle\\
\langle 0101110000101001\rangle\\
\langle 1011100001010010\rangle\\
\langle 0111000010100101\rangle\\
\langle 1110000101001010\rangle\\
\langle 1100001010010101\rangle\\
\langle 1000010100101011\rangle\\
\langle 0000101001010111\rangle\\
\langle 0001010010101110\rangle\\
\langle 0010100101011100\rangle\\
\langle 0101001010111000\rangle\\
\langle 1010010101110000\rangle\\
\langle 0100101011100001\rangle
\end{array}\right. \cdots w_{1:15}^{5}:\left\{\begin{array}{l}
\langle 0101000111001100\rangle\\
\langle 1010001110011000\rangle\\
\langle 0100011100110001\rangle\\
\langle 1000111001100010\rangle\\
\langle 0001110011000101\rangle\\
\langle 0011100110001010\rangle\\
\langle 0111001100010100\rangle\\
\langle 1110011000101000\rangle\\
\langle 1100110001010001\rangle\\
\langle 1001100010100011\rangle\\
\langle 0011000101000111\rangle\\
\langle 0110001010001110\rangle\\
\langle 1100010100011100\rangle\\
\langle 100010100011100\rangle\\
\langle 000101000111001\rangle
\end{array}\right. \cdots w_{1:15}^{15}:\left\{\begin{array}{l}
\langle 1001011101000100\rangle\\
\langle 0010111010001001\rangle\\
\langle 0101110100010010\rangle\\
\langle 1011101000100100\rangle\\
\langle 0111010001001001\rangle\\
\langle 1110100010010010\rangle\\
\langle 1101000100100101\rangle\\
\langle 1010001001001011\rangle\\
\langle 0100010010010111\rangle\\
\langle 1000100100101110\rangle\\
\langle 0001001001011101\rangle\\
\langle 0010010010111010\rangle\\
\langle 0100100101110100\rangle\\
\langle 1001001011101000\rangle\\
\langle 0010010111010001\rangle
\end{array}\right.$$

$$w_{1:31}^{1}:\left\{\begin{array}{l}
\langle 00100000100001011011101111000101\rangle\\
\langle 01000001000010110111011110001010\rangle\\
\langle 10000010000101101110111100010100\rangle\\
\langle 00000100001011011101111000101001\rangle\\
\langle 00001000010110111011110001010010\rangle\\
\langle 00010000101101110111100010100100\rangle\\
\langle 00100001011011101111000101001000\rangle\\
\langle 01000010110111011110001010010000\rangle\\
\langle 10000101101110111100010100100000\rangle\\
\langle 00001011011101111000101001000001\rangle\\
\langle 00010110111011110001010010000010\rangle\\
\langle 00101101110111100010100100000100\rangle\\
\langle 01011011101111000101001000001000\rangle\\
\langle 10110111011110001010010000010000\rangle\\
\langle 01101110111100010100100000100001\rangle\\
\langle 11011101111000101001000001000010\rangle\\
\langle 10111011110001010010000010000101\rangle\\
\langle 01110111100010100100000100001011\rangle\\
\langle 11101111000101001000001000010110\rangle\\
\langle 11011110001010010000010000101101\rangle\\
\langle 10111100010100100000100001011011\rangle\\
\langle 01111000101001000001000010110111\rangle\\
\langle 11110001010010000010000101101110\rangle\\
\langle 11100010100100000100001011011101\rangle\\
\langle 11000101001000001000010110111011\rangle\\
\langle 10001010010000010000101101110111\rangle\\
\langle 00010100100000100001011011101111\rangle\\
\langle 00101001000001000010110111011110\rangle\\
\langle 01010010000010000101101110111100\rangle\\
\langle 10100100000100001011011101111000\rangle\\
\langle 01001000001000010110111011110001\rangle
\end{array}\right. \cdots w_{1:31}^{11}:\left\{\begin{array}{l}
\langle 10100001011000011101000000101111\rangle\\
\langle 01000010110000111010000001011111\rangle\\
\langle 10000101100001110100000010111110\rangle\\
\langle 00001011000011101000000101111101\rangle\\
\langle 00010110000111010000001011111010\rangle\\
\langle 00101100001110100000010111110100\rangle\\
\langle 01011000011101000000101111101000\rangle\\
\langle 10110000111010000001011111010000\rangle\\
\langle 01100001110100000010111110100001\rangle\\
\langle 11000011101000000101111101000010\rangle\\
\langle 10000111010000001011111010000101\rangle\\
\langle 00001110100000010111110100001011\rangle\\
\langle 00011101000000101111101000010110\rangle\\
\langle 00111010000001011111010000101100\rangle\\
\langle 01110100000010111110100001011000\rangle\\
\langle 11101000000101111101000010110000\rangle\\
\langle 11010000001011111010000101100001\rangle\\
\langle 10100000010111110100001011000011\rangle\\
\langle 01000000101111101000010110000111\rangle\\
\langle 10000001011111010000101100001110\rangle\\
\langle 00000010111110100001011000011101\rangle\\
\langle 00000101111101000010110000111010\rangle\\
\langle 00001011111010000101100001110100\rangle\\
\langle 00010111110100001011000011101000\rangle\\
\langle 00101111101000010110000111010000\rangle\\
\langle 01011111010000101100001110100000\rangle\\
\langle 10111110100001011000011101000000\rangle\\
\langle 01111101000010110000111010000001\rangle\\
\langle 11111010000101100001110100000010\rangle\\
\langle 11110100001011000011101000000101\rangle\\
\langle 11101000010110000111010000001011\rangle
\end{array}\right.$$

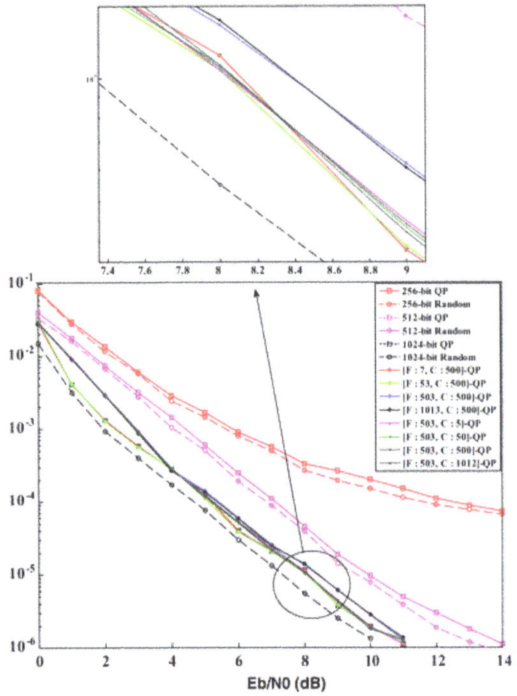

Fig. 7 *Comparison of BER curves for QP-/random-based interleaved wireless systems with 200 $\simeq$ <N $\simeq$ <1000; comparison of gain gaps between QP-based systems for selected cycle shifts (Cs) and multiplicative factors (Fs)*

tified for modern wireless applications, the fundamental design issues of channel codes directed a great deal of efforts for designing short-length FECs; that is, adopting short-frame system structures at expense of BER degradations. Hence, this design guideline limits our investigation to be conducted for range $N \simeq <1000$ bits; that is, for the comparative investigation between the QP and the random permutations' gain gaps for different lengths because of the reason discussed in (4) of the outlined list in Section 2. Fig. 7 further shows the performance curves as a function of interleaver size, which presents how performance improves as the interleaver length increases for $N_{\mathrm{INT}} \simeq <1000$. The following important observations can be made from the obtained results: (i) As far as the QP- and random-based systems are operating in close SNRs, that is, gain gaps for yielding similar BER are not high for the three possible interleaver lengths in $200 \simeq <N \simeq <1000$ range (i.e. $N \in \{256, 512, 1024\}$) the research suggests to employ QPs for modern short-frame applications which also require lower implementation complexities. (ii) For the three possible and selected N_s which satisfy both QP construction conditions and $N \simeq <1000$, 256 bit scheme is not capable of operating without an error floor. Whereas the 512 bit scheme is provisioning the BER performance in both waterfall and flare regions, and reduces the further complexities associated with the high gain $N = 1024$ bit scheme.

5 Conclusion

In this paper, in order to contrive an attractive design tradeoff for the QP-based STTCs, algebraic permutation and SRCC component code design issues and FEC-rate flexibility, which have dramatic

$$\cdot w^{21}_{1:31}: \left\{ \begin{array}{l}
\langle 00011010111000010111000110001100 \rangle \\
\langle 00110101110000101110001100011000 \rangle \\
\langle 01101011100001011100011000110000 \rangle \\
\langle 11010111000010111000110001100000 \rangle \\
\langle 10101110000101110001100011000001 \rangle \\
\langle 01011100001011100011000110000011 \rangle \\
\langle 10111000010111000110001100000110 \rangle \\
\langle 01110000101110001100011000001101 \rangle \\
\langle 11100001011100011000110000011010 \rangle \\
\langle 11000010111000110001100000110101 \rangle \\
\langle 10000101110001100011000001101011 \rangle \\
\langle 00001011100011000110000011010111 \rangle \\
\langle 00010111000110001100000110101110 \rangle \\
\langle 00101110001100011000001101011100 \rangle \\
\langle 01011100011000110000011010111000 \rangle \\
\langle 10111000110001100000110101110000 \rangle \\
\langle 01110001100011000001101011100001 \rangle \\
\langle 11100011000110000011010111000010 \rangle \\
\langle 11000110001100000110101110000101 \rangle \\
\langle 10001100011000001101011100001011 \rangle \\
\langle 00011000110000011010111000010111 \rangle \\
\langle 00110001100000110101110000101110 \rangle \\
\langle 01100011000001101011100001011100 \rangle \\
\langle 11000110000011010111000010111000 \rangle \\
\langle 10001100000110101110000101110001 \rangle \\
\langle 00011000001101011100001011100011 \rangle \\
\langle 00110000011010111000010111000110 \rangle \\
\langle 01100000110101110000101110001100 \rangle \\
\langle 11000001101011100001011100011000 \rangle \\
\langle 10000011010111000010111000110001 \rangle \\
\langle 00000110101110000101110001100011 \rangle
\end{array} \right.$$

$$\cdots w^{31}_{1:31}: \left\{ \begin{array}{l}
\langle 101100010000000111001111001100100 \rangle \\
\langle 011000100000001110011110011001001 \rangle \\
\langle 110001000000011100111100110010010 \rangle \\
\langle 100010000000111001111001100100101 \rangle \\
\langle 000100000001110011110011001001011 \rangle \\
\langle 001000000011100111100110010010110 \rangle \\
\langle 010000000111001111001100100101100 \rangle \\
\langle 10000001110011110011001001011000 \rangle \\
\langle 00000011100111100110010010110001 \rangle \\
\langle 00000111001111001100100101100010 \rangle \\
\langle 00001110011110011001001011000100 \rangle \\
\langle 00011100111100110010010110001000 \rangle \\
\langle 00111001111001100100101100010000 \rangle \\
\langle 01110011110011001001011000100000 \rangle \\
\langle 11100111100110010010110001000000 \rangle \\
\langle 11001111001100100101100010000001 \rangle \\
\langle 10011110011001001011000100000011 \rangle \\
\langle 00111100110010010110001000000111 \rangle \\
\langle 01111001100100101100010000001110 \rangle \\
\langle 11110011001001011000100000011100 \rangle \\
\langle 11100110010010110001000000111001 \rangle \\
\langle 11001100100101100010000001110011 \rangle \\
\langle 10011001001011000100000011100111 \rangle \\
\langle 00110010010110001000000111001111 \rangle \\
\langle 01100100101100010000001110011110 \rangle \\
\langle 11001001011000100000011100111100 \rangle \\
\langle 10010010110001000000111001111001 \rangle \\
\langle 00100101100010000001110011110011 \rangle \\
\langle 01001011000100000011100111100110 \rangle \\
\langle 10010110001000000111001111001100 \rangle \\
\langle 001011000100000001110011110011001 \rangle
\end{array} \right.$$

effect on the free distance of the resultant STTC method, have been all addressed and characterised when they are all amalgamated. The results of this paper can be extended in a number of ways which are of great theoretical and practical interest, including investigation on the potential enhancements as a result of deploying directional antennas at one or both ends of the MISO radio link, and investigation over mobile-to-mobile scattering propagation environments.

6 References

[1] Bauch G.: 'Concatenation of space-time block codes and turbo-TCM'. IEEE Int. Conf. Communications (ICC), June 1999, vol. 2, pp. 1202–1206

[2] Mehran F., Maunder R.G.: 'Wireless MIMO systems employing joint turbo-like STBC codes with bit-level algebraically-interleaved URSCs'. IEEE Int. Wireless Symp. (IWS), April 2013, pp. 1–4

[3] Mehran F., Rahimian A.: 'Physical layer performance enhancement for femtocell SISO/MISO soft real-time wireless communication systems employing serial concatenation of quadratic interleaved codes'. 20th Iranian Conf. Electrical Engineering (ICEE), May 2012, pp. 1188–1193

[4] Rahimian A., Mehran F.: 'BEP enhancement for semi-femtocell MIMO systems employing SC-QICs and OSTBCs', *Int. J. Electron. Commun. Comput. Technol.*, 2013, **3**, (1), pp. 329–332

[5] Rahimian A., Mehran F.: 'Short-length FSU-SCQICs over coherent and incoherent stochastic aeronautical MISO channels'. 19th Asia-Pacific Conf. Communications (APCC), August 2013, pp. 655–656

[6] Rahimian A., Mehran F., Maunder R.G.: 'Serial concatenation of quadratic interleaved codes in different wireless Doppler environments'. IEEE Fourth Int. Conf. Electronics Information and Emergency Communication (ICEIEC), November 2013, pp. 94–101

[7] Hanzo L., Woodard J.P., Robertson P.: 'Turbo decoding and detection for wireless applications', *Proc. IEEE*, 2007, **95**, (6), pp. 1178–1200

[8] Rahimian A., Mehran F., Maunder R.G.: 'Joint space-time algebraically-interleaved turbo-like coded incoherent MIMO systems with optimal and suboptimal MAP probability decoders'. IEEE Fourth Int. Conf. Electronics Information and Emergency Communications (ICEIEC), November 2013, pp. 1–8

[9] Nguyen T.M.: 'Phase-ambiguity resolution for QPSK modulation systems'. JPL Publication 89–4, Part I: A Review, May 1989, pp. 1–25

[10] Jung P., Nasshan M.: 'Dependence of the error performance of turbo-codes on the interleaver structure in short frame transmission systems', *Electron. Lett.*, 1994, **30**, (4), pp. 287–288

[11] Dogandzic A., Zhang B.: 'Estimating Jakes' Doppler power spectrum parameters using the Whitte approximation', *IEEE Trans. Signal Process.*, 2005, **53**, (3), pp. 987–1005

[12] Takeshita O.Y., Costello D.J.: 'New classes of algebraic interleavers for turbo-codes'. IEEE Int. Symp. Information Theory (ISIT), August 1998

[13] Takeshita O.Y., Costello D.J.: 'New deterministic interleaver designs for turbo codes', *IEEE Trans. Inf. Theory*, 2000, **46**, (6), pp. 1988–2006

[14] Tarokh V., Jafarkhani H., Calderbank A.: 'Space-time block codes from orthogonal designs', *IEEE Trans. Inf. Theory*, 1999, **45**, pp. 1456–1467

[15] Tarokh V., Jafarkhani H., Calderbank A.: 'Space-time block coding for wireless communications: performance results', *IEEE J. Sel. Areas Commun.*, 1999, **17**, pp. 451–460

[16] Clarkson K.L., Sweldens W., Zheng A.: 'Fast multiple-antenna differential decoding', *IEEE Trans. Commun.*, 2001, **49**, (2), pp. 253–261

[17] Jakes W.C.: 'Microwave mobile communications' 2nd edition (Wiley, New York, 1974)

Compact U-shape radiating patch with rectangular ground planar monopole antenna

Vijay Kisanrao Sambhe, Rahul Narayanrao Awale, Abhay Wagh

Department of Electrical Engineering, Veermata Jijabai Technological Institute, Mumbai, India
E-mail: vksambhe@vjti.org.in

Abstract: A compact U-shape radiating patch with rectangular ground planar monopole antenna is proposed. Antenna is fabricated on FR4 substrate with permittivity 4.4 and loss tangent 0.02 with dimension $75(LR) \times 48(WR) \times 1.6(h)$ mm^3. Measured return loss is ≤ -10 dB for the entire impedance bandwidth (800–3500 MHz). In addition, different key parameters which affect the impedance bandwidth are analysed and results discussed. Moreover, antennas have acceptable gain flatness with good omnidirectional radiation patterns. Its ease of fabrication, compatibility with other electronic devices, and radiation pattern make it a competent candidate for global system of mobile (890–960 MHz), digital communication system (1700–1900 MHz) and Bluetooth (2.45 GHz) cellular communication applications.

1 Introduction

Planar rectangular monopole antennas (RMAs) have become increasingly common because of the increasing boom in wireless communications. Narrow impedance bandwidth (typically a few percents) is the most serious disadvantage of microstrip patch antennas. In modern wireless communication systems, the required operating bandwidths for antennas are about 8.1% for an advanced mobile phone system (AMPS; 824–894 MHz), 7.6% for a global system for mobile communication (global system of mobile (GSM); 890–960 MHz) and 9.5% for a digital communication system (DCS; 1710–1880 MHz). On the other hand, compact microstrip antennas are essential in personal mobile communication systems for low-cost manufacturing and high integration with other electronic devices.

Recently, several techniques have been proposed to reduce the size of wide-band microstrip antennas [1–16]. In [1–4], a U-shape radiating patch and a half U-shape patch and their analyses are presented. V-slot loaded rectangular microstrip antennas and E-shape patch antenna have been studied in [5, 6]. Furthermore, E-shape patch antennas for a 5–6 GHz wireless computer network are designed in [7]. Moreover, stack patch antennas with different shapes are introduced in [8, 14]. L-probe proximity fed annular ring microstrip antenna in [10] and compact broadband C-shaped stacked microstrip antennas for enhancing the impedance bandwidth is studied in [13]. However, in the low microwave frequency range such as AMPS, GSM and DCS bands, the sizes of microstrip antennas are usually too large to be installed into a mini system. Thus, the problem of achieving a wide impedance bandwidth for a compact microstrip antenna is becoming an important topic in modern microstrip antenna design.

Apparently, the techniques presented in [1–16] are not suitable if low profile, wide bandwidth and high gain are required simultaneously. In this paper, a simple, low profile and compact rectangular microstrip antenna is designed and developed.

2 Antenna geometry

The geometry and optimum dimensions of the proposed antenna along with rectangular monopole is shown in Fig. 1. The proposed antenna is designed from a rectangular patch with size 75 (LR) × 48 (WR) mm^2 with removal of centre portion. The dimensions of U-shape monopole antenna (USMA) are further optimised to cover 800–3500 MHz impedance bandwidth. It is seen that removing the centre portion of the RMA not only reduces the size of the antenna but also increases the impedance bandwidth. The structure is fabricated on an easily available FR4 substrate with substrate permittivity of 4.4 and loss tangent of 0.02. The antenna is fed to a 50 Ω microstrip line with 3 × 37 mm^2 dimension and terminated by subminiature version A (SMA) connector. To meet the actual design requirements, that is, operating frequency, bandwidth and radiation pattern, some approximations are considered. The calculations are based on the transmission line model [17]. The effective dielectric constant of the substrate is given as

$$\varepsilon_{\text{eff}} = \frac{\varepsilon_r + 1}{2} + \frac{\varepsilon_r - 1}{2}\left[1 + 12\frac{H}{W}\right]^{-(1/2)} \tag{1}$$

The normalised extension of the length of the patch [17] is calculated by

$$\Delta L = 0.412 * H\frac{(\varepsilon_{\text{eff}} + 0.3)\big((W/H) + 0.264\big)}{(\varepsilon_{\text{eff}} - 0.258)\big((W/H) + 0.8\big)} \tag{2}$$

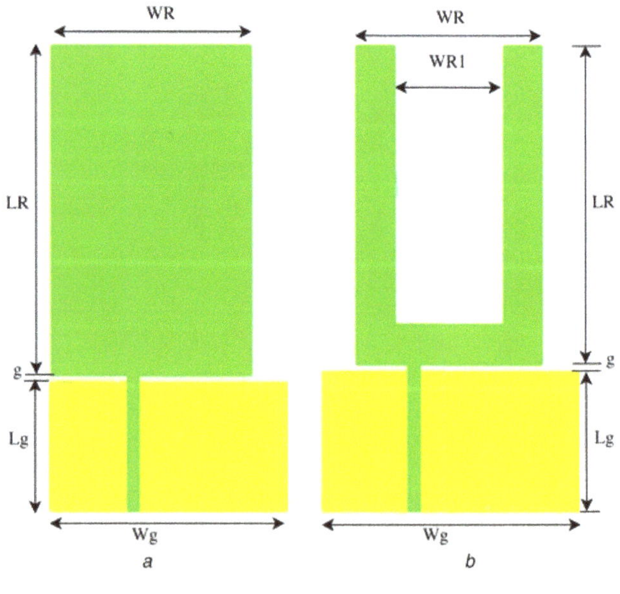

Fig. 1 *Geometry and configuration of the antennas*
a RMA with $LR = 75$ mm, $WR = 48$ mm, $Lg = 35$ mm, $Wg = 66$ mm and gap $g = 2$ mm
b Proposed antenna with $LR = 75$ mm, $WR = 48$ mm, $WR1 = 28$ mm, $Lg =$ 35 mm, $Wg = 66$ mm and gap $g = 2$ mm

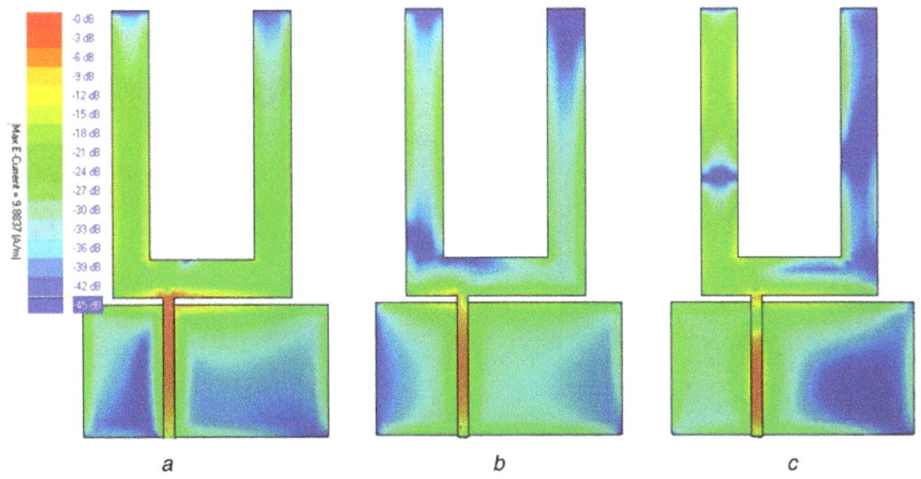

Fig. 2 *Surface current distribution of the proposed antenna*
a 1 GHz
b 2 GHz
c 3 GHz

where W is the width of the patch and H is the height of the substrate. The actual length of the patch [17] is expressed as

$$L = L_{\text{eff}} - 2\Delta L \tag{3}$$

The width and effective length of the microstrip patch are calculated by

$$W = \frac{c}{2f_0\sqrt{((\varepsilon_r + 1)/2)}} \tag{4}$$

$$L_{\text{eff}} = \frac{c}{2f_0\sqrt{\varepsilon_{\text{eff}}}} \tag{5}$$

Using the above equations and iterative trials, the dimensions of the antenna are optimised. In addition, ground plane dimensions are also optimised to achieve the optimum band of frequency as it affects the resonant frequencies and operating bandwidth. HyperLynx 3D (commercially available) software from Mentor Graphics version 15.2 is employed to perform the design, simulation and optimisation.

3 Results and discussion

The performance of USMA depends on the number of parameters. Here, some key parameters are considered for discussion which affects the impedance bandwidth, such as the gap (g), between the patch and ground plane and the length (Lg) and width (Wg) of the ground plane. Moreover, the length (LR) and width (WR) of the radiating patch are considered. In addition, antenna performance also depends on the size and shape of the ground plane and the fed point location of the microstrip line. The parameters that have a significant effect on the wide-band performance are discussed and analysed.

The surface current distribution at 900 MHz, 1.8 GHz and 2.45 GHz are shown in Fig. 2. In the monopole antenna, both the radiating patch and the ground plane act as a radiator. Fig. 2*a* shows that the current distribution is greater at the edges of the RMA and there is small Jx current component at the edges of the ground plane. The Jy current, that is, the vertical component, is greater at 1 GHz. It is less at 2 GHz as shown in Fig. 2*b*. At 3 GHz, the current distribution is greater in the lower part of the U-shape monopole antenna and Jy current is greater in the ground plane. Therefore, radiation patterns are nearly omnidirectional with negligible cross-polar component at the higher frequencies. The cross-polar component is < -20 dB.

The gap (g) between the radiating patch and the ground plane affects the impedance bandwidth as it acts as a matching

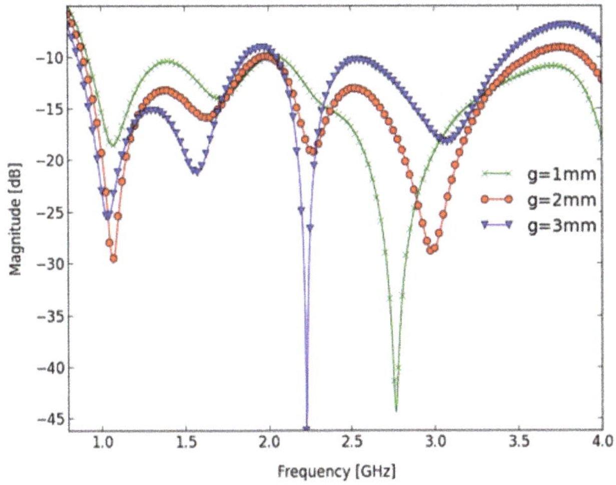

Fig. 3 *Return loss against frequency of the antenna for different gap 'g'*

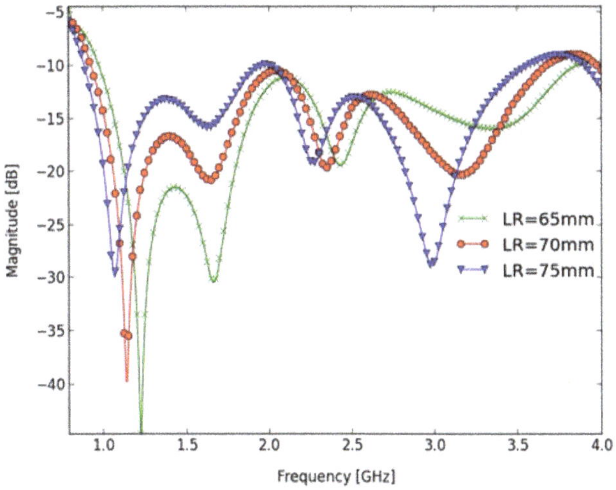

Fig. 4 *Return loss against frequency for different 'LR'*

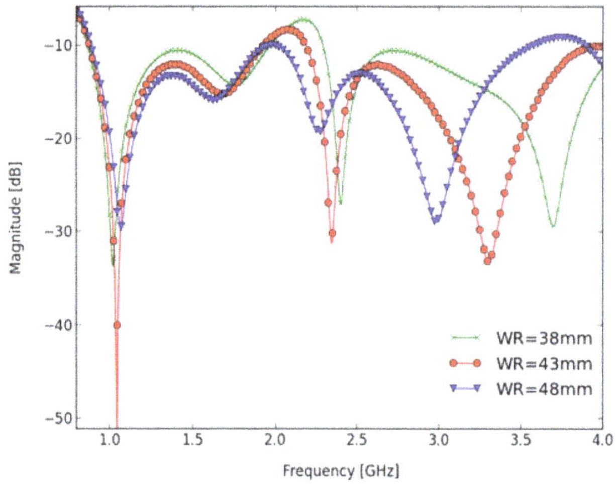

Fig. 5 *Return loss against frequency of the antenna for different 'WR'*

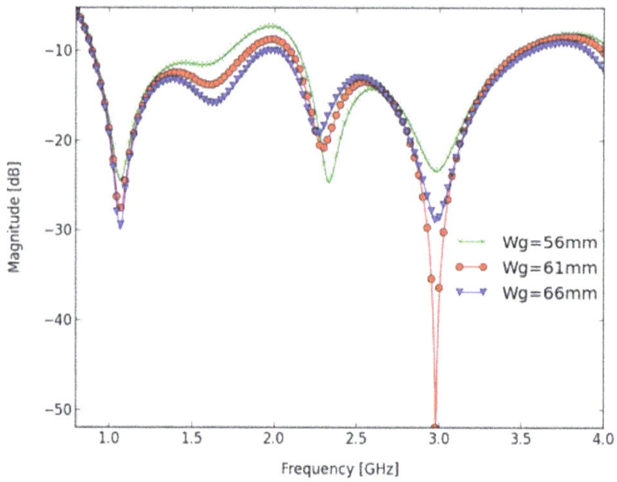

Fig. 7 *Return loss against frequency of the antenna for different 'Wg'*

network. The impedance bandwidth of the proposed antenna at different gap *g* is shown in Fig. 3. The optimum impedance bandwidth is obtained with $g = 2$ mm. At $g = 2$ mm, the capacitance that results from the spacing between the edge of the ground plane and the radiating patch reasonably balances the inductance of the antenna. The decreased gap *g* shifts the first resonant frequency 890–915 MHz and improves the impedance bandwidth. Moreover, increased *g* decrease lower band frequency from 890 to 850 MHz with impedance mismatch between 1.82 and 2.05 GHz.

Furthermore, the length *LR* of the radiating patch decreases from 75 to 70 mm, then 70 to 65 mm and the results are shown in Fig. 4. It is noted from the plot that the entire impedance bandwidth is shifted towards higher-frequency ends. In this structure, the first resonant frequency depends on the length of the rectangular patch. At $L = 75$ mm, first resonant frequency is 888 MHz and it is shifted towards high-frequency end when *L* is decreased. At $L = 70$ mm, and $L = 65$ mm, overall bandwidth remains same but GSM (890–960 MHz) band suppressed at $L = 65$ mm. The return loss against frequency plot of the antenna for different widths *WR* is shown in Fig. 5. It is clearly noted from the plot that the impedance mismatch occurs by decreasing *WR* from its optimum width.

In addition the length (*Lg*) and width (*Wg*) of the ground plane affect the impedance bandwidth. For clear understanding, the length of the ground plane is varied first, and then the width and the other parameters kept constant. The simultaneous effect on the impedance bandwidth is observed for different lengths *Lg* of the ground plane and shown in Fig. 6. The plot indicates, decreased length *Lg* shifts the bandwidth towards higher-frequency ends such as the length of the radiating patch. Moreover, decreased width *Wg* of the ground plane is shown in Fig. 7 and has negligible effect on the impedance bandwidth. Impedance mismatch only occurs when *Wg* is reduced from 66 to 61 mm and then 56 mm. Finally, microstrip fed line location is shifted towards right and left by 5 mm and simultaneous effect on impedance bandwidth is observed. Return loss against frequency plot of the feed location is shown in Fig. 8. The plot clearly indicates how crucial the fed location is as far as the proposed antenna is concerned. The impedance bandwidth is decreased when the fed location is changed to either the right or left side of the optimum location. Furthermore, the comparison between the RMA and USMA is shown in Table 1.

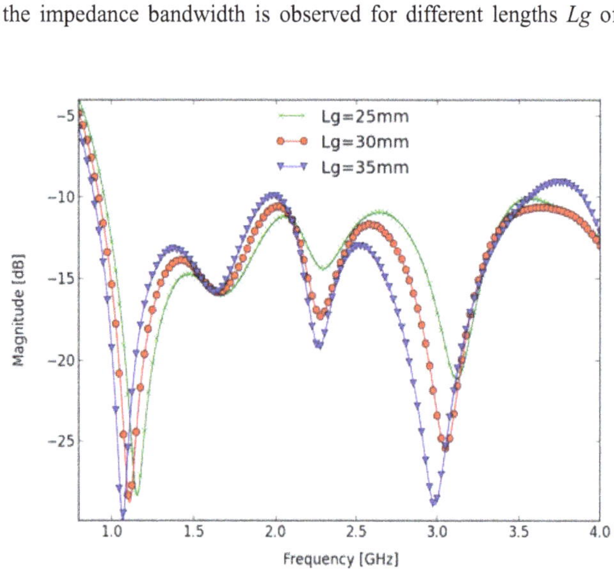

Fig. 6 *Return loss against frequency of the antenna for different 'Lg'*

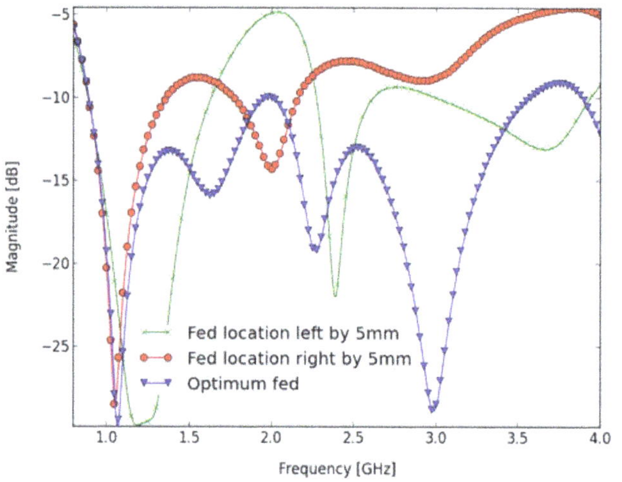

Fig. 8 *Return loss against frequency plot for different fed location of the antenna*

Table 1 Comparison between RMA and USMA

Antenna structure	*LR*, mm	*WR*, mm	Simulated bandwidth, MHz	Measured bandwidth, MHz	Gain, dBi
RMA	80	48	890–2348	818–2272	2–3.4
USMA	75	48	890–3557	818–3485	2–3.5

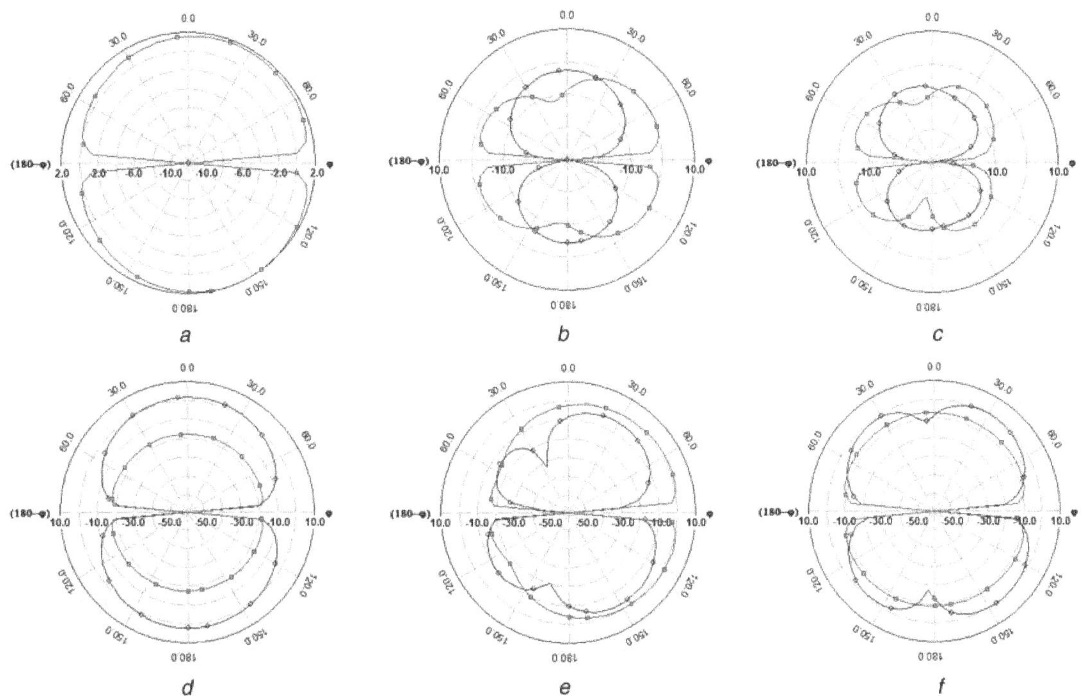

Fig. 9 *Radiation patterns of the proposed antenna for different frequencies*
a 1 GHz at $\phi = 0°$
b 2 GHz at $\phi = 0°$
c 3 GHz at $\phi = 0°$
d 1 GHz at $\phi = 90°$
e 2 GHz at $\phi = 90°$
f 3 GHz at $\phi = 90°$

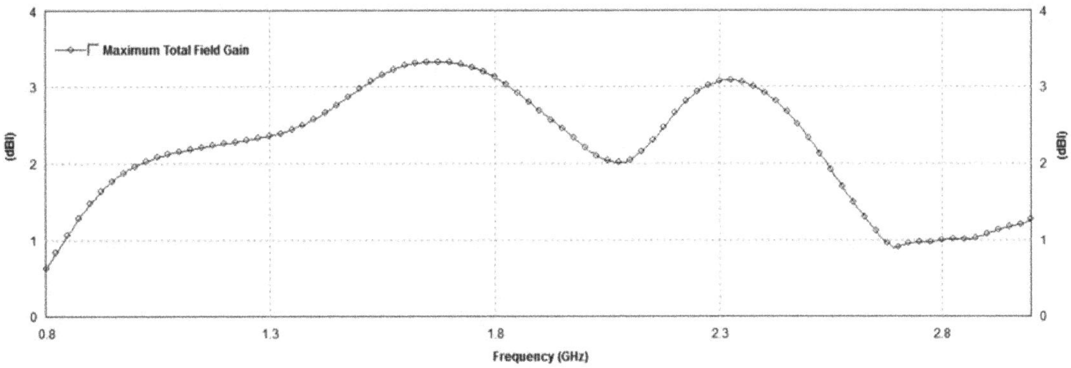

Fig. 10 *Gain against frequency of the proposed antenna*

The antenna radiation pattern of the proposed antenna is shown in Fig. 9. At each frequency, the radiation patterns in the *E* planes and *H* planes are normalised with respect to the maximum at their crossing points. The antenna exhibits stable omnidirectional radiation patterns over the entire bandwidth. At higher frequency, the radiation patterns deteriorate because the equivalent radiating area changes with frequency over the wide-band. Unequal phase distribution and significant magnitude of higher-order mode at higher frequencies also play a part in the deterioration of radiation pattern at higher frequencies. The measured return loss against the frequency plot of the proposed antenna is shown in Fig. 10. The graph shows that antenna impedance is within the standing wave ratio circle and there is good impedance matching for the entire frequency band. It is < -10 dB from 834 MHz to 3.5 GHz and there is good agreement between the measured and simulated impedance bandwidths. The mismatch between the measured and simulated one occurs because of fabrication and microstrip line fed to radiating patch. Moreover, there is acceptable gain from 2 to 3.5 dBi as shown in Fig. 11. Omnidirectional radiation patterns, good gain and compact size of the proposed structure make it a competent candidate in wireless communication applications (Fig. 12).

4 Conclusion

A compact U-shape radiating patch with rectangular ground planar monopole antenna is designed. The proposed antenna covers GSM (890–960 MHz), DCS (1710–1880 MHz) and 2.4 GHz (wireless fidelity and Bluetooth) frequency bands. Moreover, antennas have effective control over the entire frequency range. The structure is fabricated on an easily available FR4 substrate of permittivity 4.4 with loss tangent 0.02 and tested by using vector network analyser. The measured return loss is ≤ -10 dB for the entire impedance bandwidth (800–3500 MHz). The antenna structure is fed to 50 Ω microstrip line and terminated by a SMA connector. Proposed

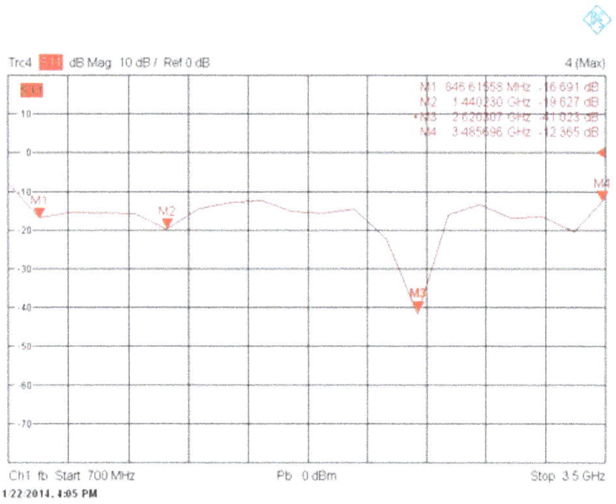

Fig. 11 *Measured return loss against frequency of the proposed antenna*

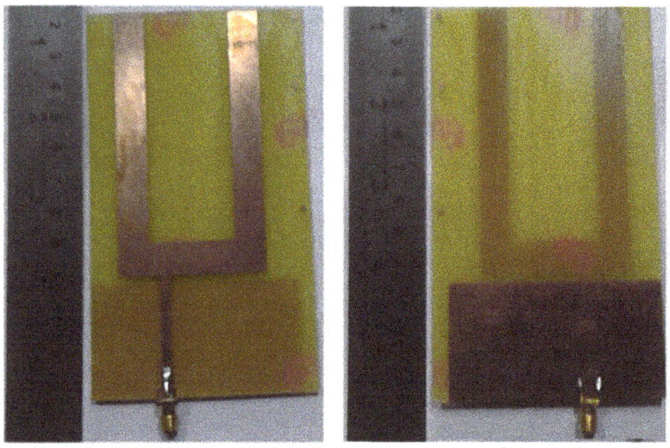

a *b*

Fig. 12 *Photograph of the proposed antenna*
a Front view
b Back view

antennas have acceptable gain flatness from 1–3.5 dBi with good omnidirectional radiation patterns. In addition, different key parameters which affect the impedance bandwidth are analysed and results are presented. Its easy fabrication and compatibility make it a competent candidate for different cellular phone applications.

5 Acknowledgment

The author acknowledges the All India Council of Technical Education (AICTE) New Delhi, an autonomous body of the Government of India for financial support under the Research Promotion Scheme (RPS) with reference no. 8023/RID/RPS-112/2011-12.

6 References

[1] Weigand S., Huff G.H., Pan K.H., Bernhard J.T.: 'Analysis and design of broad-band single-layer rectangular U-slot microstrip patch antennas', *IEEE Trans. Antennas Propag.*, 2003, **51**, (3), pp. 457–468

[2] Deshmukh A.A., Kumar G.: 'Half U-slot loaded rectangular microstrip antenna'. Proc. of IEEE Int. Symp. on Antennas and Propagation Society (APS), June 2003, vol. **2**, pp. 876–879

[3] Lee K.-F., Luk K.M., Tong K.F., Yung Y.L., Huynh T.: 'Experimental study of the rectangular patch with a U-shaped slot'. Proc. of IEEE Int. Symp. on Antennas and Propagation Society (APS), 1996, vol. **1**, pp. 10–13

[4] Chair R., Mak C.-L., Lee K.-F., Luk K.-M., Kishk A.A.: 'Miniature wide-band half U-slot and half E-shaped patch antennas', *IEEE Trans. Antennas Propag.*, 2005, **53**, (8), pp. 2645–2652

[5] Deshmukh A.A., Kumar G.: 'Broadband compact V-slot loaded RMSAs', *IEEE Electron. Lett.*, 2006, **42**, (7), pp. 951–952

[6] Ge Y., Esselle K.P., Bird: 'E-shaped patch antennas for high-speed wireless networks', *IEEE Trans. Antennas Propag.*, 2004, **52**, (12), pp. 3213–3219

[7] Ge Y., Esselle K.P., Bird T.S.: 'Broadband E-shaped patch antennas for 5-6 GHz wireless computer networks'. Proc. of IEEE Int. Symp. on Antennas and Propagation Society (APS), June 2003, vol. **2**, pp. 22–27

[8] Ooi B.-L., Qin S., Leong M.-S.: 'Novel design of broad-band stacked patch antenna', *IEEE Trans. Antennas Propag.*, 2002, **50**, (10), pp. 1391–1395

[9] Yang F., Zhang X.-X., Ye X., Rahmat-Samii Y.: 'Wide-band E-shaped patch antennas for wireless communications', *IEEE Trans. Antennas Propag.*, 2001, **49**, (7), pp. 1094–1100

[10] Guo Y.-X., Luk K.-M., Lee K.-F.: 'L-probe proximity-fed annular ring microstrip antennas', *IEEE Trans. Antennas Propag.*, 2001, **49**, (1), pp. 19–21

[11] Deshmukh A.A., Kumar G.: 'Compact broadband shorted square microstrip antenna'. Proc. of IEEE Int. Symp. on Antennas and Propagation Society (APS), June 2003, vol. **2**, pp. 872–875

[12] Wang Y.J., Lee C.K., Koh W.J.: 'Single-patch and single-layer square microstrip antenna with 67.5 percent bandwidth'. Proc. of IEE Int. Conf. on Microwaves, Antennas and Propagation, December 2001, vol. **148**, no. 6, pp. 418–422

[13] Deshmukh A.A., Kumar G.: 'Compact broadband S-shaped microstrip antennas', *IEEE Electron. Lett.*, 2006, **42**, (5), pp. 260–261

[14] Deshmukh A.A., Kumar G.: 'Compact broadband C-shaped stacked microstrip antennas'. Proc. of IEEE Int. Symp. on Antennas and Propagation Society (APS), 2002, vol. **2**, pp. 538–541

[15] Sambhe V.K., Awale R.N., Wagh A.: 'Compact inverted a shape radiating patch with rectangular ground wide band monopole antenna'. Proc. of Elsevier Science & Technology Third Int. Conf. on Recent Trends in Engineering & Technology (ICRTET), March 2014, pp. 210–214

[16] Sambhe V.K., Awale R.N., Wagh A.: 'Dual band inverted L-shape monopole antenna for cellular phone applications', *Microw. Opt. Technol. Lett.*, 2014, **56**, (12), pp. 2751–2755

[17] Kumar G., Ray K.P.: 'Broadband microstrip antennas' (Artech house, Norwood, MA, 2003)

[18] Balanis C.A.: 'Antenna theory and design' (John Wiley and sons Inc. USA, 2005, 3rd edn.)

[19] Kraus J.D., Marhefka R.J.: 'Antennas' (McGraw-Hill, 1988, 3rd edn.)

24

Minimum jitter-based adaptive decision feedback equaliser for giga-bit-per-second serial links

Alaa Rahman Al-Taee, Fei Yuan, Andy Ye

Department of Electrical and Computer Engineering, Ryerson University, Toronto, ON, Canada
E-mail: fyuan@ryerson.ca

Abstract: This study presents a minimum jitter-based adaptive decision feedback equaliser (DFE) for giga-bit-per-second (Gbps) serial links. The adaptation in search for the optimal tap coefficients of DFE is carried out with the objective to minimise data jitter at the edge of data eyes. Jitter minimisation is achieved by adjusting the slope of the DFE that counteracts that of the channel. The effectiveness of the proposed adaptive DFE is evaluated by embedding the DFE in a 2 Gbps serial link. The data link is analysed using Spectre from Cadence Design Systems with BSIM4 device models. Simulation results demonstrate that the proposed adaptive DFE is capable of opening closed data eyes with 83% vertical opening, 68% horizontal opening and 16% data jitter over 1 m FR4 channel while consuming 15.45 mW.

1 Introduction

The explosive growth of data processed by digital systems demands that data be transmitted over wire channels at giga-bit-per-second (Gbps). Increasing the number of parallel links undoubtedly improves effective channel bandwidth, its effectiveness, however, is severely undermined by high routing cost and deteriorating data skew [1]. Inter-chip and point-to-point on-chip data links over long interconnects are most achieved using serial links where data and clock are transmitted simultaneously via a single channel. The data rate of serial links is limited by inter-symbol interference (ISI) arising from channel imperfections including finite bandwidth, reflection and cross-talk. ISI manifests itself as both pre-cursors and post-cursors that span beyond the temporal boundary of one symbol time with latter typically dominating. To remove post-cursors, the most widely used method is decision feedback equalisation (DFE) introduced by Austin in 1967 [2]. Fig. 1 illustrates the operation of a 3-tap DFE. The slicer slices v_s at the main cursor and the result of the slicer passes through three delay stages of unit-interval (UI) delay. The output of the jth delay stage is multiplied by weight factor c_j such that $v_{\text{in},j} = c_j v_{k-j}$ where $v_{\text{in},j}$ is the jth post-cursor and v_{k-j} is the jth past decision of the slicer is satisfied. The temporally distributed feedback signal $v_f = v_{\text{in},1}u(k-1) + v_{\text{in},2}u(k-2) + v_{\text{in},3}u(k-3)$ where $u(k-r) = 1$ if $k = r$ and 0 otherwise is subtracted from v_{in} such that $v_s[k] = 0$ at $k = 1, 2, 3$. As a result, the first three post-cursors of v_s are removed ideally. The uncertainty of the characteristics of channels requires that weight factor c_j be set in accordance of the characteristics of the channels such that a complete removal of the post-cursors can be achieved [3]. In [4], the optimal tap coefficients are obtained by minimising the power of the difference (error) between the output and input of the slicer, that is, the least mean square (LMS) of the error. The jth optimal tap coefficient is obtained using iterative operation $c_{j,k+1} = c_{j,k} + h\varepsilon_k v_{k-j}$ where $j = 1, 2, \ldots, N$, N is the number of the taps of DFE, $c_{j,k+1}$ and $c_{j,k}$ are the coefficients of tap-j in steps $k+1$ and k, respectively, and h is the step size used to adjust the tap coefficients in each iteration step. LMS-based DFE is difficult to implement because of the need for ε_k and v_{k-j}, which can only be obtained using analogue-to-digital converters. Sign–sign LMS (SS-LMS)-based DFE that only needs the signs of ε and v_{k-j} and obtains the optimal tap coefficients using iterative algorithm $c_{j,k+1} = c_{j,k} + h\ \text{sign}[\varepsilon_k]\text{sign}[v_{k-j}]$ where $\text{sign}[x] = 1$ if $x \geq 0$ and -1 otherwise is proven to be an efficient and yet effective alternative of LMS-based DFE [5].

Since when data eye, that is, the waveform of the input of the slicer, is fully open, the difference between the output and input of the slicer is minimised. This observation suggests that the eye-opening of received data can also be used to guide the search for optimal DFE tap coefficients. As eye-opening is directly related to bit error rate (BER), for example, the larger the eye-opening, the better the BER, eye-opening-monitor (EOM)-based DFE provides an explicit link between DFE operation and BER. Both one-dimensional (1D) [6–10] and 2D EOMs [11–14] have been reported. A common drawback of EOM-based DFE is the lack of a tight constraint on data jitter, arising from duty-cycle distortion, switching noise and ISI, and having a detrimental effect on both the vertical and horizontal openings of data eyes. For example, 1D EOMs only focus on the vertical opening at the centre of data eye and do not have any constraint on data jitter. Although 2D EOMs do have a constraint on data jitter, it is lower-bound by the vertical eye-opening at the centre of the data eye, as illustrated

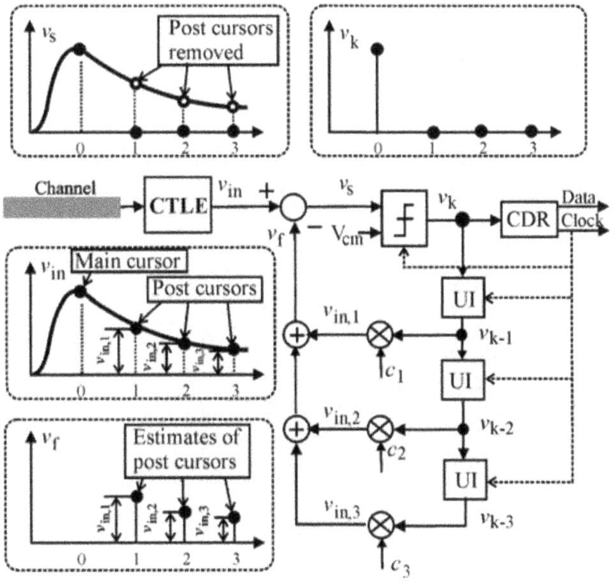

Fig. 1 *3-tap DFE. Continuous-time liner equaliser is used to boost data symbols prior to slicing*
This step is necessary to ensure that v_s is large enough such that the decision made by the slicer is most likely correct

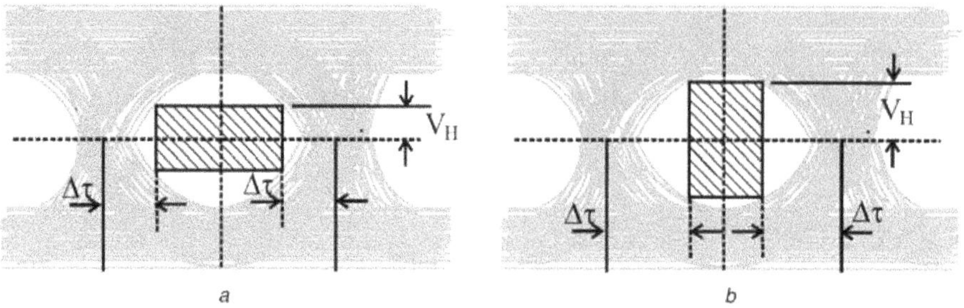

Fig. 2 *Jitter constraint is lower bound by the vertical eye-opening at the centre of data eye*
It is seen that the smaller the vertical opening V_H, the tighter the jitter constraint
a smaller vertical eye-opening, better jitter constraint
b Larger vertical eye-opening, poorer jitter constraint

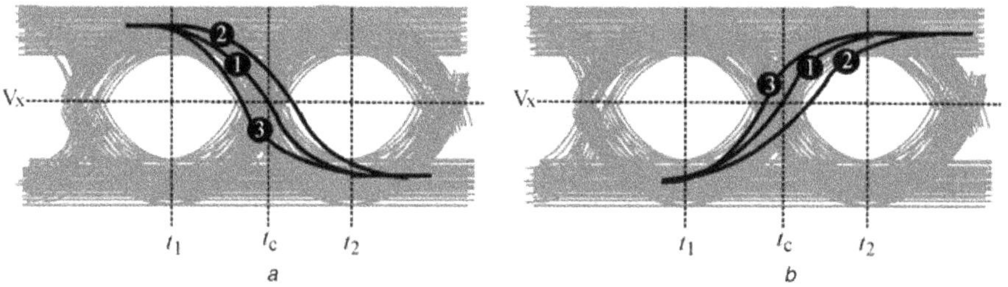

Fig. 3 *Jitter-based error detection*
a High-to-low transitions
b Low-to-high transitions

graphically in Fig. 2. It is highly desirable from a clock recovery point of view to minimise data jitter without sacrificing eye-opening. In [15], jitter information at the edges of data eyes is obtained by employing a set of temporally spaced samplers that sample the transition edge of the eyes at uniformly incremented time instants. Although effective, this approach is power hungry especially when the number of samplers is large. The dual-mode adaptive DFE proposed in [16] consists of a data DFE for maximising the vertical opening of data eyes and a jitter DFE for minimising the jitter at the edges of data eyes.

This paper presents a jitter-based adaptive DFE for Gbps serial links. The adaptation in search for the optimal tap coefficients of DFE is carried out with the objective to minimise data jitter at the edge of data eyes. Jitter minimisation is achieved by adjusting the slope of the DFE that counteracts that of the channel. The rest of this paper is organised as follows: Section 2 details the principle and implementation of the proposed maximum jitter adaptive DFE and simulation results that validates the effectiveness of the DFE. This paper is concluded in Section 3.

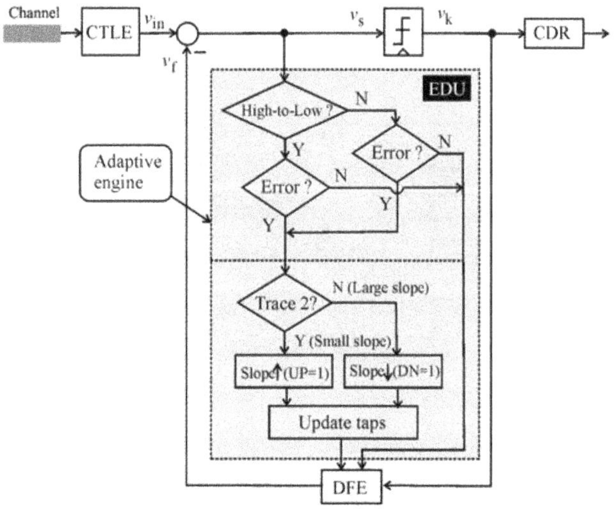

Fig. 4 *AE of maximum jitter adaptive DFE*

2 Jitter-based adaptive DFE

2.1 Algorithm

The detection of whether the minimum data jitter at the edges of received data is achieved or not is performed by comparing the voltage of received data with their common-mode voltage V_x at t_1, t_c and t_2 where $t_1 = t_c - T_s/2$, $t_2 = t_c + T_s/2$, t_c is threshold-crossing time instant and T_s is the symbol time, as shown in Fig. 3:

(1) If $V_s(t_1) > V_x$ and $V_s(t_2) < V_x$, it is a high-to-low transition (Fig. 3*a*).

- If $V_s(t_1) > V_x$ and $V_s(t_c) = V_x$ (Trace 1), the desirable trace is obtained and no error is flagged.
- $V_s(t_1) > V_x$ and $V_s(t_c) > V_x$ (Trace 2), the slope of the rising edge is too small. An error is flagged.
- If $V_s(t_1) > V_x$ and $V_s(t_c) < V_x$ (Trace 3), the slope of the rising edge is too large. An error is flagged.

(2) If $V_s(t_1) < V_x$ and $V_s(t_2) > V_x$, it is a low-to-high transition (Fig. 3*b*).

- If $V_s(t_1) < V_x$ and $V_s(t_c) = V_x$ (Trace 1), the desirable trace is obtained. No error is flagged.
- If $V_s(t_1) < V_x$ and $V_s(t_c) < V_x$ (Trace 2), the slope of the rising edge is too small. An error is flagged.
- $V_s(t_1) < V_x$ and $V_s(t_c) > V_x$ (Trace 3), the slope of the rising edge is too large. An error is flagged.

The preceding analysis reveals that the error signal indicating whether the slope of the transition is too small or too large can be generated using elementary logic. The tap coefficients of DFE can therefore be updated using an iterative algorithm similar to SS-LMS: $c_{j,k+1} = c_{j,k} + h\varepsilon_k \text{ sign}[v_{k-j}]$ where ε_k is the error in step k. The adaptive engine (AE) shown in Fig. 4 first determines the

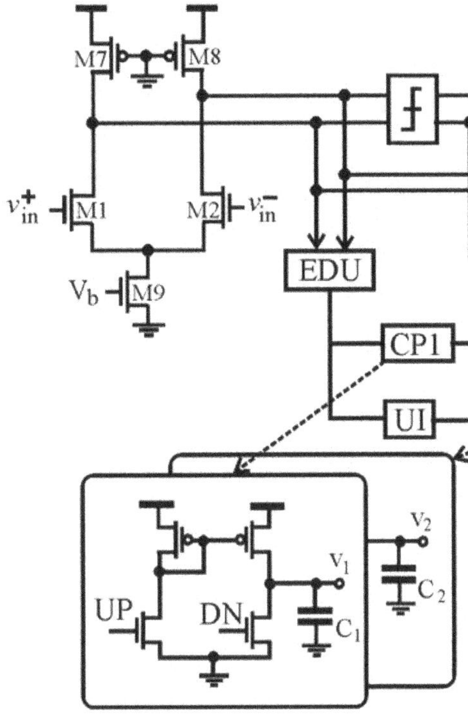

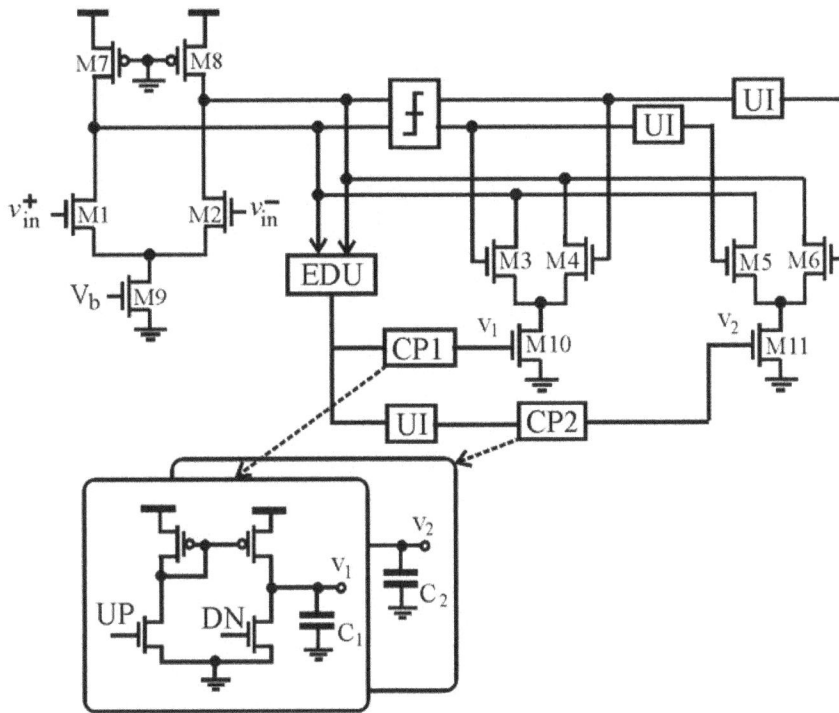

Fig. 5 *Configuration of jitter-based adaptive DFE*

type of the transition, it then rules whether the slope of the transition is too large or too small. If it is too large, DFE counteracts it by discharging the capacitors subsequently v_1 and v_2. This in turn lowers the tap coefficients and weakens DFE equalisation. Otherwise, it increases the tap coefficients and strengthens DFE equalisation.

The amount of the adjustment of v_1 and v_2 subsequently that of tap coefficients is set with the following considerations: if the adjustment is too small, the equalisation action of the DFE will be too weak to be effective. On the other hand, if the adjustment is too large, the equalisation action of the DFE will be too strong, resulting in undesirable oscillation subsequently a long adaption process. The value of the capacitors must also be chosen with care. If the capacitances are too large, the output voltage will not change fast enough to provide an appropriate voltage to DFE, resulting in a slow adaptation process. On the other hand, if the capacitances are too small, a large voltage fluctuation will exist, which will in turn have a negative effect on the search for the optimal coefficients of DFE. The voltage incremental step for tuning the coefficient of tap-1 is set to be five times that for tuning the coefficient of tap-2. This is achieved by setting $C_1 = 1$ pF and $C_2 = 5$ pF. Note that although $C_2 = 5C_1$ and the same error ε_k is used for tap-1 and tap-2, the iterative adjustments of tap-1 and that of tap-2 are independent of each other. This is because $c_{1,k+1} = c_{1,k} + h\varepsilon_k$ sign$[v_{k-1}]$ and $c_{2,k+1} = c_{2,k} + h\varepsilon_k$ sign$[v_{k-2}]$. The difference of v_{k-1} and v_{k-2} ensures that c_1 and c_2 are adjusted independently.

2.2 *Implementation*

The configuration of the proposed jitter adaptive DFE with two taps is shown in Fig. 5. The DFE consists of a differential amplifier to boost received signal prior to slicing, a slicer implemented using a three-stage clocked comparator capable of minimising duty-cycle distortion and kick-back [17–19], several clocked delay units, two

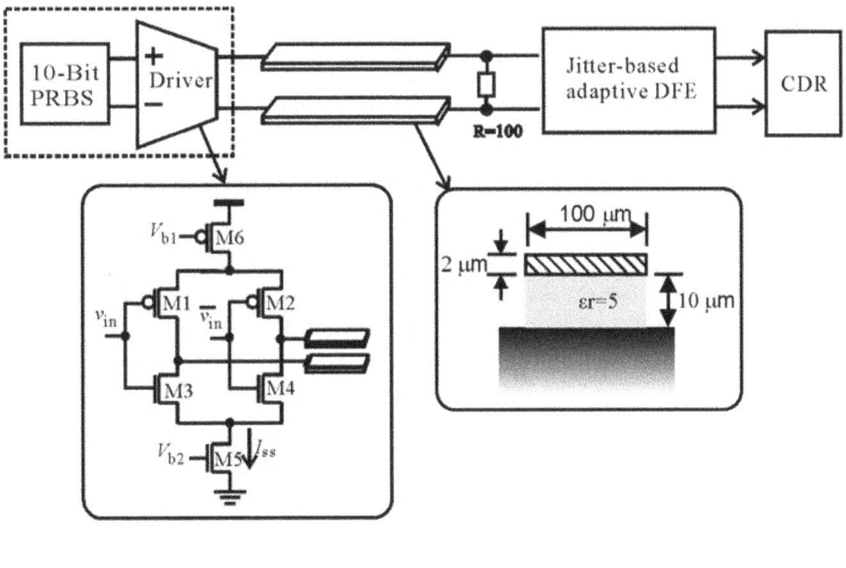

Fig. 6 *Test bench*

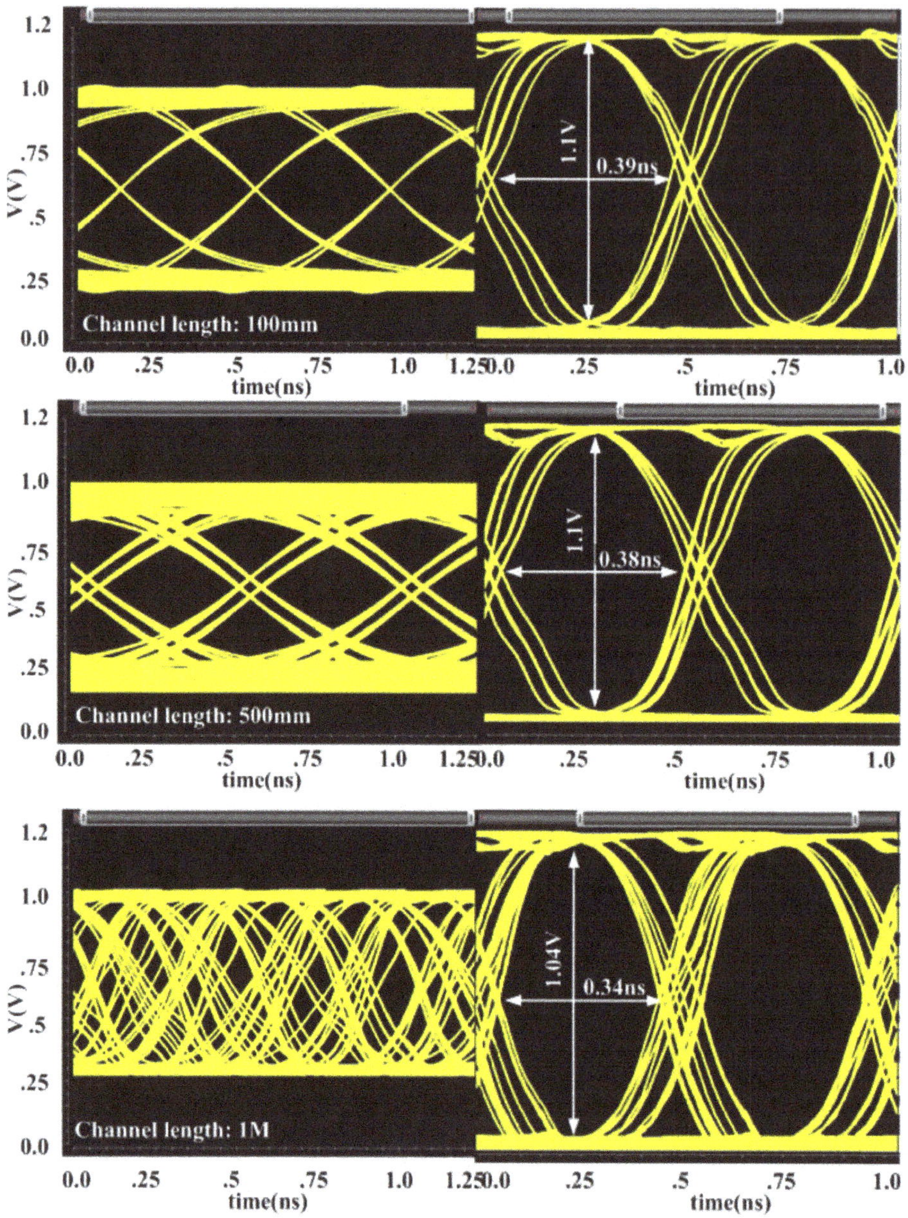

Fig. 7 *Simulated eye diagram*
Left: without DFE, and right: with DFE

current-steering tap generators with tap coefficients set by the tail currents, an error detect unit (EDU) consisting of three comparators and some logic gates, and two charge pumps that adjust the bias voltages of the tail current sources of the tap generators. Tap-1 takes the output of the slicer directly, whereas tap-2 has explicit delay units. The implementation of tap-1 is the same as that in [20]. Together with the delay of the second tap generator, the output of tap-2 has the delay of two unit intervals. It should be emphasised that the error signal from the EDU to the second tap generator needs to be delayed by one UI because of the temporal displacement of tap-1 and tap-2.

2.3 Simulation results

The proposed DFE is embedded in a 2 Gbps serial link shown in Fig. 6. The low-voltage differential-signalling driver conveys a 2 mA current to the channel that is terminated with a 100 Ω resistor at the far end of the channel. The data conveyed to the channel is a 10 bit pseudo-random bit sequence of full voltage swing. The serial link is analysed using Spectre from Cadence Design Systems with BSIM4 device models.

Fig. 7 shows the simulated eye diagram of the signal at the input of the slicer without and with the proposed DFE for various channel lengths. It is seen that when the DFE is absent, the eye-opening is reduced with the increase in channel length and the eye is completely closed when channel length is 1 m. When the proposed DFE is activated, the data eye is open nicely. The vertical eye-opening for channel length 0.1, 0.5 and 1 m is found to be 1.1, 1.1 and 1.0 V, respectively. The horizontal eye-opening for channel lengths 0.1, 0.5 and 1 m is found to be 0.39, 0.83 and 0.34 ns, respectively. With 0.5 ns symbol time, the data jitter for channel length 0.1, 0.5 and 1 m is found to be 55, 60 and 80 ps, respectively.

Fig. 8 shows the adaptation process with channel length 60 mm. It is observed v_1 and v_2 rise rapidly and reach their steady-state values in ~20 ns. The fluctuations of v_1 and v_2 in their steady-state agree well with the bang-bang nature of the proposed DFE. If one wants to remove the fluctuations of v_1 and v_2, a jitter dead zone can be introduced, as shown in Fig. 9. Traces 1 and 2 set the lower and

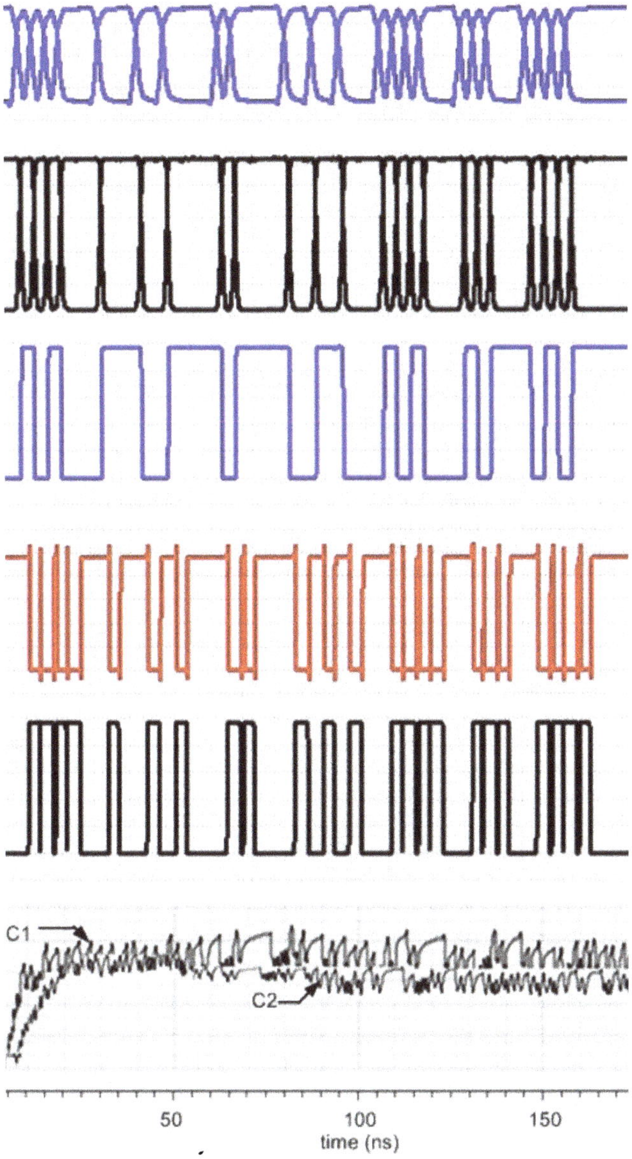

Fig. 8 *Top figure: waveform of unequalised data symbols (channel length: 60 mm), second figure: waveform of equalised data symbols, third figure: output of slicer, fourth figure: error signal (low-to-high transition), fifth figure: error signal (high-to-low transition) and last figure:* v_1 *and* v_2 *of tap generators*

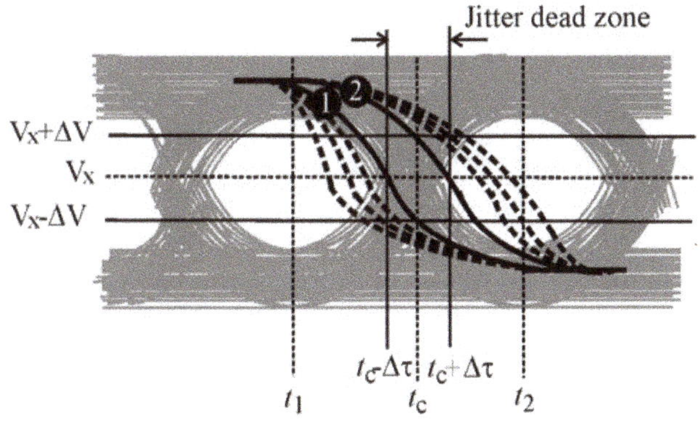

Fig. 9 *Jitter dead zone*

Table 1 Performance of the proposed jitter adaptive DFE (1 m channel length)

Tech.	IBM 130 nm
Supply voltage	1.2 V
Data rate	2 Gbps
Power of charge pumps	3.41 mW
Power of EDUs	6.33 mW
Power of DFE core (pre-amp., tap generators, delay units, ...)	5.71 mW
Total power	15.45 mW
Vertical eye-opening	1.0 V (83%)
Horizontal eye-opening	0.34 ns (68%)
Jitter	80 ps (16%)

upper boundaries. The corresponding voltages at t_c are $V_x - \Delta V$ and $V_x + \Delta V$, respectively. The corresponding jitter dead zone is given by $[t_c - \Delta\tau,\ t_c + \Delta\tau]$. Traces that fall into the dead zone will not flag an error. As a result, no action will be taken by DFE and v_1, v_2 will remain unchanged. Traces that fall outside the dead zone (dotted traces) will flag an error, which will in turn activate DFE and adjust v_1 and v_2 accordingly. Table 1 tabulates the performance of the proposed DFE at 1 m channel length.

3 Conclusions

A minimum jitter-based adaptive DFE for Gbps serial links was presented. The adaptation in search for the optimal tap coefficients of DFE is carried out with the objective to minimise data jitter at the edge of data eyes. Jitter minimisation is achieved by adjusting the slope of the DFE that counteracts that of the channel. The effectiveness of the proposed adaptive DFE is evaluated by embedding the DFE in a 2 Gbps serial link. The data link is analysed using Spectre from Cadence Design Systems with BSIM4 device models. Simulation results demonstrate that the proposed adaptive DFE is capable of providing 83% vertical eye-opening, 68% horizontal eye-opening and 16% data jitter over 1 m FR4 channel while consuming only 15.45 mW. Although the example used to validate the proposed DFE contains only two DFE taps, the underlining principle is applicable to DFE with an arbitrary number of taps. The method presented here cannot change the number of DFE taps adaptively. If one wants to have the number of taps also to be adaptive to the characteristics of channels, an algorithm that predetermines the number of taps based on the characteristics of the channel is needed.

4 Acknowledgments

Financial support from the Natural Science and Engineering Research Council (NSERC) of Canada, the Ontario Graduate Scholarship (OGS) programme and the computer-aided design tools provided by CMC Microsystems, Kingston, ON, Canada are gratefully acknowledged by the authors.

5 References

[1] Hu A., Yuan F.: 'Inter-signal timing skew compensation of parallel links with voltage-mode incremental signaling', *IEEE Trans. Circuits Syst. I.*, 2009, **56**, (4), pp. 773–783
[2] Austin M.: 'Decision-feedback equalization for digital communication over dispersive channels.' IEEE Int'l Research Laboratory of Electronics Technical Report, 461, August 1967
[3] Yuan F., Al-Taee A., Ye A., Sadr S.: 'Design techniques for decision feedback equalization of multi-Gbps serial data links: a state-of-the-art review', *IET Circuits Devices Syst.*, 2014, **8**, (2), pp. 118–130
[4] Winters J., Gitlin R.: 'Electrical signal processing techniques in long-haul fiber-optic systems', *IEEE Trans. Commun.*, 1990, **38**, (9), pp. 1439–1453

[5] Cui H., Lee J., Jeon S., *ET AL.*: 'A single-loop SS-LMS algorithm with single-ended integrating DFE receiver for multi-drop DRAM interface', *IEEE J. Solid-State Circuits*, 2011, **46**, (9), pp. 2053–2063
[6] Bien F., Kim H., Hur Y., *ET AL.*: 'A 10 Gb/s reconfigurable CMOS equalizer employing a transition detector-based output monitoring technique for band-limited serial links', *IEEE Trans. Microw. Theory Tech.*, 2006, **54**, (12), pp. 4538–4547
[7] Hong D., Cheng K.: 'An accurate jitter estimation technique for efficient high speed I/O testing'. Proc. Asian Test Symp., October 2007, pp. 224–229
[8] Chen L., Zhang X., Spagna F.: 'A scalable 3.6–5.2 mW 5-to-10 Gb/s 4-tab DFE in 32 nm'. IEEE Int'l Solid-State Circuits Conf. Digest of Technical Papers, February 2009, pp. 180–181
[9] Spagna F., Chen L., Deshpande M., *ET AL.*: 'A 78 mw 11.8 Gb/s serial link transceiver with adaptive RX equalization and baud-rate CDR in 32 nm CMOS'. IEEE Int'l Solid-State Circuits Conf. Digest of Technical Papers, February 2010, pp. 366–367
[10] Pozzoni M., Erba S., Viola P., *ET AL.*: 'A multi-standard 1.5 to 10 Gb/s latch-based 3-tap DFE receiver with a SSC tolerant CDR for serial backplane communication', *IEEE J. Solid-State Circuits*, 2009, **44**, (4), pp. 1306–1315
[11] Ellermeyer T., Langman U., Wedding B., Pohlmann W.: 'A 10 Gb/s eye-opening monitor IC for decision-guided adaptation of the frequency response of an optical receiver', *IEEE J. Solid-State Circuits*, 2000, **35**, (12), pp. 1958–1963
[12] Analui B., Rylyakov A., Rylov S., Meghelli M., Hajimiri A.: 'A 10-Gb/s two-dimensional eye-opening monitor in 0.13-µm standard CMOS', *IEEE J. Solid-State Circuits*, 2005, **40**, (12), pp. 2689–2699
[13] Noguchi H., Yoshida N., Uchida H., Ozaki M., Kanemitsu S., Wada S.: 'A 40-Gb/s CDR circuit with adaptive decision-point control based on eye-opening monitor feedback', *IEEE J. Solid-State Circuits*, 2008, **43**, (12), pp. 2929–2938
[14] Bhatta D., Kim K., Gebara E., Laskar J.: 'A 10 Gb/s two dimensional scanning eye opening monitor in 0.18 µm CMOS process'. Proc. IEEE Int. Microwave Symp. Digest, June 2009, pp. 1141–1144
[15] Gerfers F., Besten G., Petkov P., Conder J., Koellmann A.: 'A 0.2–2 Gb/s 6x OSR receiver using a digitally self-adaptive equalizer', *IEEE J. Solid-State Circuits*, 2008, **43**, (6), pp. 1436–1448
[16] Wong K., Rylyakov A., Yang C.: 'A 5 mW 6 Gb/s quarter-rate sampling receiver with a 2-tap DFE using soft decisions', *IEEE J. Solid-State Circuits*, 2007, **42**, (4), pp. 881–888
[17] Fayed A., Ismail M.: 'A low-voltage, low-power CMOS analog adaptive equalizer for UTP-5 cables', *IEEE Trans. Circuits Syst. I.*, 2008, **55**, (2), pp. 480–495
[18] Marcu M., Durbha S., Gupta S.: 'Duty-cycle distortion and specifications for jitter test-signal generation'. Proc. IEEE Int. Symp. Electromagnetic Compatibility, 2008, pp. 1–4
[19] Carusone T., Johns D., Martin K.: 'Analog integrated circuit design' (John Wiley and Sons, New York, 2012, 2nd edn.)
[20] Payne R., Bhakta B., Ramaswamy S., *ET AL.*: 'A 6.25 Gb/s binary adaptive DFE with first post-cursor tap cancellation for serial backplane communications'. IEEE Int. Solid-State Circuits Conf. Digest of Technical Papers, February 2005, pp. 68–69

Stochastic channel model for simulation of mobile *ad hoc* networks

Gunnar Eriksson, Karina Fors, Kia Wiklundh, Ulf Sterner

Swedish Defence Research Agency (FOI), Box 1165, SE-58111 Linkoping, Sweden
E-mail: kia.wiklundh@foi.seulf.sterner@foi.se

Abstract: Channel models, applicable to mobile *ad hoc* network (MANET) simulations, need to be both accurate and computationally efficient. It has been shown that inaccuracies in the channel model can seriously affect various network performance measures. It is essential that the model gives realistic time and spatial variability as the terminals move. Furthermore, the frequency selective effects from multipath propagation must be realistically modelled, so that the effects of different signalling bandwidths are captured correctly. However, to meet the necessary low-complexity constraints current commonly used channel models for network simulations are very simplified. In this study, the authors propose a model structure that is able to capture the essence of the channel characteristics, and to cope with the constraint of low computational complexity. The model describes the channel time and frequency variability between nodes in a MANET. It models the large- and small-scale fading, where the correlation between the fading parameters as well as the spatial correlation is considered. Furthermore, the study presents parameters for the proposed model based on wideband peer-to-peer channel measurements in an urban environment at 300 MHz. When analysing the link and network performance, they show that the proposed channel model describes the channel dynamics appropriately.

1 Introduction

Mobile *ad hoc* networks (MANETs) consist of mobile devices connected by wireless links. The network is characterised by peer-to-peer communication between the nodes without any centralised base station. For example, MANETs are used in vehicle communication applications and are under introduction for military use. Simulations of MANETs can be very demanding from a computational point of view. A large number of nodes, which are typical for military networks, stress the need for a channel model with low computational complexity. If we consider that the number of links between n nodes grows as $n(n-1)/2$ (assuming reciprocal channels), and that the link qualities need to be updated frequently because of mobility, the simulation time may soon get unmanageable for large networks. Hence, traditionally, very simplistic channel models – for example, two-ray models – have been used to describe the radio channel between the nodes. A summary of propagation models, used in network simulator tools, is presented in [1, Table 1]. The network performance is, however, very dependent on the dynamic behaviour of the individual links. This is a well-known problem that has been recognised by many researchers within the *ad hoc* network community, see, for example, [2–4], to give a more realistic description of the channel dynamics, the model must be able to generate both slow and fast fading; the first is mainly caused by shadowing, and the latter by that different propagation paths interfere constructively or destructively. Furthermore, the frequency selectivity of the channel (which is caused by multipath propagation) must be realistically modelled so that the effects on systems with different signal bandwidths can be investigated.

A large number of ideas have emerged on how to introduce the variability of the channels in network simulations. For example, in [5] a 'double-ring with a line-of-sight (LOS) component' model is proposed to incorporate both LOS and scattering effects for mobile scenarios. The model exhibits the statistical properties of a Rician fading channel and considers the small-scale fading for such a scenario. However, the large-scale fading and the correlation between the large-scale fading and other channel parameters are not addressed. Although the autocorrelation function of the fading envelope is derived, it is determined from the assumed theoretical model and it is not compared with real channel realisations. In [3], a different approach is adopted and the link stability and availability are modelled by using a distance transition probability matrix. The underlying channel modelling considers distance-dependent path loss, multipath and shadowing. The dynamic channel variation is generated from a semi-Markov smooth (SMS) mobility model. This link model will, to some extent, incorporate spatial correlation. However, this correlation will be determined from the SMS model and not from a certain terrain or scenario.

In recent years, there has been extensive research on the subject of channel modelling for vehicular *ad hoc* networks (VANETs). In [6], a survey of existing channel models for different scenarios in vehicular applications is presented. Several of these incorporate both small- and large-scale fading. This paper points out the key characteristics needed of channel models in VANETs and divides

Table 1 Model parameters

Description	Parameter	Value	Unit
distance dependent	n	4.15	—
	G_0	−4.8	dB
	d_{ref}	1	M
global means	$\mu_{G_{\mathrm{LS}}}$	0	dB
	μ_K	−3.3	dB
	μ_{σ_τ}	−66.4	dBs[a]
standard deviations and corr. coeff. for $\boldsymbol{C}_A$	$\sigma_{\bar{G}_{\mathrm{LS}}}$	6.6	dB
	$\sigma_{\bar{K}}$	3	dB
	$\sigma_{\bar{\sigma}_\tau}$	1.6	dB
	$\rho_{\bar{G}_{\mathrm{LS}}\bar{K}}$	0.74	—
	$\rho_{\bar{G}_{\mathrm{LS}}\bar{\sigma}_\tau}$	−0.39	—
	$\rho_{\bar{K}\bar{\sigma}_\tau}$	−0.45	—
standard deviations and corr. coeff. for $\boldsymbol{C}_B$	$\sigma_{\tilde{G}_{\mathrm{LS}}}$	3.1	dB
	$\sigma_{\tilde{K}}$	3.2	dB
	$\sigma_{\tilde{\sigma}_\tau}$	1.4	dB
	$\rho_{\tilde{G}_{\mathrm{LS}}\tilde{K}}$	0.5	—
	$\rho_{\tilde{G}_{\mathrm{LS}}\tilde{\sigma}_\tau}$	−0.22	—
	$\rho_{\tilde{K}\tilde{\sigma}_\tau}$	−0.18	—
correlation distances	$(\Delta d)_{\mathrm{c,\,LS}}$	20.2	m
	$(\Delta d)_{\mathrm{c},\,K}$	5.5	m
	$(\Delta d)_{\mathrm{c},\,\sigma_\tau}$	14.6	m

[a] Decibels relative to 1 s.

the existing models in three different types, tap-delay models, ray-based models and geometry-based stochastic models. The authors highlight that tap-delay models do not directly take into consideration the time variations of the channel typical for vehicular scenarios. Furthermore, ray-based models often lead to time-consuming simulations, when many rays are taken into account. The authors claim that a statistical approach is often needed to yield a representative behaviour of the dynamics in VANETs and that it has the potential to yield reasonable simulation times. This is one reason for why a geometry-based stochastic channel model is often to prefer in network analysis. In [7], a summary of existing channel modelling and measurements especially for vehicle-to-vehicle (V2 V) applications is also presented.

In [8], the authors discuss the need of a more detailed description of the physical layer to be used for network simulations. The purpose is to improve the quality of the results from network simulations and they claim that the basic threshold reception model in order to simulate the carrier sense functionality and determine successful reception is not adequate. It is especially interference because of collisions that is difficult to take into consideration without a more detailed description of the physical layer. Furthermore, the NS-2 simulator that the authors refer to as not having a proper threshold reception model is using relatively simple channel models, that is, the simulator uses either a free-space model, a two-ray ground reflection model or a shadowing model. On the basis of these reasons the authors suggest network simulations on bit level. To improve the quality of the network results with reasonable computational complexity, another approach is to develop a channel model with appropriate behaviour that provides accurate estimates of the signal-to-noise ratio (SNR) to be used in the network simulations.

In this paper, we propose a model structure that is able to capture the essence of the channel characteristics and that copes with the constraint of low computational complexity when used in network simulations. For network performance analysis, the particular geographical location is generally not interesting, as long as the statistical description meets the considered environment to be used in. The proposed model is a scenario-based stochastic channel model and it is similar to a geometry-based stochastic channel model in the sense that the channel statistics varies over positions. The model includes different environment-dependent channel parameters that control the distributions as the large-scale fading, the Rician K-factor and the delay spread. In this paper, model parameters are exemplified for an urban peer-to-peer scenario at 300 MHz. However, the model framework can be used for other environments and other frequency bands as long as the statistics for the channel parameters are available. Also results from a deterministic channel model are possible to use for derivation of the needed channel parameters. The proposed model has similarities with other existing models, as, for example, the one proposed in [9]. However, our model also considers the correlation between the channel parameters, and a first-order autoregressive (AR) filter is used to maintain accurate spatial correlation during terminal movement. We also demonstrate the behaviour of the proposed stochastic channel model in terms of the statistical behaviour of the channel parameters and by showing two typical examples of results at link and network levels.

The remainder of this paper is organised as follows. Section 2 describes the proposed channel model structure, including the distance-dependent path gain, large-scale and small-scale fading. It also describes the incorporation of correlation between large-scale parameters and the spatial correlation. Section 3 describes an example of parameter generation based on channel measurements. Section 4 exemplifies the impact of using the proposed channel model, both on link level and on network level. Comparisons are performed with measurement data and a plain two-ray ground model. Furthermore, the statistical characteristics of the channel parameters are examined. Finally, Section 5 concludes this paper.

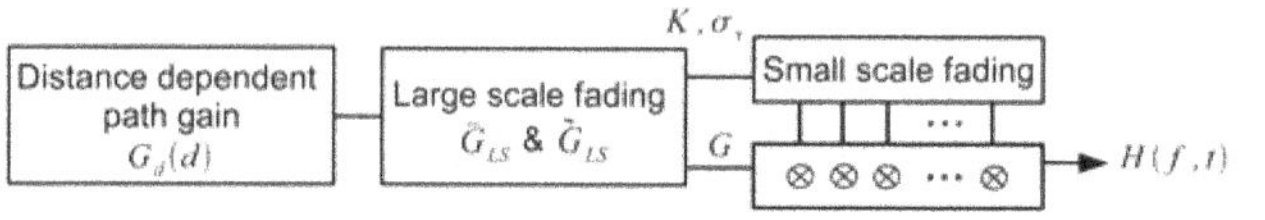

Fig. 1 *Stochastic channel model structure*

2 Channel model

To evaluate multiple access control (MAC) solutions and routing protocols for MANETs by use of simulations, the dynamics of the radio channels must be properly modelled. Current channel models for network evaluations are in their most simple form based on the distance between transmitter and receiver (e.g. a plain-earth model) to more thorough models with ray-tracing wave propagation simulations for each link. The former ones do not consider environmental effects such as fading properties, whereas the later models, in their most sophisticated form, can become very computationally demanding. In [10, 11], the small- and large-scale fading have been shown to significantly affect the performance of *ad hoc* networks. Furthermore, the correlation in time, space and frequency is often important when studying network performance. For example, when analysing the link availability in time for network protocols with fast acknowledgements, the behaviour of the channel dynamics is of great importance.

2.1 Proposed channel model structure

To address a dynamic channel behaviour (the large- and small-scale fading and the correlation properties), we propose the channel model structure shown in Fig. 1; the parameters are explained later in this section. The model is divided in three different blocks, where each block determines one layer of the total channel fading process.

- The first block includes the distance-dependent path gain $G_d(d)$, where d is the geometrical distance between two MANET nodes. Here, $G_d(d)$ is computed from an empirically developed link attenuation model.
- The second block creates the large-scale channel variations. This includes the large-scale (slow) fading G_{LS}, which are the path gain variations relative to $G_d(d)$. In addition, other large-scale parameters are generated to control the small-scale fading process in the third block. Examples of such parameters are the root-mean-square (RMS) delay spread σ_τ and the Rician K-factor, which are related to the coherence bandwidth and the amplitude distribution of the small-scale fading, respectively.
- The third block considers the small-scale fading process with a Rician amplitude distribution. The small-scale variations are created for multiple subchannels, whose mutual correlation depends on the coherence bandwidth. Finally, the small-scale fading process is multiplied with the large-scale fading obtained in block two to conduct the total fading.

These three blocks constitutes the base of the model. The radio channel in position $\boldsymbol{r}$ can be characterised by the frequency selective transfer function as

$$H(f, \boldsymbol{r}) = \sqrt{G(\boldsymbol{r})}Y(f, \boldsymbol{r}) \tag{1}$$

where $G(\boldsymbol{r})$ (expressed in linear scale in (1)) is the link's composite path gain, which can be obtained from

$$G(\boldsymbol{r}) = G_d(d) + G_{LS} \quad [\text{dB}] \tag{2}$$

and where f and $Y(f, \boldsymbol{r})$ are the frequency of interest and the complex-valued small-scale fading in position $\boldsymbol{r}$, respectively; the latter is described in Section 2.4. Furthermore, in (2), $G_d(d)$ is the distance-dependent path gain, see Section 2.2, and G_{LS} is the large-scale fading, see Section 2.3.

A similar approach to model the channel fading, with a similar block structure, has been proposed in [9] for wireless personal area networks for indoor use. However, we impose the small-scale fading in the frequency-transfer function instead of adding a time-domain component. Moreover, the incorporation of the correlation between large-scale channel parameters differs from what is proposed in [9].

2.2 Distance-dependent path gain

The distance-dependent path gain $G_d(d)$, generated in the first block shown in Fig. 1, is modelled to be a function of the geometrical distance d as

$$G_d(d) = G_0 - 10n\log_{10}\left(\frac{d}{d_{ref}}\right) \text{[dB]} \tag{3}$$

where G_0 is the path gain at the reference distance d_{ref} and n is the path-gain exponent.

2.3 Large-scale fading

For the large-scale channel variation, three important parameters are modelled. The parameter G_{LS} is the large-scale fading, K and σ_τ are the K-factor and the channels RMS delay spread, respectively. The large-scale parameters are defined as

$$\begin{pmatrix} G_{LS} \\ K \\ \sigma_\tau \end{pmatrix} = \begin{pmatrix} \bar{G}_{LS} \\ \bar{K} \\ \bar{\sigma}_\tau \end{pmatrix} + \begin{pmatrix} \tilde{G}_{LS} \\ \tilde{K} \\ \tilde{\sigma}_\tau \end{pmatrix} + \begin{pmatrix} \mu_{G_{LS}} \\ \mu_K \\ \mu_{\sigma_\tau} \end{pmatrix} \tag{4}$$

All parameters in (4) are expressed in logarithmic units. The two components $\bar{(\cdot)}$ and $\tilde{(\cdot)}$ are the local mean and the superimposed variation around the local mean, respectively. The last term $\boldsymbol{\mu} = [\mu_{G_{LS}} \;\; \mu_K \;\; \mu_{\sigma_\tau}]^T$ is the global mean for each channel parameter and is valid for a whole scenario and can be seen as an offset factor. The local mean $\bar{(\cdot)}$ on the contrary varies at a quite large spatial scale and can typically be considered constant over one block in an urban scenario. The superimposed variation $\tilde{(\cdot)}$ depends mainly on the local environment in the vicinity of the radio nodes; its spatial scale is therefore considerably smaller.

The proposed structure in (4) is based on the findings in [12], where it was stated that $\bar{(\cdot)}$ and $\tilde{(\cdot)}$ can be considered as approximately normal distributed. By dividing each parameter's variation into the two separate components $\bar{(\cdot)}$ and $\tilde{(\cdot)}$, we can handle the different spatial scales of the variation more easily. Furthermore, the analysis of the results from the urban measurements in [12] showed that there is a dependency between the channel parameters. To consider the correlation between the large-scale parameters, covariance matrices are introduced. If we let $\bar{\boldsymbol{\Omega}} = [\bar{G}_{LS} \; \bar{K} \; \bar{\sigma}_\tau]^T$ and $\tilde{\boldsymbol{\Omega}} = [\tilde{G}_{LS} \; \tilde{K} \; \tilde{\sigma}_\tau]^T$, where $(\cdot)^T$ denotes the transpose operator, we can generate correlated realisations of these parameters as

$$\bar{\boldsymbol{\Omega}} = \boldsymbol{C}_A^{1/2}\boldsymbol{x} \tag{5}$$

$$\tilde{\boldsymbol{\Omega}} = \boldsymbol{C}_B^{1/2}\boldsymbol{y} \tag{6}$$

where $\boldsymbol{C}_A$ and $\boldsymbol{C}_B$ are the covariance matrices for parameter vectors $\bar{\boldsymbol{\Omega}}$ and $\tilde{\boldsymbol{\Omega}}$, respectively, and the matrix square root is defined so that $\boldsymbol{C} = \boldsymbol{C}^{1/2}\boldsymbol{C}^{1/2}$. The elements of $\boldsymbol{x}$ and $\boldsymbol{y}$ are independent normal-distributed variables with zero mean and unit variance. As $\boldsymbol{x}$ and $\boldsymbol{y}$ are modelled as zero-mean processes, the global mean $\boldsymbol{\mu}$ must be added in (4) to obtain the correct level of the large-scale parameters.

On the basis of measurement analysis, Gudmundson [13] proposed the autocorrelation properties of the large-scale fading process to be modelled by a simple exponential function. This is attractive from a computational point of view because the fading process can be generated by filtering a white Gaussian noise process through an AR filter. In our proposed model, we assume that the autocorrelation functions for all large-scale parameters can be modelled by an exponential function. Hence, for a generic element $\tilde{\Omega}$ of the mutually correlated parameters in vector $\tilde{\boldsymbol{\Omega}}$ in (6), a sequence of spatially correlated realisations of that channel parameter $\tilde{\Omega}_F(\boldsymbol{r}_p)$, $p = 1, \ldots, P$, is generated as

$$\tilde{\Omega}_F(\boldsymbol{r}_p) = \begin{cases} \alpha_p \tilde{\Omega}_F(\boldsymbol{r}_{p-1}) + \sqrt{1-\alpha_p^2}\,\tilde{\Omega}_p, & \text{if } 2 \le p \le P \\ \tilde{\Omega}_p, & \text{if } p = 1 \end{cases} \tag{7}$$

where $\boldsymbol{r}_p$ is the pth position, $\tilde{\Omega}_p$ is a realisation from (6) and α_p is a coefficient that determines the statistical dependency between $\tilde{\Omega}_F(\boldsymbol{r}_p)$ and $\tilde{\Omega}_F(\boldsymbol{r}_{p-1})$. Under the assumption of local wide-sense stationarity, α_p can be expressed as

$$\alpha_p = \rho_p(\Delta d) \tag{8}$$

where $\Delta d = |\boldsymbol{r}_p - \boldsymbol{r}_{p-1}|$ is the distance between two adjacent positions and ρ_p is an autocorrelation function that is valid within a local area around $\boldsymbol{r}_p$. On the basis of the assumption of an exponential autocorrelation, we let

$$\rho_p(\Delta d) = \left(\frac{1}{c}\right)^{-\Delta d/(\Delta d)_c} \tag{9}$$

where $(\Delta d)_c$ is the correlation distance at correlation level c. In (9), the correlation distance is implicitly dependent on $\boldsymbol{r}_p$.

To summarise the methodology, the large-scale parameters can be generated as follows:

(1) Generate the covariance matrices $\boldsymbol{C}_A$ and $\boldsymbol{C}_B$ based on parameters from a reference scenario.
(2) Generate the local mean values $\bar{\boldsymbol{\Omega}}$ for G_{LS}, K and σ_τ according to (5).
(3) Generate the superimposed process $\tilde{\boldsymbol{\Omega}}$, (6), and filter it according to (7) to incorporate the spatial correlation.
(4) Determine the scenario dependent global mean value $\boldsymbol{\mu}$.
(5) Compute the combined large-scale result according to (4).

From (4), the G_{LS} is used in (2), whereas K and σ_τ are used in the derivation of the small-scale fading.

2.4 Small-scale fading

In general, the small-scale fading process $Y(f, \boldsymbol{r})$ is characterised by its distribution and by its frequency and spatial correlation properties. In our model, we assume that $Y(f, \boldsymbol{r})$ is Rician distributed and that the time dispersion of the channel can be described as a superposition of a dense multipath component (with an exponentially decaying delay power spectrum), and a specular component. To simplify the model, we make the approximation that the realisations are spatially independent if $|\boldsymbol{r}_p - \boldsymbol{r}_{p-1}| \ge d_\varepsilon$ (a certain threshold distance); if the movement is $< d_\varepsilon$, we assume that $Y(f, \boldsymbol{r}_p) = Y(f, \boldsymbol{r}_{p-1})$. In this way, the effects of static nodes are captured.

Following the modelling approach for dense multipath components in [14–16], we let the dense multipath components have the delay power spectrum

$$S_{\mathrm{dm}}(\tau) = \begin{cases} 0, & \text{if } \tau < \tau_0 \\ \frac{a_1}{b_1}\exp\left(-\frac{\tau-\tau_0}{b_1}\right), & \text{if } \tau \geq \tau_0 \end{cases} \tag{10}$$

where τ_0 is the delay of the first arriving multipath component, b_1 is the decay constant of the power spectrum and a_1 is the total dense multipath power; for the brevity of notation, we have dropped the explicit dependency on $\boldsymbol{r}$ in the equation. The specular component (which is assumed to be statistically independent of the dense multipath components) has the delay power spectrum

$$S_{\mathrm{sc}}(\tau) = a_0\,\delta(\tau - \tau_0) \tag{11}$$

where a_0 is the power of the specular component, and again the dependency on $\boldsymbol{r}$ is dropped for brevity.

The frequency autocorrelation function for the dense multipath components $\psi_{\mathrm{dm}}(\Delta f)$ is obtained by a Fourier transformation of (10) as

$$\psi_{\mathrm{dm}}(\Delta f) = \frac{a_1}{1 + \mathrm{j}2\pi\Delta f b_1}\exp(-\mathrm{j}2\pi\Delta f\,\tau_0) \tag{12}$$

where a_1 and b_1 are determined by the Rician K-factor and the delay spread. Under the power constrain $a_0 + a_1 = 1$, the parameters a_0, a_1 and b_1 in (10)–(12) can be expressed, respectively, as

$$a_0 = \frac{K}{K+1} \tag{13a}$$

$$a_1 = \frac{1}{K+1} \tag{13b}$$

$$b_1 = \frac{\sigma_\tau}{\sqrt{1-a_0^2}} \tag{13c}$$

Finally, with the explicit dependency on $\boldsymbol{r}$ reintroduced, we can now write the small-scale fading process $\boldsymbol{y}(\boldsymbol{r}_p) = [Y(f_1, \boldsymbol{r}_p)Y(f_2, \boldsymbol{r}_p), \ldots, Y(f_{n_f}, \boldsymbol{r}_p)]^{\mathrm{T}}$ at discrete frequency points $f_1, f_2, \ldots, f_{n_f}$ as

$$\boldsymbol{y}(\boldsymbol{r}_p) = \sqrt{\frac{K(\boldsymbol{r}_p)}{K(\boldsymbol{r}_p)+1}} + \sqrt{\frac{1}{K(\boldsymbol{r}_p)+1}}\boldsymbol{C}_f^{1/2}(\boldsymbol{r}_p)\boldsymbol{w}_p \tag{14}$$

where the elements of $\boldsymbol{w}_p$ are independent identically distributed complex Gaussian stochastic variables with zero mean and unit variance, and $\boldsymbol{C}_{\mathrm{f}}(\boldsymbol{r}_p)$ is the frequency correlation matrix with elements

$$[\boldsymbol{C}_{\mathrm{f}}]_{kl} = \frac{1}{a_1}\psi_{\mathrm{dm}}(f_k - f_l) \tag{15}$$

The two large-scale parameters $K(\boldsymbol{r}_p)$ and σ_τ $(\boldsymbol{r}_p)$ that govern the statistics of the small-scale fading are obtained from (7), and $\tau_0(\boldsymbol{r}_p)$ is computed from the geometrical path length of the link.

2.5 Model simplifications

Modern wideband radio systems usually apply advance modulation and coding schemes. Such systems can take advantage of multipath propagation, and obtain diversity gains on frequency selective radio channels. Therefore, the performance of such systems mainly depends on the instantaneous SNR, averaged over the operating bandwidth. Under such assumptions, it is possible to reduce the computational complexity of the proposed model even further by simplifications to be used in small-scale fading calculations and hence reduce the complexity and simulation time, as well.

The diversity order that can be extracted by an idealised receiver system can be approximated as $m = W_{\mathrm{s}}/W_{\mathrm{coh}}$, where W_{s} and W_{coh} are the system bandwidth and the coherence bandwidth of the channel, respectively.

Hence, m can be viewed as the number of independently fading subchannels, which makes the frequency correlation matrix $\boldsymbol{C}_{\mathrm{f}}$ in (14) to become an identity matrix $\boldsymbol{I}_m$. Then, the instantaneous SNR is obtained from the wideband path-gain $G_{\mathrm{wb}}(\boldsymbol{r})$, which is computed as

$$G_{\mathrm{wb}}(\boldsymbol{r}_p) = G(\boldsymbol{r}_p) + 10\log_{10} \| \boldsymbol{y}(\boldsymbol{r}_p) \|^2 \ [\mathrm{dB}] \tag{16}$$

where $\|\cdot\|$ denotes the Euclidean norm and $G(\boldsymbol{r}_p)$ is the composite path gain, according to (2).

3 Example of parameterisation of the model by using an urban scenario

Parameters for the proposed model have in this paper been extracted from a peer-to-peer measurement campaign at 300 MHz in an urban environment; for further details and results, see [12]. However, the model framework can be used for other environments and other frequency bands as long as such channel measurements for such conditions are available. The measurement campaign was conducted in the city centre of the fifth largest city in Sweden, which with international standards is a rather small town. Fig. 2 shows an aerial photo of the measurement area. This part of the town, which typically has three- to six-storey buildings, is rather flat but slopes gently towards the river on the east side. The transmitter Tx and the receiver Rx were both placed on vehicles with the antenna arrays mounted on top of each vehicle. The antenna heights were ~1.8 and 2.1 m above the ground for the Tx and Rx, respectively. The scenario consists of 3 different Tx locations (Tx1, Tx2, Tx3) and from each Tx location, 25 Rx routes (Rx1–Rx25) were conducted. During the measurements the Tx was stationary at each site, whereas the Rx was driven along the measurement routes. The measurements were performed at a centre frequency of 285 MHz with a 20 MHz wide probing signal.

On the basis of the large set of measured channel transfer functions derived from the measurement campaign, the parameters of the proposed model can be generated in accordance with Section 2, see Table 1.

We can, for example, see that the distance-dependent path-gain exponent n for this scenario is 4.15. Furthermore, in the same table, we show the extracted parameters for the standard deviations and cross-correlation coefficients of our two three-dimensional large-scale processes $\bar{\boldsymbol{\Omega}}$ and $\tilde{\boldsymbol{\Omega}}$; that is, the local mean and the superimposed process, respectively. From the parameters in the table, we can compute the covariance matrices $\boldsymbol{C}_A$ and $\boldsymbol{C}_B$, used in (5) and (6), respectively, according to

$$\boldsymbol{C}_A = \begin{bmatrix} \sigma^2_{\bar{G}_{\mathrm{LS}}} & \rho_{\bar{G}_{\mathrm{LS}}\bar{K}}\sigma_{\bar{G}_{\mathrm{LS}}}\sigma_{\bar{K}} & \rho_{\bar{G}_{\mathrm{LS}}\bar{\sigma}_\tau}\sigma_{\bar{G}_{\mathrm{LS}}}\sigma_{\bar{\sigma}_\tau} \\ \rho_{\bar{G}_{\mathrm{LS}}\bar{K}}\sigma_{\bar{G}_{\mathrm{LS}}}\sigma_{\bar{K}} & \sigma^2_{\bar{K}} & \rho_{\bar{K}\bar{\sigma}_\tau}\sigma_{\bar{K}}\sigma_{\bar{\sigma}_\tau} \\ \rho_{\bar{G}_{\mathrm{LS}}\bar{\sigma}_\tau}\sigma_{\bar{G}_{\mathrm{LS}}}\sigma_{\bar{\sigma}_\tau} & \rho_{\bar{K}\bar{\sigma}_\tau}\sigma_{\bar{K}}\sigma_{\bar{\sigma}_\tau} & \sigma^2_{\bar{\sigma}_\tau} \end{bmatrix} \tag{17}$$

and

$$\boldsymbol{C}_B = \begin{bmatrix} \sigma^2_{\tilde{G}_{\mathrm{LS}}} & \rho_{\tilde{G}_{\mathrm{LS}}\tilde{K}}\sigma_{\tilde{G}_{\mathrm{LS}}}\sigma_{\tilde{K}} & \rho_{\tilde{G}_{\mathrm{LS}}\tilde{\sigma}_\tau}\sigma_{\tilde{G}_{\mathrm{LS}}}\sigma_{\tilde{\sigma}_\tau} \\ \rho_{\tilde{G}_{\mathrm{LS}}\tilde{K}}\sigma_{\tilde{G}_{\mathrm{LS}}}\sigma_{\tilde{K}} & \sigma^2_{\tilde{K}} & \rho_{\tilde{K}\tilde{\sigma}_\tau}\sigma_{\tilde{K}}\sigma_{\tilde{\sigma}_\tau} \\ \rho_{\tilde{G}_{\mathrm{LS}}\tilde{\sigma}_\tau}\sigma_{\tilde{G}_{\mathrm{LS}}}\sigma_{\tilde{\sigma}_\tau} & \rho_{\tilde{K}\tilde{\sigma}_\tau}\sigma_{\tilde{K}}\sigma_{\tilde{\sigma}_\tau} & \sigma^2_{\tilde{\sigma}_\tau} \end{bmatrix} \tag{18}$$

where σ^2_X is the variance of X and ρ_{XY} is the cross-correlation

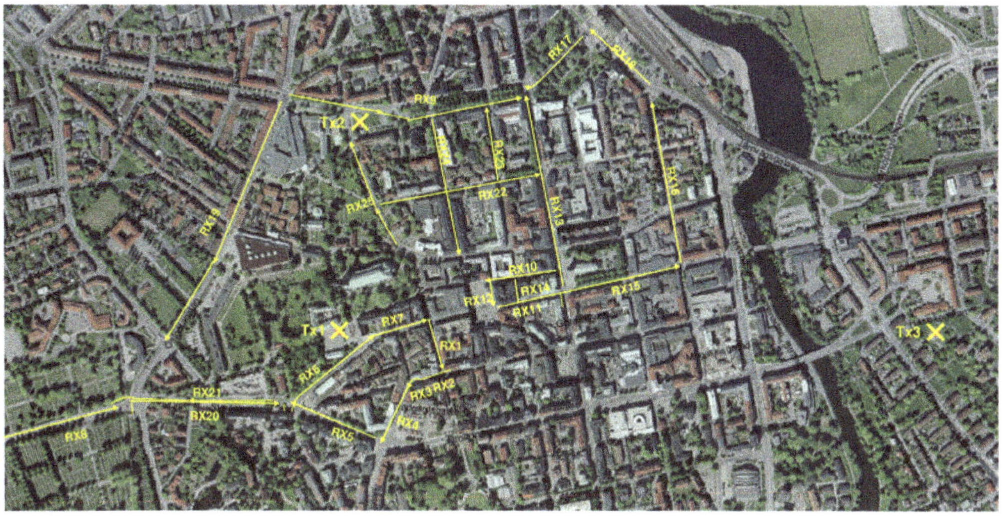

Fig. 2 *Measurement area in the city centre of Linkoping, Sweden. Transmitter sites (crosses) and receiver routes*

coefficient defined as

$$\rho_{XY} = \frac{\mathrm{cov}(X,\ Y)}{\sigma_X \sigma_Y} = \frac{E((X - \mu_X))(Y - \mu_Y))}{\sigma_X \sigma_Y} \tag{19}$$

in which $\mathrm{cov}(X, Y)$ denotes the covariance between X and Y.

From the urban measurements, the mean value of the correlation distance $(\Delta d)_c$ in (9) has been calculated for respective large-scale channel parameter at a correlation level $c = 0.5$. The correlation distances are also given in Table 1.

4 Example of the channel model behaviour

In the following, we will exemplify the behaviour of the proposed stochastic channel model. Firstly, we compare the distributions of model-generated large-scale parameters with the parameters derived from measurements directly. Secondly, we study the performance on link level in an *ad hoc* network in terms of probability that the SNR exceeds a certain threshold, which corresponds to the SNR requirement of a certain service. The performance results obtained for the proposed stochastic channel model are compared with the results for a simple two-ray model as well as results based on channel measurements. Finally, we will study the network performance of a simulated *ad hoc* network in an urban environment. The proposed stochastic channel model is used to generate the maximum data rate possible for the instantaneous channel conditions of the links. Hence, different kinds of routing algorithms can be evaluated. With this approach, the network performance in terms of the probability of packet delivery ratio is analysed for networks of different connectivities.

4.1 *Statistical properties of the channel parameters*

To verify the statistics of the modelled large-scale channel variations, we compare the cumulative distribution functions (CDFs) of the generated channel parameters in (4) – that is, large-scale fading G_{LS}, Rician K-factor and RMS delay spread σ_τ – with the CDFs that were computed from the measured data. In Figs. 3–5, the CDFs of the large-scale channel parameters are shown for measurements and for the proposed channel model. The curves denoted 'Measurements' are derived from the measurements directly, considering all combinations of transmitter and receiver positions, see [12]. The corresponding curves denoted 'Model' are derived from simulations with the proposed channel model, for the same

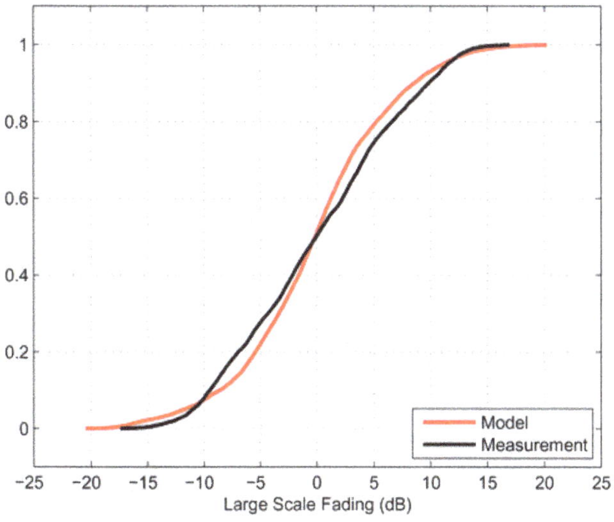

Fig. 3 *CDFs for the large-scale fading G_{LS} in dB*

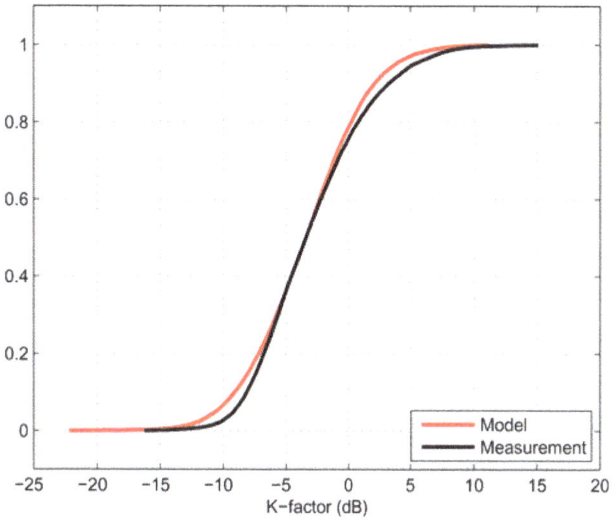

Fig. 4 *CDFs for the K-factor in dB*

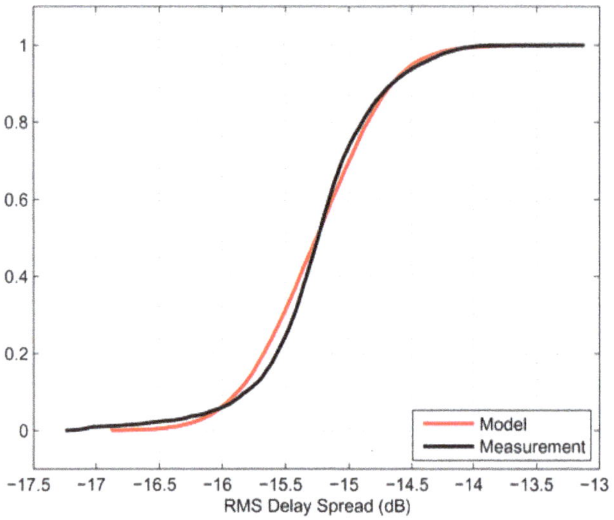

Fig. 5 *CDFs for the RMS delay spread σ_τ. (Expressed as $\log_e(\sigma_\tau)$)*

distributions of the distances. In these figures, we can see that the agreement between the CDFs for the parameters generated by the model and the CDFs from the measured data is quite good for all three parameters. This justifies the proposed distributions in the second block of the model structure, see Fig. 1.

4.2 Channel model behaviour on link level in an ad hoc *network*

To study the link performance in a MANET, channel realisations are obtained from the proposed channel model, and compared with the performance obtained when directly measured channels are used. For this purpose, a network is created from a subset of the measurement routes [17], with the consequence that the mobility is determined from the available measurement positions. To be specific, the node positions are the same as the Tx and Rx positions in the measurements.

We use a link model that have two different states, that is, if the SNR is larger than a certain SNR threshold, the link is considered to function, otherwise to fail. Generally, the SNR threshold depends on the specific radio communication system and service requirement. This is a reasonable approximation when considering a system that utilises strong channel coding. Such a system will have a sharp threshold behaviour, where the message error probability goes from one to a negligible level within a narrow SNR

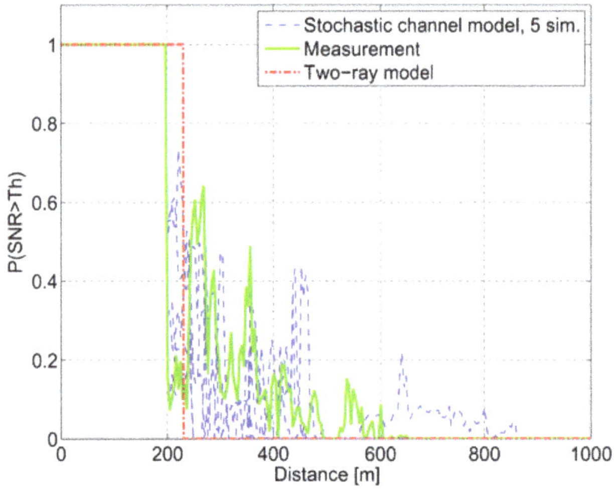

Fig. 6 *Probability that SNR exceeds a certain threshold as a function of distance*

range. The use of this link model is a common approach to determine link performance in network simulations [1] and is usually needed to limit the computational time. The proposed stochastic channel model is used to derive the probability that the SNR will exceed the threshold for the same node movements as for the created network based on the measured links. The probability that the SNR exceeds the threshold is shown in Fig. 6 for the created network with the measured links and with the proposed channel model. In addition, the result for a simple two-ray model [18] is included in this figure, for comparison. The two-ray model is an example of a simple channel model, which is commonly used in many network simulation tools. We can see that the results with the simple two-ray model exhibit a typical threshold behaviour with transition at a certain distance. This distance corresponds to the SNR threshold, where the transition from acceptable link performance to a not acceptable link performance appears. In contrast to the results of the two-ray model, the probability that the SNR will exceed the threshold for the measurements and the proposed channel does not exhibit a distinct transition and assumes values between zero and one for a large range of distances. Furthermore, we can see that the overall behaviour of the probability that the SNR will exceed the threshold derived from the proposed channel model and the measurements have a similar variability. However, the results with the stochastic channel model and the measurements differ, since the results for the measurements are based on a subset of the measurement campaign, whereas the results for the stochastic channel model are based on the whole measurement campaign results.

4.3 Channel model behaviour on network level

To exemplify how the choice of channel model, and its degree of details, can affect the network performance, we have simulated a mobile scenario with 64 nodes. In that scenario, we assume that the nodes are moving around for 400 s in an 8 × 8 km square area at a speed of 50 km/h. To model the movements of the nodes, we use the random walk model in [17]. According to the mobility model, all nodes move independently of each other and, if a node hits the boundary of the square, it bounces back like a ball. The user traffic is modelled as broadcast transmissions of packets. A source is randomly selected among the 64 nodes to send one packet. Thereafter, a new source is randomly selected and so on. A basic time-division multiple access MAC protocol is used for the simulations. Therefore no robustness issues have to be addressed at the MAC layer because of packet collisions. For such protocols, the time is divided into time slots that are grouped into repeating frames. Each node is assigned one time slot in each frame and the traffic in the network is kept sufficiently low to avoid congestion in the network. To route the packets in the network, we use the multi-point-relay (MPR) method according to the simplified multicast forwarding framework [19], and the MPR selection mechanisms are the optimised link state routing protocol [20]. Both user and overhead traffic are transmitted in the network.

For the network simulations, we use an in-house developed radio network simulator. On the basis of the channel realisation, when a packet is sent, the SNR and the instantaneous channel capacity is computed and used to decide whether a packet can be correctly received in a node or not. If the experienced channel capacity is higher than the data rate, the packet is assumed to be correctly received. Note, that in reality no system will reach the channel capacity; there will always be some implementation losses. However, such losses are neglected in these simulations as our focus is merely on the channel behaviour and its influence on the network performance.

The performance is studied in terms of delivery ratio, defined as the fraction of packets that reach the destinations. A packet that is not reaching the destination is lost either because no route exists or that a link deteriorates so that a transmitted packet cannot be

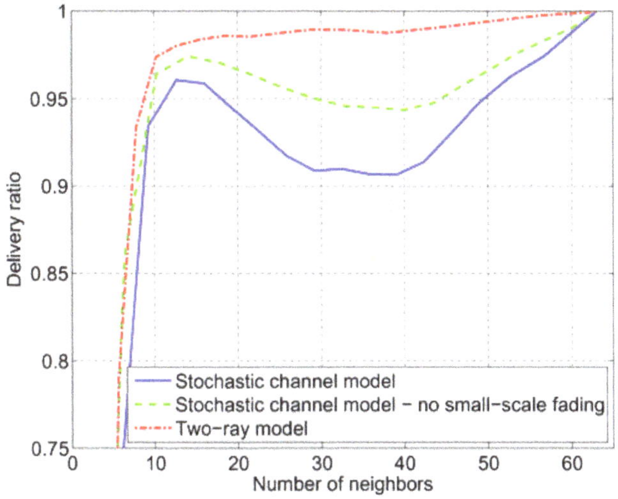

Fig. 7 *Delivery ratio as a function of the network connectivity*

received. No acknowledgments or automatic repeat request mechanisms are used on the link level, that is, between a specific transmitter and receiver. The delivery ratio should be high (say, over 95%) to assure a satisfactory quality of service. The delivery ratio is calculated for different network connectivities. As a measure of the network connectivity, we use the number of one-hop-neighbours any given node has in average measured at MAC layer. The average value is taken over all the transmissions from all the nodes during one simulation run. To obtain different network connectivities, we adjust the output power of the nodes in the scenario and hence different network situations are created from a sparse to a dense network.

To illustrate how the choice of channel model, and how its degree of details can affect the result, we compare the network performance when using different channel models. In Fig. 7, the delivery ratio is shown as a function of the network connectivity, measured as the average number of one-hop-neighbours. The proposed channel model and the two-ray model are used for the network simulations. To further investigate the effects of the channel dynamics, the proposed model without small-scale fading is included as an intermediate detailed model. Unfortunately, measurements of the actual links are generally not possible in reality to use in the network simulations with random movement and a large number of the nodes, as was possible in the link level example in previous section. For example, our scenario with 64 nodes would imply 2016 instantaneous links that are moving randomly and need to be updated with a high rate. There are some of the reasons of why a stochastic channel model is urgent for network simulations.

This figure shows results for the different channel models, the two-ray model, the proposed channel model without small-scale fading and the proposed channel model with small-scale fading, which are models with an increasing degree of detailed channel description. In the simulations, the delivery ratio is between 90 and 100% for networks with over ten node neighbours. This is in the critical region of acceptable performance (a critical value of delivery ratio is often near 95%). Hence, it is particularly important with proper modelling of the channel to obtain accurate estimates of the delivery ratio. From the simulations, we can see that the values of the delivery ratio, for the three channel descriptions, do not differ substantially. Furthermore, when the level of model details increases, the delivery ratio decreases. The simple two-ray model deceptively yields a higher delivery ratio than what the proposed channel model gives. Additionally as can be seen, the largest differences in the results between the proposed channel model and the two-ray ground model occur when the nodes have 30–40 neighbours. This difference is a consequence of the inherent robustness of the used broadcasting method, that is, the number of redundant MPRs is reduced when the number of neighbours increases. For these network types, the probability decreases of having an alternative route when a link disappears. This leads to a larger impact from the fading on the delivery ratio. The overall conclusion is that the network simulation results benefit from a more detailed channel description than the two-ray model can offer. Furthermore, the small-scale fading needs to be incorporated in the model when analysing the network performance.

5 Conclusions

There is a need for more realistic channel models for analysis of *ad hoc* network performance. The channel model must include both large-scale and small-scale fading as an effect from the mobility of the network terminals. In this paper, a model structure is proposed that captures the essence of the channel characteristics and copes with the constraint of low computational complexity. The fading statistics of the model are determined by a number of parameters that describe the distributions of the large-scale fading, the Rician K-factor and the delay spread. The model considers the mutual correlation between the channel parameters and the spatial correlation of the parameters. Examples justify that the proposed channel model, with model parameters estimated from urban peer-to-peer scenario at 300 MHz, models the channel dynamics appropriately when analysing the link and network performance.

6 References

[1] Stanica R., Chaput E., Beylot A.-L.: 'Simulation of vehicular ad-hoc networks: challenges, review of tools and recommendations', *Comput. Netw.*, 2011, **55**, (14), pp. 3179–3188

[2] Zang L.F., Rowe G.B.: 'Improved modelling for ad-hoc networks', *Electron. Lett.*, 2007, **43**, (21), pp. 1156–1157

[3] Zhao M., Wang W.: 'The impacts of radio channel and node mobility on link statistics in mobile ad hoc networks'. Proc. IEEE Globecom 2007, Washington, USA, November 2007

[4] Chapin J., Chan V.: 'The next 10 years of DOD wireless networking research'. 2011–Milcom 2011 Military Communications Conf., November 2011, pp. 2238–2245

[5] Wang L.-C., Liu W.-C., Cheng Y.-H.: 'Statistical analysis of a mobile-to-mobile Rician fading channel model', *IEEE Trans. Veh. Technol.*, 2009, **58**, (1), pp. 32–38

[6] Mecklenbrauker C., Molisch A., Karedal J., ET AL.: 'Vehicular channel characterization and its implications for wireless system design and performance', *Proc. IEEE*, 2011, **99**, (7), pp. 1189–1212

[7] Wang C.-X., Cheng X., Laurenson D.: 'Vehicle-to-vehicle channel modeling and measurements: recent advances and future challenges', *IEEE Commun. Mag.*, 2009, **47**, (11), pp. 96–103

[8] Mittag J., Papanastasiou S., Hartenstein H., Strom E.: 'Enabling accurate cross-layer PHY/MAC/NET simulation studies of vehicular communication networks', *Proc. IEEE*, 2011, **99**, (7), pp. 1311–1326

[9] Karedal J., Johansson A.J., Tufvesson F., Molisch A.F.: 'A measurement-based fading model for wireless personal area networks', *IEEE Trans. Wirel. Commun.*, 2008, **7**, (11), pp. 4575–4585

[10] Gray R.S., Kotz D., Newport C., ET AL.: 'Outdoor experimental comparison of four ad hoc routing algorithms'. Proc. of the ACM/IEEE Int. Symp. on Modeling, Analysis and Simulation of Wireless and Mobile Systems (MSWiM), 2004, pp. 220–229

[11] Kiess W., Mauve M.: 'A survey on real-world implementations of mobile ad-hoc networks', *Ad Hoc Netw.*, 2007, **5**, (3), pp. 324–339 [Online]. Available at http://www.sciencedirect.com/science/article/pii/S1570870505001149

[12] Eriksson G., Linder S., Wiklundh K., ET AL.: 'Urban peer-to-peer MIMO channel measurements and analysis at 300 MHz'. Proc. IEEE Milcom 2008, San Diego, CA, USA, November 2008

[13] Gudmundson M.: 'Correlation model for shadow fading in mobile radio systems', *Electron. Lett.*, 1991, **27**, (23), pp. 2145–2146

[14] Erceg V., Greenstein L.J., Tjandra S.Y., ET AL.: 'An empirically based path loss model for wireless channels in suburban environments', *IEEE J. Sel. Areas Commun.*, 1999, **17**, (7), pp. 1205–1211

[15] Pedersen K.I., Mogensen P.E., Fleury B.H.: 'A stochastic model of the temporal and azimuthal dispersion seen at the base station in outdoor propagation environments', *IEEE Trans. Veh. Technol.*, 2000, **49**, (2), pp. 437–447

[16] Cassioli D., Win M.Z., Molisch A.F.: 'The ultra-wide bandwidth indoor channel: From statistical model to simulation', *IEEE J. Sel. Areas Commun.*, 2002, **20**, (6), pp. 1247–1257
[17] Nilsson J., Sterner U.: 'Robust MPR-based flooding in mobile ad-hoc networks'. 2012–Milcom 2012 Military Communications Conf., 2012, pp. 1–6
[18] Ahlin L., Zander J.: 'Principles of wireless communications' (Studentlitteratur, Lund, 1998, 2nd edn.)
[19] Macker J.: 'Simplified multicast forwarding (SMF)'. IETF, Network Working Group, Internet-Draft, January 2012
[20] Clausen T., Jacquet P.: 'Optimized link state routing protocol (OLSR)'. IETF, Network Working Group, RFC 3626, October 2003

26

Design of coplanar waveguide band-pass filter for S-band application

Pratik Mondal, Amit Ghosh, Susanta Kumar Parui

Department of Electronics and Telecommunication Engineering, Indian Institute of Engineering Science and Technology, Shibpur, Howrah 711 103, India
E-mail: pratik.mondal.1987@ieee.org

Abstract: Coplanar waveguide (CPW) has a huge demand for designing band-pass filter (BPF). In this study, the filter designed by open-ended CPW series stub which acts as a resonant circuit thus giving a band-pass response. As the number of open stub discontinuity is increased, the frequency response and roll-off rate of the proposed BPF has improved gradually. Electromagnetic simulated and measured results show a very good agreement with each other. The proposed filter is designed to obtain a frequency range of 1.97–4 GHz (S-band) having rising edge and falling edge selectivities of 35.9 and 45.7 dB/GHz, respectively.

1 Introduction

Band-pass filter (BPF) is an important component of every trans-receiver. It is a passive component which selects certain bands of frequencies and rejects frequencies outside of the specified range, especially those frequencies which have the potential to interfere with the information signal. Previously, many research papers showed the design mechanism of compact BPF, parallel coupled BPF of image parameter method [1]. Coplanar waveguide (CPW) technology has a major advantage of easy integration with lumped elements as well as active components. Moreover, in CPW technology, the characteristic impedance is determined by the ratio of $w/(w+s)$, so size reduction is possible without limit. The realisation of CPW BPF by the discontinuity of the open stub has been proposed previously [2–4].

In this paper, an S-band BPF is designed by utilising CPW technology, having a major advantage of easy integration with lumped elements as well as active components, which is strongly desired in communication systems. The S-band spectrum is a leading application in the field of satellite mobile services, and is also used by weather radar, surface ship radar and by the National Aeronautics and Space Administration to communicate with the space shuttle and the International Space Station [5]. Our filter has not only considered the length of the open-ended series stub to form the equivalent resonant circuit at any particular resonant frequency, but has also optimised the width of the slots to form the discontinuity, such as the frequency response is achieved to obtain the bandwidth of 2–4 GHz, and by multiple series stubs arranged accordingly to achieve sharp selectivity and low insertion loss. Circuit cross talk parasitic radiation is minimised by series stubs being arranged in the central strip, and by fields being confined to the central strip only. Furthermore, air bridges are not required because of the symmetry of discontinuities [3].

The BPF is constructed using flame retardant 4 (FR4) substrate material of thickness 1.59 mm, relative dielectric constant of 4.4 and loss tangent of 0.02, which is a single side plated copper coated board (CCB). The software used for simulation of designs is carried out by the computer simulation technology (CST) software which computes by the field solving technique method of moments.

2 Design of open-ended series stub

In CPW technology, a series stub is designed by creating a discontinuity in the central strip by implementing two slots originating from the edge of the central strip on both sides of the ground, such that the slots are connected to each other as shown in Fig. 1*a*. At the end of the discontinuity or series stub, an open circuit is created which gives a short circuit at the input port for a length of quarter wavelength of the stub, that is, $\lambda_g/4$ where λ_g is the guided wavelength and thus giving a band-pass response [6].

The characteristic of this CPW circuit primarily depends on the length 'L' as depicted in Fig. 1. Now if the length is near to the quarter wavelength ($\sim\lambda_g/4$), then it will behave as a resonant circuit of resonating frequency of which the quarter wavelength is considered. Otherwise, if the length is considered to be too small ($<\lambda_g/10$), then the same component will behave as an equivalent capacitance [7].

Thus, we will focus on the length of the stub to be quarter wavelength

$$L = \lambda_g/4 \quad (1)$$

where λ_g is the guided wavelength and is given by

$$\lambda_g = \lambda_0/\sqrt{\varepsilon_{eff}} \quad (2)$$

where ε_{eff} is the effective permittivity and is approximately given by

$$\varepsilon_{eff} = (\varepsilon_r + 1)/2$$

The discontinuity formed will be that of a quarter wave open-ended CPW series stub. Therefore the equivalent transmission line model as shown in Fig. 1*b*, which shows an open-circuited stub of length $\lambda_g/4$ which transforms a short circuit at the starting terminal of discontinuity which is responsible for a band-pass response.

3 BPF using oppositely oriented open-ended stub

The basic filter design is constructed by implementing two $\lambda_g/4$ open stub oppositely oriented such that the discontinuities of each stub are close to each other as shown in Fig. 2*a*. The pattern formed with these slots is like a metal I-shape at the central strip, keeping in mind that the discontinuities must be kept closer to each others which provides a greater field confinement.

The length of the stub is chosen to have a centre frequency near to 3 GHz and the gaps have been optimised to have a good response [8]. This schematic diagram is then simulated using CST software and the result obtained is that of BPF having a 3 dB pass band from 1.87 to 4.04 GHz, thus covering the S-band with low insertion loss of −0.35 dB but the roll-off rate is observed to be poor as

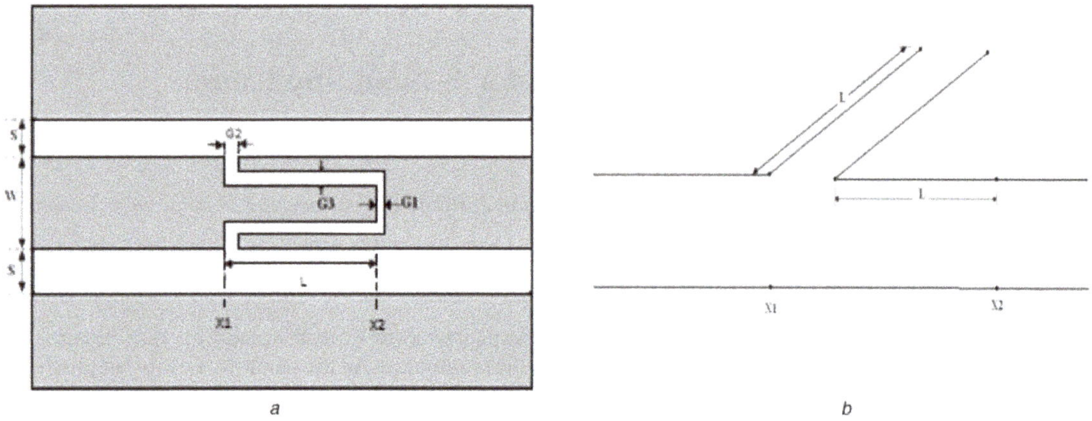

Fig. 1 *Central strip on both sides of the ground such that the slots are connected to each other*
a Open-ended series CPW stub
b Equivalent transmission line model

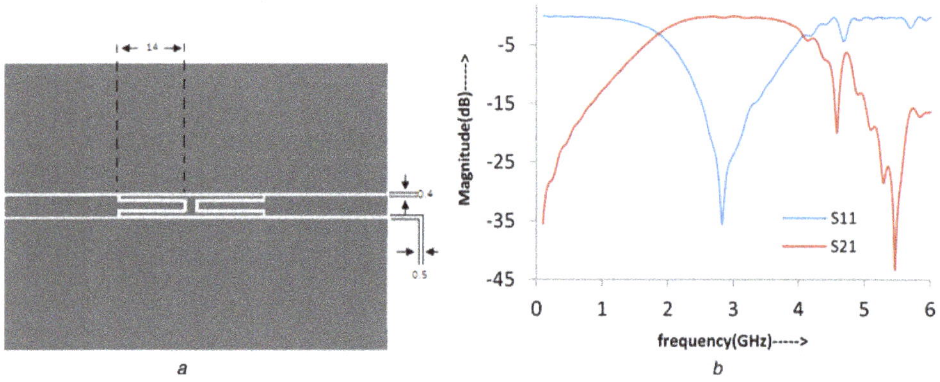

Fig. 2 *Basic filter design is constructed by implementing two oppositely oriented $\lambda_g/4$ open stub*
a Schematic of two oppositely oriented open-ended series stub (all dimension in millimetres)
b EM-simulates *S*-parameter responses

shown in Fig. 2*b* having rising edge selectivity of 12 dB/GHz and falling edge selectivity of 14.65 dB/GHz.

As open-ended series stub of length $\lambda_g/4$ transforms a short circuit at the starting terminal of discontinuity, the equivalent circuit model can be realised by the parallel inductance and capacitance (LC) resonator circuit as shown in Fig. 3*a*. The simulated results of the equivalent circuit model as described in Fig. 3*a* are compared with electromagnetic (EM)-simulated results and good agreement is obtained as shown in Fig. 3*b*.

4 Design of the proposed BPF

Now in order to increase the sharpness of the filter, the order of the filter needs to be increased thereby its component in the design.

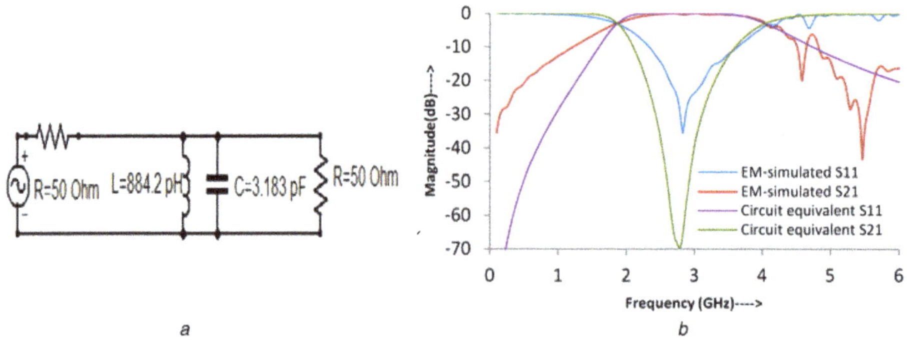

Fig. 3 *Simulated results of the equivalent circuit model*
a Equivalent circuit modelling
b Comparative study of circuit simulated and EM-simulated response

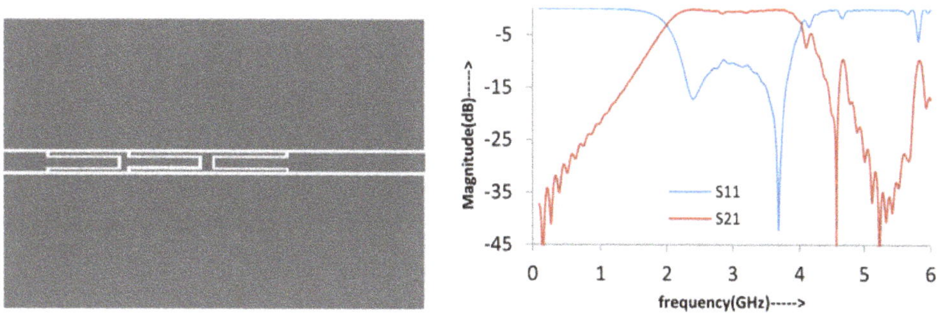

a b

Fig. 4 *Open-ended stub in series*
a Schematic of BPF using three open-ended series stub
b EM-simulates *S*-parameter responses

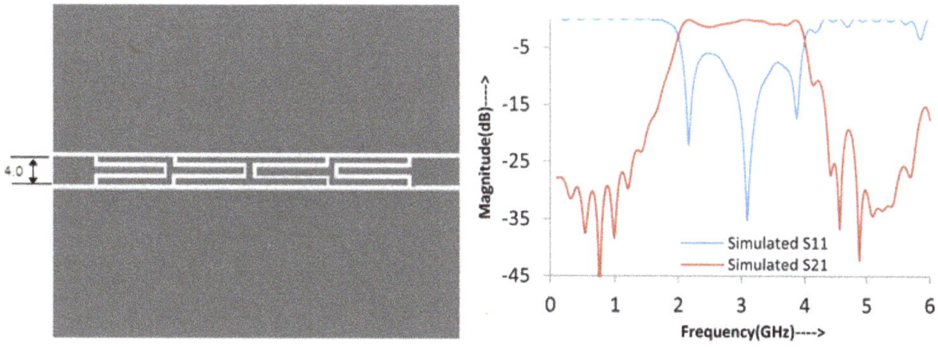

a b

Fig. 5 *Layout design of the proposed CPW BPF is given along with its precise dimensions*
a Schematic of the proposed BPF using four open-ended series stub in symmetric manner (all in millimetres)
b EM-simulates *S*-parameters

Hence we add another open stub in series with the previous figure as seen in Fig. 4*a*. It is then simulated to observe the result as shown in Fig. 4*b* having a bandwidth of 2 GHz, ranging from 2 to 4.02 GHz with an insertion loss of −0.88 dB having an improved sharpness in the band-pass response of 18.89 and 34.69 dB/GHz rising edge and falling edge selectivities, respectively.

To obtain more sharp BPF or to achieve sharp skirt-selectivity, another stub is provided on the other side to have a complete symmetric structure such that a good possible compactness is implemented to achieve a greater field confinement within the circuit than that of other straight open stub in series or shunt stubs. It is also to be noted that the order of the filter is also increased; that is, the number of poles increases as the number of series stubs are increased. Finally, a three pole BPF has been obtained with quite higher selectivity or roll off. The layout design of the CPW BPF is given along with its precise dimensions as shown in Fig. 5*a*. Now this layout design is simulated and the corresponding result shows an S-band filter response from 1.97 to 4 GHz with an insertion loss of −1.07 dB and here we observe the sharp selectivity of 35.9 dB/GHz at the rising edge and 45.7 dB/GHz at the falling edge as shown in Fig 5*b*.

Then, we made a comparative study starting from the first diagram of two oppositely oriented open stubs, then three stubs, and finally four stubs, and the response is observed. We have seen that the bandwidth for all the cases is almost the same from 2 to 4 GHz but there is a notable change in the sharpness of the filter response; as the number of stubs is increased, the roll-off rate of the BPF response improves gradually, as shown in Fig. 6. However, with the increase of stub discontinuities, the insertion loss slightly increases, although the loss is quite low.

Fig. 7*a* shows an equivalent lumped circuit model for the proposed BPF having four open stubs in a symmetric manner. This lumped model has been further simulated with quite universal circuit simulator and the results are also similar to the EM-simulated response as shown in Fig. 7*b*.

Finally, this schematic diagram of multiple stub series CPW BPF is then fabricated using a single side copper plated FR4 substrate having relative permittivity $\varepsilon_r = 4.4$ and the height of the substrate to be 1.59 mm. After proper pattern being etched on the material, it is then connected to 50 Ω SMA connectors at both the ports and the complete fabricated picture is shown in Fig. 8*a*. The scattering parameters of fabricated structure are then measured by Agilent make Vector Network Analyzer (model N5230A) and the result

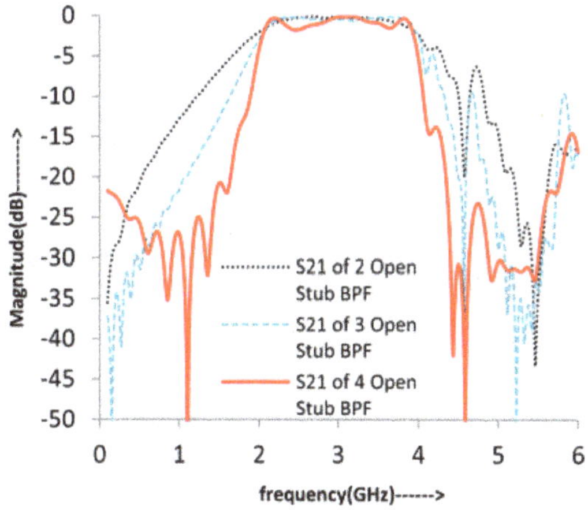

Fig. 6 *Comparative study of open-ended series stub BPF*

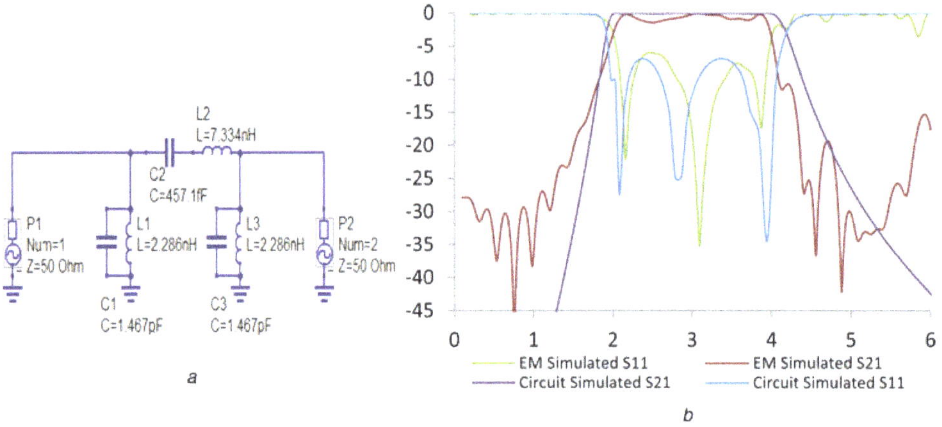

Fig. 7 *Equivalent lumped circuit model for the proposed BPF having four open stubs*
a Circuit equivalent of the proposed BPF
b Comparison of EM-simulates and circuit simulated *S*-parameter responses

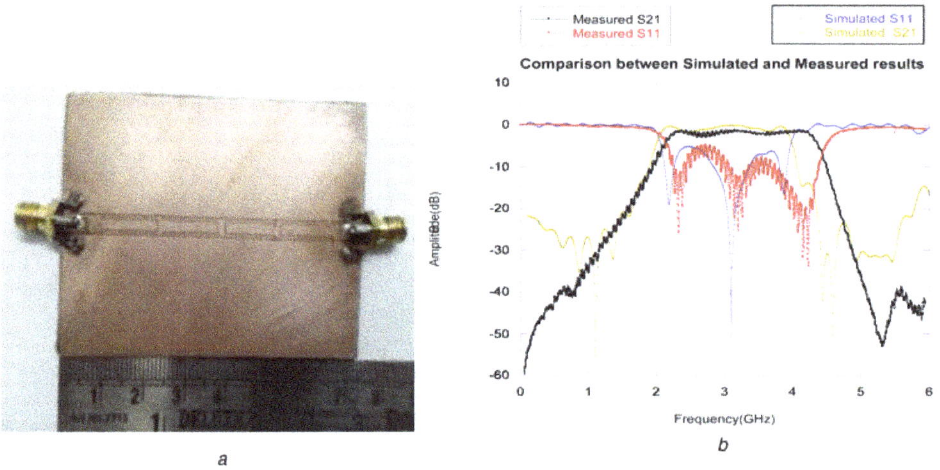

Fig. 8 *Ports and the complete fabricated picture*
a Fabricated prototype of the proposed BPF
b Comparison between simulated and measured *S*-parameters

obtained is compared with the previous simulated response, which is observed to have near agreement in the response as shown in Fig. 8*b*.

5 Conclusion

A BPF is constructed by open-ended CPW series stub which acts as a resonant circuit, thus giving a band-pass response. In this paper, the sharpness of the filter has been improved by proper arrangement and increase of series open-ended stubs, thus providing higher-order BPF. As it is designed with CPW technology having a complete uniplanar structure, there is an ease of fabrication to a single-sided CCB board and a good relaxation in design, as it primarily depends on the width to gap ratio. This multiple series stub design is easy to design and fabricate, not requiring any via holes or air bridge but providing a simple yet effective way to design a filter having a very good roll-off rate. Moreover, the simulated and measured results show a very good agreement with each other. S-band BPF is utilised over the application of satellite and radar communication.

6 Acknowledgment

This work was supported by the University Grants Commission (U.G.C.) Government of India.

7 References

[1] Dai G.L., Xia M.-Y.: 'Novel miniaturized bandpass filters using spiral-shaped resonators and window feed structures'. Progress in Electromagnetics Research, PIER 100, 1997, pp. 235–243

[2] Simons R.N., Ponchak G.E.L.: 'Modeling of some coplanar waveguide discontinuities', *IEEE Trans. Microw. Theory Tech.*, 1988, **36**, pp. 1796–1803

[3] Dib N.I., Katehi L.P.B., Ponchak G.E., Simons R.N.: 'Theoretical and experimental characterization of coplanar waveguide discontinuities for filter applications', *IEEE Trans. Microw. Theory Tech.*, 1991, **39**, pp. 873–882

[4] Hettak K., Dib N., Sheta A.-F., Toutain S.: 'A class of novel uniplanar series resonators and their implementation in original applications', *IEEE Trans. Microw.*, 1998, **46**, (9), pp. 1270–1276

[5] Rainee N.S.: 'Coplanar wave guide circuits, components, and systems' (John Wiley & Sons Inc., New York, NY, 2001), pp. 272–274

[6] Weller T.M., Katehi L.P.: 'Miniature stub and filter designs using the microshield transmission line'. IEEE MTT-S Digest, 1995, **2**, pp. 675–678

[7] Sharma A.K., Wang H.: 'Experimental models of series and shunt elements in coplanar MMIC's'. IEEE Int. Microwave Symp. Digest, Albuquerque, NM, 1–5 June 1992, pp. 1349–1352

[8] Masood R., Mohsin S.A.: 'Optimization of the S-parameter response of a coplanar waveguide series short stub for broadband applications'. IEEE, 2010, pp. 384–388

Permissions

All chapters in this book were first published in TJE, by The Institution of Engineering and Technology (TIET); hereby published with permission under the Creative Commons Attribution License or equivalent. Every chapter published in this book has been scrutinized by our experts. Their significance has been extensively debated. The topics covered herein carry significant findings which will fuel the growth of the discipline. They may even be implemented as practical applications or may be referred to as a beginning point for another development.

The contributors of this book come from diverse backgrounds, making this book a truly international effort. This book will bring forth new frontiers with its revolutionizing research information and detailed analysis of the nascent developments around the world.

We would like to thank all the contributing authors for lending their expertise to make the book truly unique. They have played a crucial role in the development of this book. Without their invaluable contributions this book wouldn't have been possible. They have made vital efforts to compile up to date information on the varied aspects of this subject to make this book a valuable addition to the collection of many professionals and students.

This book was conceptualized with the vision of imparting up-to-date information and advanced data in this field. To ensure the same, a matchless editorial board was set up. Every individual on the board went through rigorous rounds of assessment to prove their worth. After which they invested a large part of their time researching and compiling the most relevant data for our readers.

The editorial board has been involved in producing this book since its inception. They have spent rigorous hours researching and exploring the diverse topics which have resulted in the successful publishing of this book. They have passed on their knowledge of decades through this book. To expedite this challenging task, the publisher supported the team at every step. A small team of assistant editors was also appointed to further simplify the editing procedure and attain best results for the readers.

Apart from the editorial board, the designing team has also invested a significant amount of their time in understanding the subject and creating the most relevant covers. They scrutinized every image to scout for the most suitable representation of the subject and create an appropriate cover for the book.

The publishing team has been an ardent support to the editorial, designing and production team. Their endless efforts to recruit the best for this project, has resulted in the accomplishment of this book. They are a veteran in the field of academics and their pool of knowledge is as vast as their experience in printing. Their expertise and guidance has proved useful at every step. Their uncompromising quality standards have made this book an exceptional effort. Their encouragement from time to time has been an inspiration for everyone.

The publisher and the editorial board hope that this book will prove to be a valuable piece of knowledge for researchers, students, practitioners and scholars across the globe.

List of Contributors

Ahmed Dooguy Kora and Ibrahima CISSE
Department of Telecommunications, ESMT, Dakar, Senegal

Jean-Pierre Cances
Parc d'Ester, Ecole Nationale Superieure d'Ingenieurs de Limoges, Limoges Cedex, France

Kingsley Okoye, Hossein Jahankhani and Abdel-Rahman H. Tawil
School of Architecture Computing and Engineering, University of East London, London E16 2RD, UK

Faisal Al-kamali
Department of Electrical, Faculty of Engineering and Architecture, IBB University, IBB, Yemen

Luzango Pangani Mfupe
Department of Electrical Engineering, FSATI, Tshwane University of Technology, Private Bag X680, Pretoria 0001,

South Africa
Meraka Institute, Council for Scientific and Industrial Research (CSIR), PO Box 395, Pretoria 0001, South Africa

Mjumo Mzyece and Anish Mathew Kurien
Department of Electrical Engineering, FSATI, Tshwane University of Technology, Private Bag X680, Pretoria 0001, South Africa

Ming-Han Lee and Tzuu-Hseng S. Li
aiRobots Laboratory, Department of Electrical Engineering, National Cheng Kung University, 1 University Road, Tainan, Taiwan 70101

Vincent J. Urick
Naval Research Laboratory, Washington, DC, USA

Rini Akmeliawati and Sara Bilal
Faculty (Kulliyyah) of Engineering, International Islamic University Malaysia, Jl, Gombak 53100, Kuala Lumpur, Malaysia

Donald Bailey
Faculty (Kulliyyah) of Engineering, International Islamic University Malaysia, Jl, Gombak 53100, Kuala Lumpur, Malaysia
School of Engineering and Advanced Technology, Massey University, New Zealand, Private Bag 11222, Palmerston North 4442, New Zealand

Serge Demidenko
School of Engineering and Advanced Technology, Massey University, New Zealand, Private Bag 11222, Palmerston North 4442, New Zealand
Centre of Technology, RMIT University Vietnam, 702 Nguyen Van Linh Blvd, Ho Chi Minh City, HCMC, Vietnam

Nuwan Gamage, Ye Chow Kuang and Melanie Ooi
School of Engineering, Monash University Malaysia, Jl Lagoon Selatan, 46150, Selangor Darul Ehsan, Malaysia

Shujjat Khan and Gourab Sen Gupta
School of Engineering and Advanced Technology, Massey University, New Zealand, Private Bag 11222, Palmerston North 4442, New Zealand

Junjie Zhang, Wenyan Yuan and Bingyao Cao
School of Electronic Engineering, Bangor University, Bangor LL571UT, UK
Key Laboratory of Specialty Fiber Optics and Optical Access Networks, Shanghai University, Shanghai 200072, People's Republic of China

Kai Wang and Min Wang
Key Laboratory of Specialty Fiber Optics and Optical Access Networks, Shanghai University, Shanghai 200072,

People's Republic of China
Roger P. Giddings and Jianming Tang
School of Electronic Engineering, Bangor University, Bangor LL571UT, UK

Zhe-Yang Huang and Chung-Chih Hung
Department of Electrical and Computer Engineering, National Chiao Tung University, Hsinchu, Taiwan

Chun-Chieh Chen
Department of Electronics Engineering, Chung Yuan Christian University, Chungli, Taiwan
Brijesh Kumbhani, Lomada Nerusupalli Baya Reddy and Rakhesh Singh Kshetrimayum
Department of Electronics and Electrical Engineering, Indian Institute of Technology Guwahati, Guwahati, India

Prabir Saha and Anup Dandapat
Department of Electronics and Communication Engineering, National Institute of Technology, Shillong, Meghalaya 793 003, India

Deepak Kumar
Department of Computer Science and Engineering, National Institute of Technology, Shillong, Meghalaya 793 003, India

Partha Bhattacharyya
Department of Electronics and Telecommunication Engineering, Bengal Engineering and Science University, Shibpur, Howrah 711 103, India

Ge Wu, Leonid Belostotski and James W. Haslett
Department of Electrical and Computer Engineering, University of Calgary, Calgary, Alberta, Canada

Naveed Ahmed Sheikh and Ashfaq Ahmad Malik
PN Engineering College, National University of Science & Technology, Karachi, Pakistan

Athar Mahboob
Department of Electrical Engineering, DHA Suffa University, Karachi, Pakistan

Khairun Nisa
Department of Computer Science and Engineering, University of Engineering & Technology, Lahore, Pakistan

Soon-mi Hwang and Kwan-hun Lee
Reliability & Failure Analysis Center, Korea Electronics Technology Institute, Sungnam-Si, Republic of Korea

Maryam Niknamfar and Mehdi Shadaram
Department of Electrical and Computer Engineering, University of Texas at San Antonio, San Antonio, TX, USA

Dongwan Kim and Sunshin An
Department of Electric and Electronics Engineering, Korea University Anam Campus, 145, Anam-ro, Seongbuk-gu, Seoul 136-701, South Korea

Yoshiyuki Akuzawa, Yuki Ito, Toshihiro Ezoe and Kiyohide Sakai
Kamakura Office, Mitsubishi Electric Engineering Co. Ltd, 730 Kamimachiya, Kamakura, Kanagawa 247-0065, Japan

Lei Sun and Kong-Pang Pun
Department of Electronic Engineering, Chinese University of Hong Kong, Shatin, Hong Kong, N.T., People's Republic of China

Wai-Tung Ng
Department of Electrical & Computer Engineering, University of Toronto, 10 King's College Road, Toronto, Canada

Shuai Wang and Tao Jin
State Key Laboratory of Novel Software Technology, Department of Computer Science and Technology, Nanjing University, Nanjing 210046, People's Republic of China

Sweta Jain, Nikhitha Kishore and Meenu Chawla
Department of Computer Science & Engineering, Maulana Azad National Institute of Technology, Bhopal, India

Vasco N. G. J. Soares
Instituto de Telecomunicações, University of Beira Interior, Covilhã, Portugal
Superior School of Technology, Polytechnic Institute of Castelo Branco, Castelo Branco, Portugal

Szu-Wei Lee
Graphic Software Development and Validation Team, Visual and Parallel Computing Group, Intel, Santa Clara, CA, USA

Ardavan Rahimian
School of Electronic, Electrical and Computer Engineering, University of Birmingham, Birmingham B15 2TT, UK

Farhad Mehran
School of Electronic, Electrical and Computer Engineering, University of Birmingham, Birmingham B15 2TT, UK

Robert G. Maunder
School of Electronics and Computer Science, University of Southampton, Southampton SO17 1BJ, UK

Vijay Kisanrao Sambhe, Rahul Narayanrao Awale and Abhay Wagh
Department of Electrical Engineering, Veermata Jijabai Technological Institute, Mumbai, India

Alaa Rahman Al-Taee, Fei Yuan and Andy Ye
Department of Electrical and Computer Engineering, Ryerson University, Toronto, ON, Canada

Gunnar Eriksson, Karina Fors, Kia Wiklundh and Ulf Sterner
Swedish Defence Research Agency (FOI), Box 1165, SE-58111 Linkoping, Sweden

Pratik Mondal, Amit Ghosh and Susanta Kumar Parui
Department of Electronics and Telecommunication Engineering, Indian Institute of Engineering Science and Technology, Shibpur, Howrah 711 103, India

www.ingramcontent.com/pod-product-compliance
Lightning Source LLC
Chambersburg PA
CBHW080301230326
41458CB00097B/5250